VOLUME ONE HUNDRED AND EIGHTY-NINE

ADVANCES IN
IMAGING AND ELECTRON PHYSICS

EXAMPLES OF LEPTONIC MONOPOLE TRACKS ON X-RAY PHOTOGRAPHIC FILMS

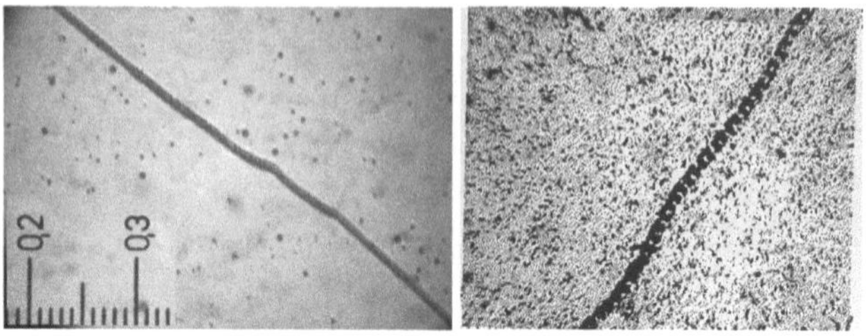

Characteristic monopole tracks in caterpillar form. We have hundreds, even thousands of such tracks.

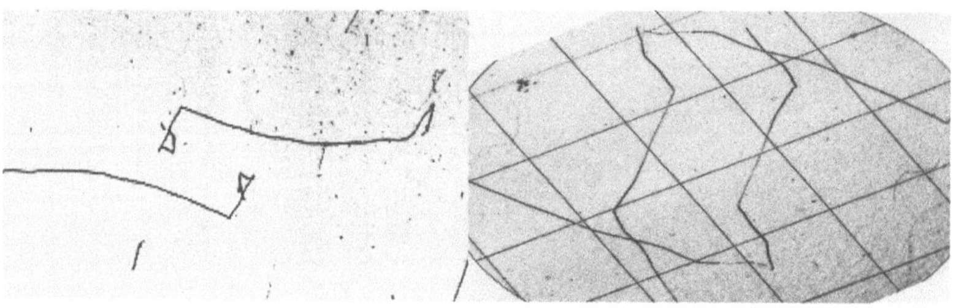

Two pairs of tracks obtained from an element, the image of which is reflected in a germanium mirror. Unlike light symmetry, we see that the symmetry is not left–right : each element of the pair is identical with the other but rotated through 180°, as predicted by theory.

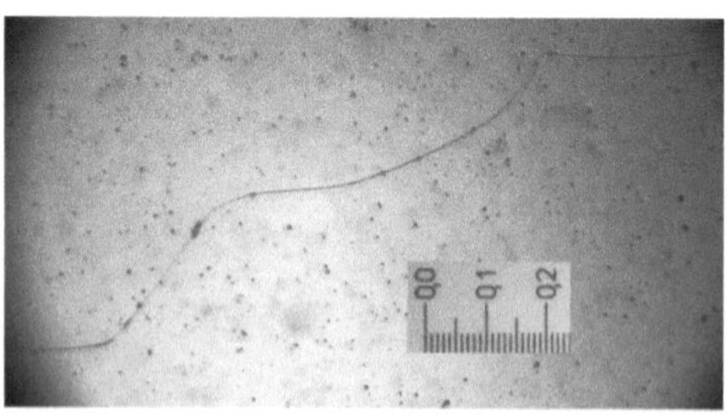

A monopole track at the Earth's North Pole. Theory predicted these tracks because of the possibility of conversion of neutrinos to monopoles caused by the action of solar magnetic fields on neutrinos. Many such tracks were observed.

VOLUME ONE HUNDRED AND EIGHTY-NINE

ADVANCES IN IMAGING AND ELECTRON PHYSICS

The Leptonic Magnetic Monopole Theory and Experiments

GEORGES LOCHAK

Fondation Louis de Broglie
23 rue Marsoulan
F-75012 Paris, France

HARALD STUMPF

Institute of Theoretical Physics
University of Tübingen
D-72076 Tübingen, Germany

EDITOR-IN-CHIEF

PETER W. HAWKES

CEMES-CNRS, Toulouse, France

ELSEVIER

AMSTERDAM • BOSTON • HEIDELBERG • LONDON
NEW YORK • OXFORD • PARIS • SAN DIEGO
SAN FRANCISCO • SINGAPORE • SYDNEY • TOKYO
Academic Press is an imprint of Elsevier

Cover photo credit:
Photo of Professor Louis de Broglie. With kind permission of Georges Lochak.
Photo of Professor Werner Heisenberg. With kind permission of Springer Science+Business Media (Hans Peter Dürr, Quanten und Felder, 1971, Vieweg).

Academic Press is an imprint of Elsevier
225 Wyman Street, Waltham, MA 02451, USA
525 B Street, Suite 1800, San Diego, CA 92101-4495, USA
125 London Wall, London, EC2Y 5AS, UK
The Boulevard, Langford Lane, Kidlington, Oxford OX5 1GB, UK

First edition 2015

ISBN: 978-0-12-802463-8
ISSN: 1076-5670

This volume is dedicated to Louis de Broglie and Werner Heisenberg, both winners of the Nobel Prize, by their former students, Georges Lochak and Harald Stumpf

CONTENTS

PREFACE

Most volumes in this series contain several chapters on a variety of subjects in the general area of imaging and electron physics. Occasionally, however, an entire volume is devoted to a single subject or to a specific theme, and this is the case here. The authors have the distinction of being former students of Louis de Broglie and Werner Heisenberg, to whom they have dedicated their book, and have spent many years wrestling with the outstanding questions of quantum mechanics. In particular, they have examined the thorny problem of the magnetic monopole and they present here their arguments and conclusions on this difficult topic.

The book is divided into two large parts. The first, by Georges Lochak, presents the theory of the leptonic monopole, a massless particle with spin ½ that travels at the speed of light. The second part, by Harald Stumpf, sets out from the suggestion that such leptonic monopoles can be created during discharges in fluids. This suggestion is pursued in great detail, with extensive theoretical background.

I have no doubt that this full presentation of ideas that depart from the mainstream of thinking in theoretical physics will keep interest in the subject alive and readers will surely be grateful to the authors for spelling out their ideas so fully. I take this opportunity to acknowledge the invaluable contribution of Madame Morénas of the Fondation Louis de Broglie in the preparation of Georges Lochak's material and of Michael Ruder, who has dealt with many queries concerning Profesor Stumpf's text.

Peter Hawkes

FOREWORD

The Past and Future of Magnetic Monopoles

G. Lochak

Contents

1. A BRIEF HISTORY OF ELECTRICITY AND MAGNETISM

The observation of electricity and magnetism began in ancient times. It is attributed to the Greeks and the Chinese.

Around 600 B.C., Thales noticed that amber had the ability to attract light bodies when rubbed with a cat skin. But the Greeks also knew about magnets, whose name is *magnes* in Greek. And thus, since amber is called *elektron,* they passed down to us not only the beginning of the study of electromagnetism, but also the etymology.

The Chinese did not know about electricity, but they knew about magnets: probably both they and the Greeks found magnetite mines (i.e., a magnetic iron oxide). But the Chinese also discovered the Earth's magnetism. They invented the compass (1000 B.C.), and merchant ships brought it to the rest of the world.

Nevertheless, the observation of things and the use of them was not yet a science, which is the explanation of *the behavior of things and their inclusion in an image of the world.* The Greeks had done this in astronomy, but not in electricity and magnetism.

The science of electricity and magnetism started with André-Marie Ampère, and was developed by James Clerk Maxwell, who considered Ampère the greatest physicist since Isaac Newton. But then Ampère lost that distinction when Einstein declared that the greatest was actually Maxwell. And Maxwell lost that title to Einstein. Records are fragile!

Now, it may be said that, despite the contributions from many others in these fields, the science of electricity and magnetism owes its fundamental ideas to three men: Ampère, Maxwell, and Michael Faraday. Consider the following points:

1. Starting from experiments of Oersted and others, Ampère and Faraday gathered electricity and magnetism in a single doctrine that claims that:
   a. An electric current produces magnetism (Ampère).
   b. A variation of magnetism produces electricity (Faraday).
   These two statements are not reciprocal because in the first case, the electric current may be static, while in the second case, magnetism must vary.
2. The next discovery was the notion of *field,* born in the same years. The idea appeared in two steps:
   a. At first, Ampère's and Faraday's attention was drawn to the lines of magnetic or electric forces. Ampère based his "Electrodynamics" on the interactions between small elements of currents and introduced a vector, known in French as the *directrice*, which today is called the *magnetic field.*
   Faraday observed figures drawn by iron filings scattered on a sheet of paper above a magnet. He interpreted them as lines of force and then as *lines of field,* a concept that is now taught in every textbook.
   b. The second step is due to Faraday alone. He introduced the entirely new idea that an electric or a magnetic force acts owing to a new property of the surrounding space: the *electromagnetic field.*

This idea required a new type of mathematics that was capable of gathering electric and magnetic properties in a new image of the world. This was the achievement of Maxwell's equations, which are engraved into the pantheon of science and are the object of universal admiration, as expressed by Ludwig Boltzmann in his *Treatise of Electromagnetism,* quoting a verse of

Goethe's *Faust*: *"War es ein Gott der diesen Zeilen schrieb ?"* (Was it a god who wrote down these signs?)[1].

Thus, beside the empty world inherited from Democritus, filled with material points obeying Newton's laws of mechanics, appeared another physical world: a resurgence of the world of Anaxagoras, filled with fields. Maxwell's equations became the counterpart of Newton's mechanics. But these contradictory worlds were destined to concur as follows:

- In 1905, the Einstein photon brought the corpuscles back into the theory of light, from which they had been expelled by the wave theory of Maxwell, Christiaan Huygens, and Augustin-Jean Fresnel. Einstein proved that the still-mysterious photoelectric effect was a consequence of his hypothesis.
- Conversely, in 1923, Louis de Broglie discovered that material corpuscles have wave properties. He gave the first formulas for matter waves, interpreted the spectrum of Niels Bohr's atom in terms of stationary electron waves, and predicted the diffraction of electrons. *Matter and Light,* the title of one of his books, became the symbol of the new quantum world: wave-particle dualism.

2. THE FATHERS OF THE MAGNETIC MONOPOLE

2.1 Maxwell (1873)

We have already noted the absence of symmetry between the Ampère and Faraday laws, which favors electricity, since we can create magnetism with a stationary electric current (the electromagnet), whereas we need a variation of magnetism to create an electric current (Faraday's induction). This asymmetry occurs because we can produce an electric current, but not a magnetic current.

[1] Concerning Maxwell's equations, there is an interesting story. When Faraday announced his *law of induction,* nobody believed in it because, unlike Ampère's law, it was not between time-independent physical quantities, but rather included the time variation of a magnetic field. This was complete nonsense: how could the variation of a field produce an effect? It took a long time before the young Maxwell brought the laws of Ampère, Faraday, and Coulomb into a set of equations, and he discovered that something was inconsistent, not in Faraday's law but in Ampère's law—namely, that a term depending on time was missing. He created this term (the *displacement current*), and his equations became the celebrated Maxwell equations, which led to the electromagnetic theory of light and, later, to radio waves. The second result of this story will be discussed elsewhere in this book: much later, in 1934, Louis de Broglie published the equation of the photon, considered as the fusion of two Dirac particles, and he derived from quantum mechanics the Maxwell equations without adding any displacement current, which appears automatically as a consequence of relativistic symmetry.

Nevertheless, in 1785, Charles-Augustin de Coulomb measured his law of force in $1/r^2$, not only for electric charges but for magnetic charges even though he possessed only what he called "electric poles " (small, charged objects), but no "magnetic poles." So he used long magnetic wires, whose extremities could not interact, as Maxwell explained in his *Treatise on Electricity and Magnetism.*

Maxwell gave a central place to Coulomb's law not only for electricity, but also for magnetism. Magnetic poles are described at the beginning of volume 2 of his *Treatise*, which shows that electric and magnetic charges have the same physical dimensions and are thus expressed in the same units. But neither in the electric case nor in the magnetic case are particles mentioned in the sense that we now understand them (as localized microobjects). The experimental discovery of the electron as a particle—the fruit of numerous works—was announced by J. J. Thomson in 1897. A first theory came from Hendrik Lorentz at this time, but the true theory came later, with the Dirac equation in 1928.

The Maxwell poles were only a rough draft, but Maxwell was the first to understand that vectors representing electricity and magnetism are of a different nature. The first is a polar vector, of the same kind as a velocity, while the second is an axial vector of the same kind as a rotation axis.

Despite some analogies, electricity and magnetism are fundamentally different; this difference appears in their symmetry properties. The image of a magnetic field in a mirror perpendicular to it is the field itself, while the image of an electric field perpendicular to a mirror is inverted. That is, the image is the mirror image of the object. Conversely, the image of a magnetic field parallel to a mirror is parallel to the field, but inverted, while the image of an electric field parallel to a mirror is the field itself.

In other words, there is no exact analogy between electricity and magnetism, contrary to what is often claimed and to the analogy falsely attributed to Maxwell's equations. These errors arise because polar and axial vectors are represented by the same symbols. Maxwell knew this, and Pierre Curie tried to impose different notations for these reasons, but in vain.

Through the discovery (quickly forgotten!) of the difference between polar and axial vectors, Maxwell was the second (after Louis Pasteur) to approach the fundamental property of *enantiomorphism*, or *chirality;* i.e., the difference between left and right, such as the left and right hands. *Chiral* comes from the Greek *kheir* (*hand*).

Actually, Pasteur was the first to discover this phenomenon in another field, not in electromagnetism but in crystallography. He discovered that there are two kinds of crystals of tartaric acid: left and right. Each of them is not its own mirror image, but the image of the other in a mirror, like two hands. When Pasteur made this discovery, he was still a student at the Ecole Normale Supérieure. In a state of great excitement, he immediately went out of the room where he worked, declaring with enthusiasm in the corridor, "The Universe is not symmetric to itself !" He was immediately convinced of the universal character of his result. And when he said that to his old master, Jean-Baptiste Biot, the latter replied: "I am moved to tears by what you are telling me !"

This discovery lay at the origin of the biological achievements of Pasteur, and it must be remembered that a few years before these lines were written, a Nobel Prize was awarded to William Knowles, Ryoji Noyori and Barry Sharpless for a discovery of great importance for medicine: the chiral catalysts, able to separate left and right synthetic molecules. A whole generation remembers the drama of a mildly sedative medicine, thalidomide, prescribed to pregnant women, who gave birth to children without arms or legs as a result of the drug. The reason for this was that the synthetic molecules were a mixture of left and right molecules, one group of which (and only one) being teratogenic. And we shall see what followed from the discovery of chirality in physics: it happened that beta radioactivity is chiral too, and it introduces enantiomorphism just like tartaric acid and magnetism.

In my opinion, it may be asserted that Pasteur and Maxwell, in two different fields of science, discovered in enantiomorphism one of the most important results of the 19th century. The following step was the discovery, by Lee and Yang, in 1956, of the "non-conservation of parity" (a somewhat barbaric name for chirality) of weak interactions. This was the last physical discovery to change our physical image of the world. But between Pasteur and Maxwell and this discovery, there were several great contributors: Pierre Curie, Henri Poincaré, and Paul Dirac.

2.2 Pierre Curie (1894)

Pierre Curie was one of the great successors to Maxwell in the field of electromagnetism. He knew even better than Maxwell the difference between electricity and magnetism, and he proved that this difference is of a geometric nature and concerns general symmetry laws. He published a paper titled "On the Symmetry of Physical Phenomena, the Symmetry of Electric and Magnetic Fields," which was the first general paper on a subject that would dominate physics in the 20th century.

This paper was thought to be a classic, but it fell into oblivion. This happened probably in part because it was written in French, which is no longer a leading language, and also because it was written in the language of crystallography, which gave rise to the science of symmetry in physics. French physicists of the 19th century were fascinated by crystallography, under the strong influence of Romé de l'Isle and René Just Haüy. Nowadays, however, crystallography is a normal branch of science—it has lost its dominant position, and few physicists know that "the spherical enantiomorphic group (18) ∞L_{∞}" is simply a pseudoscalar symmetry. Reading Pierre Curie, I was grateful to my masters in mineralogy at the Sorbonne, who taught me these things (among many others).

In this book, we shall explore different problems of symmetry, but in a modern language, adding many questions that were unknown in the time of Curie (particles, transformations of charges, *CPT* theorem, etc.). A special point in Curie's paper will be particularly important: it is a kind of postscript to the principal paper, namely, "On the Possibility of the Existence of Magnetic Conductibility and of Free Magnetism", in other words, magnetic monopoles, and Curie announced the symmetry laws of such phenomena, if they indeed existed. He looked for the condition of their observation, and he showed that a magnetically charged sphere must have a pseudoscalar symmetry (i.e., (18) $\infty L_{\infty}\ldots$).

Such a magnetic sphere is not to be superposed on its image in a mirror because the North and South poles are mutual images (right and left) in a mirror, unlike the electric charges + and −, which are mutual images too, not in space but in time reversal, as was shown by Richard Feynman. The chiral property of magnetism will be described more precisely later in this book. The fact that the North and South monopoles are antiparticles is a consequence of the theory of monopoles, but in the frame and language of quantum mechanics.

In particular, it will be found that this chiral symmetry cannot be represented by a pseudoscalar constant of charge, which leads to contradictions, as will be shown later in this foreword. This fact is easy to understand even before seeing any formal proof because all the physical constants are pure scalars: otherwise, the Planck constant would have to be transformed into a kinetical moment and the velocity of light would be reversed by the P and T transformations. In the theory of monopoles, the sign of charge will not be affected by P or T transformations, but a P transformation will be the transition between two eigenstates of a charge operator; i.e., between two quantum states.

This difference between the classical and quantum theory is fundamental because the classical theory is subject to an objection, which is invalidated in quantum theory. The objection asserts that speaking of electric or magnetic charges does not correspond to different physical objects, but only to an arbitrary choice (see Jackson, 1975 and many others). We address this issue shortly.

Let us now introduce densities of electric and magnetic currents and charges in the Maxwell equations[2]:

$$\begin{aligned} &\nabla.\mathbf{E} = \rho_e; \quad \nabla\times\mathbf{H} = \frac{1}{c}\frac{\partial\mathbf{E}}{\partial t} + \frac{1}{c}\mathbf{J}_e \\ &\nabla.\mathbf{H} = \rho_m; \quad -\nabla\times\mathbf{E} = \frac{1}{c}\frac{\partial\mathbf{H}}{\partial t} + \frac{1}{c}\mathbf{J}_m. \end{aligned} \tag{1}$$

The system [Eq. (1)] remains invariant under the following transformation, where α is an arbitrary angle:

$$\begin{aligned} &\mathbf{E} = \mathbf{E}'\cos\alpha + \mathbf{H}'\sin\alpha; \quad \mathbf{H} = -\mathbf{E}'\sin\alpha + \mathbf{H}'\cos\alpha \\ &\rho_e = \rho'_e\cos\alpha + \rho'_m\sin\alpha \quad \rho_m = -\rho'_e\sin\alpha + \rho'_m\cos\alpha \\ &\mathbf{J}_e = \mathbf{J}'_e\cos\alpha + \mathbf{J}'_m\sin\alpha \quad \mathbf{J}_m = -\mathbf{J}'_e\sin\alpha + \mathbf{J}'_m\cos\alpha. \end{aligned} \tag{2}$$

The primed and unprimed variables $(\mathbf{E}', \mathbf{H}', \rho'_e, \rho'_m, \mathbf{J}'_e, \mathbf{J}'_m)$ and $(\mathbf{E}, \mathbf{H}, \rho_e, \rho_m, \mathbf{J}_e, \mathbf{J}_m)$ obey the same equation [Eq. (1)]. Now, suppose that $\rho_m = \mathbf{J}_m = 0$, so that Eq. (1) represents a purely electric particle. Here, α can be chosen freely so that we can put $\alpha = \frac{\pi}{2}$ and find, according to Eq. (2), the following equalities:

$$\mathbf{H}' = \mathbf{E}, \quad \mathbf{E}' = -\mathbf{H}, \quad \rho'_m = \rho_e, \quad \mathbf{J}'_m = \mathbf{J}_e, \quad \rho'_e = 0, \quad \mathbf{J}'_e = 0. \tag{3}$$

The primed system becomes

$$\begin{aligned} &\nabla.\mathbf{E}' = 0; \quad \nabla\times\mathbf{H}' = \frac{1}{c}\frac{\partial\mathbf{E}'}{\partial t} \\ &\nabla.\mathbf{H}' = 4\pi\rho'_m; \quad -\nabla\times\mathbf{E}' = \frac{1}{c}\frac{\partial\mathbf{H}'}{\partial t} + \frac{4\pi}{c}\mathbf{J}'_m. \end{aligned} \tag{4}$$

This is an electromagnetic field interacting with a purely magnetic particle, and it seems that we have changed the electric case into a magnetic one by a simple change of variables, so that the difference seems purely formal.

[2] We use the rationalized Gauss system of Heaviside-Lorentz, and we suppose that we are in a vacuum, which means that: (1) The factor 4π disppears, and (2) we have $\varepsilon = \mu = 1$, so that $\mathbf{D} = \mathbf{E}$, $\mathbf{B} = \mathbf{H}$.

This change may be formal, but it is physically wrong because not only the equality of values [Eq. (3)], but already the transformations [Eq. (2)] are not covariant with respect to the symmetry laws: they equate or add vectors and pseudovectors. It must be emphasized once more that despite the fact that there are strong links between electricity and magnetism, they are profoundly different in their symmetry properties.

Such an error is impossible in our theory, in which the difference of affine transformation between electricity and magnetism is evident because $(\rho_e, \mathbf{J}_e)$ will automatically appear as a polar quadrivector, while $(\rho_m, \mathbf{J}_m)$ is an axial space-time quadrivector, orthogonal to the first one, by virtue of the Clifford algebra of Dirac's theory of the electron. Let us recall that the polar and axial transformations that will be deduced from our equations had already been proved by Maxwell and Curie. Electric and magnetic particules are really different and the choice between them cannot be free.

2.3 Poincaré (1896)

Two years after Curie, but independently of him, Poincaré wrote a short paper about an experiment of Kristian Birkeland, which he characterized as some "remarks." The paper apparently had no connection with the problem of the magnetic monopole, but we shall return to this point later because actually, Poincaré gave the first differential equation describing, in classical physics, the interaction between an electron and a magnetic monopole. The paper was an important contribution to the theory of the monopole for three reasons:

1. The problem of Poincaré was the collision between a moving electron and a motionless magnetic pole. But the equation is exactly the same for a moving magnetic monopole and a motionless electric pole.
2. We shall prove later in this book that the geometrical optics approximation of my quantum equation is exactly the Poincaré equation.
3. The Poincaré equation gave a theoretical explanation of the Birkeland effect. As a result of points 1 and 2, we can assert that reciprocally, the quantum equation of the monopole was also confirmed by the classical approximation, even before any new experiment had been carried out.

2.4 Dirac (1931)

In a series of famous papers, Dirac described, if not really a quantum theory of a magnetic monopole, at least a theory of the interaction between an electric charge and a fixed magnetic Coulomb pole. Despite the fact that Dirac did not speak of symmetry laws, the latter were implicitly introduced.

Dirac used a gauge reasoning—the first of this kind—and he obtained a remarkable law that claims that the product (*eg*) of the electric charge (*e*) of the incident particle and the magnetic charge of the motionless target is quantized. Therefore, for a given magnetic charge, an electric charge must be a multiple of an elementary charge in accordance with the experimental facts: it was Dirac's goal to prove it.

Further, it will be shown that our theory of leptonic monopoles gives the Dirac law in a more precise form. The fact that we find the same kind of law is not surprising because this law is a consequence of the discrepancy between electric and magnetic symmetries, and it does not depend on a particular model.

Nevertheless, it must be acknowledged that there has been, up to now, only one experimental proof of this law given in a series of experiments by Alexander Mikhailov based on the method initiated by Robert Millikan for the measurement of the electron charge. Unfortunately, Mikhailov later gave another series of results based on a totally different method, and he found a charge that was more than 10,000 times smaller. I think that the first result was the correct one, but we obviously need more experiments.

Dirac's paper originated a development in physics : namely, the investigation of magnetic monopoles. Many papers appeared, two of which were very successful. In 1974, Gerard 't Hooft and Alexander Polyakov showed that the Great Unification Theory (GUT) involves the existence of magnetic monopoles. It gave confidence in the whole monopole theory despite the fact that no new effect was predicted and there was no proposal about the observation of monopoles. It may even be asserted that the converse was true because this theory predicted a mass of the order of 10^{16} GeV for this monopole. If this were true, there would be no hope of ever creating such a particle. It must also be added (but in such circumstances, it is a small detail), that this theory says nothing about symmetries.

In conclusion, it may be said that after Dirac, the theory of monopoles was more or less put under the umbrella of the hypothesis of giant masses, which is the simplest way of explaining why they were not observed. Needless to say, this is not our problem because we shall speak of a massless monopole that is confirmed by observable phenomena, while the rival theories are unable to find any experimental confirmation, precisely on account of the predicted giant masses.

3. SOME INTRODUCTORY WORDS ABOUT THE LEPTONIC MONOPOLE

The theory that we shall develop here is quite different. First, it is now time to say how this theory is indebted to Dirac. Curiously, it owes nothing to his work on the magnetic monopole except for his authorship of the law of charges, which is found in our theory in another way, thus providing a mutual confirmation.

Actually, this theory is indebted to the Dirac equation of the electron. I worked on his equation in 1954, at the Institut Henri Poincaré, and it all seemed to me a rotating of the Dirac equation, obviously because of the spin. This is why I suggested in 1956 with a coworker, Jakobi, a representation of the Dirac equation through a density and seven angles: six relativistic Euler angles (three real rotations in $\mathbb{R}^3$ and three imaginary angles for the velocity) and a small ugly duckling: the angle A. This strange angle was defined in the Dirac equation by Yvon and Takabayasi, who discovered its central role in the theory of the electron.

I knew Takabayasi quite well at that time, late in the 1950s, when he worked in the de Broglie group at the Institut Henri Poincaré: everybody in our group (including de Broglie himself) was aware of the importance of this angle but failed to give it any physical meaning. Only 30 years later did I understand that it is related to magnetism, as will be discussed in Chapter 2.

The angular representation so obtained was a complicated system of equations, but two equations stand out owing to their formal simplicity and their resemblance. Both implied classical Poisson brackets. The first equation seemed quite simple:

$$\left[\frac{\varphi}{2}, J_4\right] = \delta(\mathbf{r} - \mathbf{r}').$$

This canonical conjugation between the Euler angle of proper rotation and the time component of the electric current J_μ suggested the conservation of electricity with the phase $\varphi/2$.

The second equation was

$$\left[\frac{A}{2}, \Sigma_4\right] = \delta(\mathbf{r} - \mathbf{r}').$$

This is the same conjugation, but between the angle A and the time component of the pseudovector Σ_μ, defined by the Dirac equation just like the polar current vector J_μ. We shall examine this point again later, and it has already apppeared here in the form $(\rho_m, \mathbf{J}_m)$, but I didn't notice, at that time, that it was the same vector. The reason was that everybody considered the space part Σ_k of Σ_μ as the spin vector because Σ_k appears in the first integral of the linear momentum in Dirac's theory of the electron. But nobody said anything about the time component: Σ_4. The second Poisson bracket was mysterious because neither the angle A nor Σ_4 was understandable.

Many years later, in 1983, I suddenly realized that A must be a pseudoscalar phase and that this relation must be the counterpart of the first one and represent the conservation of magnetism. I rapidly deduced the equation of a magnetic monopole, which appeared so to say, as the hidden second slope of the equation of the electron.

This theory is completely different from the other theories of the monopole: The new monopole is a massless fermion, which is able to participate in low-energy interactions and which automatically obeys the Curie symmetry laws and the Dirac law between the electric and magnetic charges. We shall see that the new monopole is a magnetically excited neutrino, a massless leptonic monopole with a quantized magnetic charge, a fundamental state of which has zero charge: the neutrino.

More than 25 years of theoretical work and 10 years of experiments were performed, the latter, essentially in Russia in Moscow and Kazan. The results were confirmed more recently in Nantes, France. In view of the results obtained, the reality of the new particle is now more than probable despite the fact that much remains to be done.

Having initiated the idea, I carried out a great part of the theoretical work, but an important role was played by Dr. Harald Stumpf of Tübingen University, a former coworker of Werner Heisenberg (as I was of Louis de Broglie). Many of our ideas may be found in previous results from our mentors and gathered in, the book by T. Borne, G. Lochak and H. Stumpf, *Nonperturbative Field Theory and the Structure of Matter* (Springer, Heidelberg 2001).

With respect to experiments, the initiator was Leonid Urutskoiev, who inspired and to a large extent performed the experiments on the magnetic monopole at the Kurchatov Institute of Moscow. One should add the important results of his remarkable coworker, Nicolas Ivoilov of Kazan University.

It gives me great pleasure to express my gratitude to all my eminent friends for their collaboration during many years and for their valuable results.

4. CHARACTERISTICS OF THE THEORY

Before discussing further details, let us give a review of our theory and show that it is profoundly anchored in the quantum theory and in the Curie symmetry laws of electromagnetism.

1. The monopole equation follows from a fact and a question:
   a. The fact is the following: While Dirac's equation of a massive electron has only one gauge invariance (the phase invariance) which generates the theory of the electron, the massless Dirac equation has two gauge invariances (and only two). The first is the phase invariance based on the unit matrix I and the operator $e^{iI\theta}$, while the second is based on the matrix γ_5 and on the operator $e^{i\gamma_5\theta}$ [3].
   b. The question (which has never been asked before) is: What is the electromagnetic interaction generated by the second gauge? The answer is that it generates electromagnetic interaction with a magnetic monopole.
2. The operator $e^{i\gamma_5\theta}$ entails a new covariant derivative and a new equation describing a magnetic monopole, just as the operator $e^{iI\theta}$ entailed the equation of the electron.
3. The equation automatically defines magnetic pseudopotentials, which were previously deduced from different arguments. Nevertheless, in these earlier papers, the pseudopotentials were considered for the global case in the absence of an electromagnetic field. The present theory was the first to consider the local case (the interaction with an external field) and to introduce the new covariant derivative, from which the magnetic charge follows.
4. The equation gives a renewal of Dirac's law connecting electric and magnetic charges.
5. In the geometrical optics approximation, the Poincaré equation is given. The cone defined by Poincaré in the classical theory plays the same role in the quantum case, with the same vertex angle.
6. The equation obeys the Curie laws of symmetry (the chiral character of a free magnetic charge). This is a crucial point because magnetism and electricity are conjugate in space-time in a sense that appears in the structure of the Maxwell tensor $F_{\mu\nu}$. In quantum mechanics, the electron and positron are symmetric in time: "A positron is an electron moving

[3] Concerning the role of the operator $e^{i\gamma_5\theta}$ in Dirac's theory, see Jakobi and Lochak (1956a, b).

backward in time," as Feynman said. As γ_5 defines the conjugation between space and time, it may be expected that a monopole and its antimonopole are symmetric in space. This is exactly what is deduced from the equation, and it is the translation in the quantum language of a law formulated by Curie.

7. The equation shows that the leptonic monopole is a magnetically excited state of the neutrino. We have hence predicted that this monopole is able to take part in weak interactions, taking the place of a neutrino; this has been experimentally confirmed by Urutskoiev and Ivoilov. As was predicted, these monopoles may be produced on the sun as excited neutrinos. And thus, in the vicinity of the Earth, they must follow the magnetic lines and fall on the poles, as confirmed by an expedition directed by Jean-Louis Etienne to the North Pole (Bardout, Lochak and Fargue, 2007). The link with beta radioactivity was later proved by Ivoilov.
8. The $e^{i\gamma_5\theta}$ gauge invariance induces a family of nonlinear equations (including the famous nonlinear Heisenberg equation). The interaction between a nonlinear monopole and a fixed Coulomb field admits exactly the same first integrals as in the linear case. These integrals are the quantum form of the classical Poincaré integral: this is a proof of what was asserted previously: that Dirac's law is not the consequence of a model, but of a group invariance (i.e., of a geometrical property). It was also proved that the nonlinear equations (including the one of Heisenberg) describe bradyon and tachyon states without any supplementary hypothesis. It was also proved that the magnetic charge implies the torsion of space, in accordance with the work of the Russian physicist Anton Rodichev.
9. This theory of the monopole agrees with de Broglie's Neutrino Theory of Light, in which the photon is considered as the fusion of two Dirac neutral particles. The arguments are as follows:
   a. We have proved that the theory of light contains a second photon, a magnetic photon, which is related to the electromagnetic interaction found in the equation of a monopole.
   b. The pseudopotentials, found in the theory of the monopole, now appear in the theory of light. De Broglie had met them in his own theory and asssociated them with antifields (a simple formal name). In this case, we define a second electromagnetism that is related to the monopole. The pseudopotentials occupy the role of the Lorentz potentials of the theory of the electron. Here, it must be stressed that de Broglie's theory of fusion does not gives a spin 1 particle, but a particle with a maximum spin 1 (analogous to a diatomic molecule). So

we find (as discussed in Chapter 6) a "para-state" of spin 1 and an "ortho-state" of spin 0, corresponding to de Broglie's Maxwellian and non-Maxwellian states. As we also have electric and magnetic photons, the theory of light is finally based on four different photons, instead of one as in the de Broglie theory.

c. The theory of the graviton of de Broglie and M. A. Tonnelat considers the graviton as a fusion of four Dirac particles, three photons appear simultaneously with the graviton: this theory is a unified theory in the sense of Einstein. But we shall show that only two of these photons are electric: the third one is magnetic. This intrusion of a magnetic photon in the relativistic theory of gravitation could be of great importance, for two reasons:
The first is of a fundamental character because the lack of magnetism in Einstein's Unified Theory of Fields could be the reason why this theory has not succeeded. The magnetic monopole opens a new way of thinking about this theory.
The second is practical: the possibility of weakening the gravitational field in the vicinity of the starting area of a rocket, which would allow enormous energy savings. Some strange phenomena, possibly related to monopole physics, could suggest such an idea: the fact that the very heavy lid (about 3500 tons) of the Chernobyl reactor was lifted during the catastrophe and pushed as a whole beside the reactor. The hypothesis that this phenomenon was due to an enormous gas pressure within the reactor meets the objection that such a pressure would probably had led to an explosion of the reactor itself, which makes more plausible the assumption of a weakening of gravitation. It must be added that, in the Kurchatov laboratory, a heavy source of monopoles was pushed aside in the same manner, and a gravimeter registered a variation of the gravitational field.

d. An important fact that will be further discussed in this book is the problem of the Dirac equation (with a proper mass) on the light cone, which is defined as the cone on which the electric current is isotropic. The Dirac equation splits into two equations: one for an ultrarelativistic electric particle and the other for a magnetic monopole. So, at the ultrarelativistic limit of the Dirac equation, the electron and the monopole appear simultaneously and symmetrically, without the introduction of any other hypothesis.

e. Two isotropic currents appear in the theory, which are space-symmetrical about each other. Their sum is the electric current and their difference is the magnetic one. Elementary algebraic properties automatically give all the physical properties in relativity and other symmetry laws. These properties are so striking that it seems possible to ask whether these isotropic currents have a more fundamental importance than just electricity and magnetism.

f. At the moment when these lines are written, our theory says nothing about the creation of monopoles and cannot explain the experimental fact that they appear in two circumstances: the disruptive electric phenomena and beta emitters plunged into a magnetic field (in this case, we have probably magnetic excitation of a neutrino). I have nothing to say about that only that the answer to this question is one of the brilliant results of my friend Dr. Stumpf, which may be found in his contribution to this book. It must be noted that the principal hypothesis of Stumpf is that the beta emission is the fundamental fact; this seems most probable for reasons that cannot be discussed at length in an introduction.

An aim of all these remarks was to suggest that the theory of a leptonic monopole is so strongly rooted in quantum theory that if it were contradicted by an experiment, the consequences would be so important that this eventuality seems very improbable. On the other hand, the theory has had major experimental support.

Nevertheless, it must be recognized that today, the existence and properties of the leptonic monopole still need to be proved in terms of our theoretical predictions and of the exactness of our experiments. And it must be recognized that the practical applications of these monopoles are very few and often only speculative. But it may be stated that when the electron was discovered as a truly existing particle, this was a daring conclusion. The theory came only 30 years later: in those days, "only some details were missing"[4]; namely, relativity, quantum mechanics, and atomic structure. As for the practical applications, how could it be predicted at that time that the electron would one day be at the very center of a far-distant industry?

In the present case, we have a theory which is probably not definitive, but which is embedded in the framework of quantum mechanics and which already gives an acceptable description of phenomena and has made experimentally verified predictions. On the other hand, these experiments are not

[4] Curiously, this expression was used in a satirical remark of W. Pauli concerning another case.

at all the first investigation of an elementary particle. They are based on a century of science.

The theory we are speaking of is not absolutely new because it is based on the discovery of numerous consequences of Dirac's theory of the electron, de Broglie's theory of light, and the general theory of spin particles. These forgotten consequences of old theories remained unknown because, during almost two centuries, the science of electromagnetism (except that of Maxwell) was principally based on electricity. This was perhaps Ampère's fault because of the discovery of the electromagnet, which made it possible to describe and even to create magnetism via electricity, so that attention was exclusively given to elecricity. Physicists had forgotten magnetism in itself, with the exception of Maxwell, once more, who introduced magnetism in his famous treatise on the basis of magnetic poles; and after him, Curie who, just as Maxwell considered magnetic poles, considered magnetic currents and free magnetism.

This book is an attempt to renew electromagnetism and gravitation around electric and magnetic charge and their different symmetry laws. Perhaps such an attempt has a future. I am confident in the motto of the family of my old mentor and friend Louis de Broglie: "Pour l'avenir [For the future]."

I cannot finish this foreword without warmly thanking our editorial project managers at Elsevier, Helene Kabes and Malathi Samayan, for their kind welcome, and our editor Peter Hawkes, to whom Harald Stumpf and myself are profoundly grateful.

It is well known that our celebrated mentors, Heisenberg and de Broglie, were rivals, eternally discussing the meaning of quantum mechanics, but each of them recognized the genius of the other, and their discord was probably not as great as one might think. After all, Einstein said, "If somebody tells you that he knows what $E = h.\nu$ means, tell him that he is a liar." And Bohr said, "If somebody says that he understands quantum mechanics, this means that he does not understand physics." This is why it is easy for the intellectuel sons of Heisenberg and de Broglie to be friends and coworkers.

FUTURE CONTRIBUTIONS

H.-W. Ackermann
Electron micrograph quality

S. Ando
Gradient operators and edge and corner detection

J. Angulo
Mathematical morphology for complex and quaternion-valued images

D. Batchelor
Soft x-ray microscopy

E. Bayro Corrochano
Quaternion wavelet transforms

C. Beeli
Structure and microscopy of quasicrystals

M. Berz and K. Makino, Eds (Vol. 191)
Femtosecond electron imaging and spectroscopy

C. Bobisch and R. Möller
Ballistic electron microscopy

F. Bociort
Saddle-point methods in lens design

K. Bredies
Diffusion tensor imaging

A. Broers
A retrospective

R.E. Burge (Vol. 191)
A scientific autobiography

N. Chandra and R. Ghosh
Quantum entanglement in electron optics

A. Cornejo Rodriguez and F. Granados Agustin
Ronchigram quantification

N. de Jonge, Ed. (Vol. 190)
CISCEM 2014

L.D. Duffy and A. Dragt
Eigen-emittance

J. Elorza
Fuzzy operators

A.R. Faruqi, G. McMullan and R. Henderson (Vol. 190)
Direct detectors

M. Ferroni
Transmission microscopy in the scanning electron microscope

R.G. Forbes
Liquid metal ion sources

P.L. Gai and E.D. Boyes
Aberration-corrected environmental microscopy

V.S. Gurov, A.O. Saulebekov and A.A. Trubitsyn
Analytical, approximate analytical and numerical methods for the design of energy analyzers

M. Haschke
Micro-XRF excitation in the scanning electron microscope

P.W. Hawkes (Vol. 190)
Electron optics and electron microscopy conference proceedings and abstracts, a supplement

R. Herring and B. McMorran
Electron vortex beams

M.S. Isaacson
Early STEM development

K. Ishizuka
Contrast transfer and crystal images

K. Jensen, D. Shiffler and J. Luginsland
Physics of field emission cold cathodes

M. Jourlin
Logarithmic image processing, the LIP model. Theory and applications

U. Kaiser
The sub-Ångström low-voltage electron microcope project (SALVE)

C.T. Koch
In-line electron holography

O.L. Krivanek
Aberration-corrected STEM

M. Kroupa
The Timepix detector and its applications

B. Lencová
Modern developments in electron optical calculations

H. Lichte
Developments in electron holography

M. Matsuya
Calculation of aberration coefficients using Lie algebra

J.A. Monsoriu
Fractal zone plates

L. Muray
Miniature electron optics and applications

M.A. O'Keefe
Electron image simulation

V. Ortalan
Ultrafast electron microscopy

D. Paganin, T. Gureyev and K. Pavlov
Intensity-linear methods in inverse imaging

N. Papamarkos and A. Kesidis
The inverse Hough transform

Q. Ramasse and R. Brydson
The SuperSTEM laboratory

B. Rieger and A.J. Koster
Image formation in cryo-electron microscopy

P. Rocca and M. Donelli
Imaging of dielectric objects

J. Rodenburg
Lensless imaging

J. Rouse, H.-n. Liu and E. Munro
The role of differential algebra in electron optics

J. Sánchez
Fisher vector encoding for the classification of natural images

P. Santi
Light sheet fluorescence microscopy

R. Shimizu, T. Ikuta and Y. Takai
Defocus image modulation processing in real time

T. Soma
Focus-deflection systems and their applications

I.F. Spivak-Lavrov
Analytical methods of calculation and simulation of new schemes of static and time-of-flight mass spectrometers

J. Valdés
Recent developments concerning the Système International (SI)

G. Wielgoszewski (Vol. 190)
Scanning thermal microscopy and related techniques

PART ONE

Theory of the Leptonic Monopole

Georges Lochak
Fondation Louis de Broglie 23 rue Marsoulan F-75012 Paris, France
E-mail: georges.lochak@free.fr

Contents

Advances in Imaging and Electron Physics, Volume 189
ISSN 1076-5670
http://dx.doi.org/10.1016/bs.aiep.2015.01.001

CHAPTER 1

Theoretical Background

1. THEORIES OF POINCARÉ, DIRAC, AND CURIE

1.1 The Birkeland-Poincaré effect

In 1896, Kristian Birkeland introduced a straight magnet in a Crookes tube, and he was puzzled by a convergence of the cathodic beam that did not depend on the orientation of the magnet (Birkeland, 1896). Henri Poincaré explained this effect by the action of a magnetic pole on the electric charges of the beam (such charges were only conjectured at that time); he showed that it may be due to the action of only one pole of the magnet and that, for symmetry reasons, it must be independent of the sign of the pole (Poincaré, 1896). (See Figure 1.1)

To describe this effect, Poincaré wrote down the equation of an electric charge in a coulombian magnetic field created by one end of the magnet. The magnetic field is expressed as

$$\mathbf{H} = g\frac{1}{r^2}\mathbf{r}, \tag{1.1}$$

where g is the magnetic charge. From the expression of the Lorentz force (Poincaré, 1896) the following equation results:

$$\frac{d^2\mathbf{r}}{dt^2} = \lambda\frac{1}{r^3}\frac{d\mathbf{r}}{dt} \times \mathbf{r}; \quad \lambda = \frac{eg}{mc}, \tag{1.2}$$

where e and m are the electric charge and the mass of the electron.

Poincaré found the following integrals of motion, where A, B, C, and Λ are arbirary constants:

$$\mathbf{r}^2 = Ct^2 + 2Bt + A; \quad \left(\frac{d\mathbf{r}}{dt}\right)^2 = C. \tag{1.3}$$

$$\mathbf{r} \times \frac{d\mathbf{r}}{dt} + \lambda\frac{\mathbf{r}}{r} = \Lambda \tag{1.4}$$

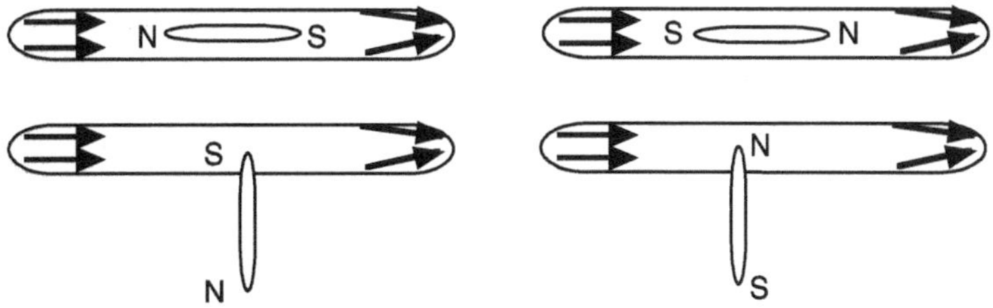

Figure 1.1 The Birkeland-Poincaré effect. When a straight magnet is introduced in a Crookes tube, the cathodic rays converge regardless of the orientation of the magnet. Above: the cases considered by Birkeland. Below: the cases corresponding to the calculations of Poincaré.

He obtained the following from Eqs. (1.4) and (1.2):

$$\boldsymbol{\Lambda}.\mathbf{r} = \lambda r; \quad \frac{d^2\mathbf{r}}{dt^2}.\mathbf{r} = \frac{d^2\mathbf{r}}{dt^2}.\frac{d\mathbf{r}}{dt} = 0. \tag{1.5}$$

This means that $\mathbf{r}$ describes an axially symmetric cone—the *Poincaré cone*, 1896—and that the acceleration is perpendicular to its surface, so that $\mathbf{r}$ follows a *geodesic line*. If the cathodic rays are emitted far from the magnetic pole, with a velocity V parallel to the z-axis, they will have an asymptote that obeys the following equations:

$$x = x_0; \quad y = y_0 \tag{1.6}$$

And we find, from Eqs. (1.3) and (1.4):

$$C = V^2; \quad \boldsymbol{\Lambda} = \{y_0 V, \ -x_0 V, \ \lambda\}. \tag{1.7}$$

Thus the z-axis is a generating line of the Poincaré cone, the half-vertex angle Θ' of which is given by

$$\sin \Theta' = \frac{V}{\lambda}\sqrt{x_0^2 + y_0^2} \tag{1.8}$$

Now, after the emission, the cathodic ray becomes a geodesic line rotating along the cone and crosses the z-axis at distances from the origin given by

$$\frac{\sqrt{x_0^2 + y_0^2}}{\sin \phi}; \quad \frac{\sqrt{x_0^2 + y_0^2}}{\sin 2\,\phi}; \quad \frac{\sqrt{x_0^2 + y_0^2}}{\sin 3\,\phi}; \ldots \quad \phi = 2\pi \sin \Theta' \tag{1.9}$$

Therefore, if the emitting cathode is a small disk of radius $\sqrt{x_0^2 + y_0^2}$, orthogonal to the z-axis and if the position of the magnetic pole is such that one of these points is on the surface of the cone, there will be a concentration of electrons emitted by the periphery of the cathode and even, approximately, by the whole disk: this is the *focusing* effect observed by Birkeland.

This is an important result because the Poincaré equation [Eq. (1.2)] and the integral of motion [Eq. (1.4)] may be seen as experimentally verified because, for electrons falling on a fixed monopole, it is proved by the Birkeland effect. Conversely, for monopoles falling on a fixed-coulomb electric charge, it is implicitly proved by the simple fact that the interacting force is the same for electricity and magnetism. Consequently, the Poincaré equation remains true.

In Eq. (1.4), the first term is clearly the orbital momentum of the electron with respect to the magnetic pole. The second term was later interpreted by J. J. Thomson (Thomson, 1904 and Lochak, 1995b), who showed that

$$\frac{eg}{c}\frac{\mathbf{r}}{r} = \frac{1}{4\pi c}\int_{-\infty}^{\infty} \mathbf{x} \times (\mathbf{E} \times \mathbf{H})\, d^3x \tag{1.10}$$

Thus, with the value of l given in Eq. (1.2), the second term of the Poincaré integral is equal to the electromagnetic momentum and Eq. (1.4) gives the *constant total angular momentum* $\mathbf{J} = m\mathbf{\Lambda}$. The presence of a nonvanishing electromagnetic angular momentum is due to the axial character of the magnetic field created by a magnetic pole and acting on a scalar electric charge.

Let us add here a remark about symmetry (Lochak, 1997a, b): the Poincaré cone is enveloped by a vector $\mathbf{r}$, which is the *symmetry axis* of the system formed by the electric and the magnetic charges, and this axis rotates (with a constant angle Θ') around the constant *angular momentum* $\mathbf{J} = \mathrm{m}\mathbf{\Lambda}$. But this is exactly the definition of the *Poinsot cone* associated with a symmetric top.

The identity of the Poincaré cone and the Poinsot cone of a symmetric top is not surprising because the system formed by electric and magnetic charges is axisymmetric and rotating around a fixed point with a constant total angular momentum, just like a top, but with a different radial motion because it is not rigid. Hence, the motion along the geodesic lines of the cone has nothing to do with a top.

Let us introduce the following definition, which has two obvious properties:

$$\mathbf{L} = \mathbf{r} \times \frac{d\mathbf{r}}{dt}; \quad L.\frac{\mathbf{r}}{r} = 0; \quad \Lambda.\frac{\mathbf{r}}{r} = \lambda. \tag{1.11}$$

Figure 1.2 summarizes all of these points.

All the calculations and interpretations of Poincaré (1896) concerning an electric charge (a cathodic ray—that is. an electron) in the field of a magnetic pole are also right for a magnetic charge (a monopole) in the field of a coulombian electric pole. The cause of this is the symmetry of Coulomb's law between electricity and magnetism. We shall see later in this chapter, that this will be true in the case of our quantum equation for a magnetic monopole, which gives, at the classical limit, the Poincaré equation.

Consider another point: All the reasonings of Poincaré concerning the convergence phenomenon of cathodic rays observed by Birkeland are independent of the sign of magnetic charges, as Poincaré claimed, because his description depends only on the half-angle Θ' of the cone, which is defined by Eq. (1.8). Actually, by virtue of Eqs. (1.2) and (1.8), this angle depends on $V/\lambda = V\, mc/eg$, but an inversion of the sign of this ratio could be compensated by an inversion of time. Therefore, the crossing points between the trajectory and the angulat momentum would be same.

Nevertheless, the sign of charges appears in the rotation sense of the spiral trajectory of an electron along the cone, because the rotation of an electron (or of a monopole) around the cone is left or right according to the sign of V/λ. This is the unique echo of the opposite variances of electric and magnetic charges, which only quantum mechanics is able to describe clearly.

1.2 P. A. M. Dirac

Dirac (1931) asked the following question: "Why are all electric charges multiples of the same unit charge?". He considered exactly the same problem as Poincaré (the interaction between an electric charge and a fixed

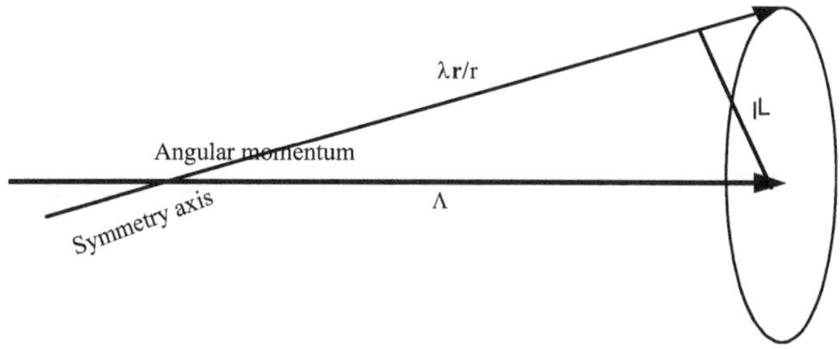

Figure 1.2 The generation of the Poincaré (or Poinsot) cone and the decomposition of the total momentum.

magnetic pole), but in quantum terms. This problem is exactly the same as the motion of a light magnetic monopole in the vicinity of a fixed electric charge. But there is a great difference: contrary to Poincaré, who knew the equation in classical mechanics, Dirac didn't know the quantum equation. We shall answer this question later in this discussion.

Here, we consider, as Dirac did, the motion of an electric charge e in the field of a fixed magnetic monopole with a charge g. The field **H** is thus defined by a vector potential **A** such that

$$curl\ \mathbf{A} = g\frac{\mathbf{r}}{r^3}. \tag{1.12}$$

It is clear that there is no continuous and uniform solution **A** of this differential equation because if we consider a surface S bounded by a loop Λ, we find according to the Stokes theorem:

$$\int_{\Sigma} \mathbf{H}.d\mathbf{S} = \int_{\Sigma} curl\mathbf{A}.d\mathbf{S} = \int_{\Lambda} \mathbf{A}.d\mathbf{l} = g\int_{\Sigma} \frac{\mathbf{r}}{r^3}.d\mathbf{S} = g\int_{\Sigma} d\Omega, \tag{1.13}$$

where d**S**, d**l**, and dΩ are elements of surface, length, and solid angle, respectively. Now, if the loop is shrunken to a point, while the pole remains inside the closed surface S, we get

$$\int_{\Lambda\to 0} \mathbf{A}.d\mathbf{l} = g\int_{\Sigma} d\Omega = 4\pi g. \tag{1.14}$$

This equality is impossible for a continuous potential **A** because the first integral vanishes. There must be a singular line around which the loop shrinks. Now, whatever the wave equation, the minimal coupling is given by a covariant derivative:

$$\nabla - i\frac{e}{\hbar c}\mathbf{A} \tag{1.15}$$

Dirac introduced into the wave function ψ a nonintegrable (nonunivalent) phase γ defining a new wave function:

$$\psi = e^{i\gamma}\psi. \tag{1.16}$$

If we apply the preceding operator [Eq. (1.15)], we know that the introduction of this phase γ is equivalent to the introduction of a new potential by a change of electromagnetic gauge:

$$\left(\nabla - i\frac{e}{\hbar c}\mathbf{A}\right)\psi = e^{i\gamma}\left(\nabla + i\nabla\gamma - i\frac{e}{\hbar c}\mathbf{A}\right)\psi. \tag{1.17}$$

We can identify the new potential with the gradient of γ, but the phase factor $e^{i\gamma}$ is admissible only if the variation of γ around a closed loop equals a multiple of 2π. Then, we must have

$$\frac{e}{\hbar c}\int_{\Lambda\to 0} \mathbf{A}\cdot d\mathbf{l} = \int_{\Lambda\to 0} \nabla\gamma\cdot d\mathbf{l} = (\Delta\gamma)_{loop} = 2\pi n. \tag{1.18}$$

Comparing Eqs. (1.14) and (1.18), we find the *Dirac condition* between electric and magnetic charges:

$$\frac{eg}{\hbar c} = \frac{n}{2}. \tag{1.19}$$

It is interesting to confirm this result on a solution of Eq. (1.12). Dirac chose the following solution:

$$A_x = \frac{g}{r}\frac{-y}{r+z}, \quad A_y = \frac{g}{r}\frac{x}{r+z}, \quad A_z = 0, \quad r = \sqrt{x^2+y^2+z^2}. \tag{1.20}$$

In polar coordinates, the solution is

$$x = r\sin\theta\cos\varphi, \quad y = r\sin\theta\sin\varphi, \quad z = r\cos\theta. \tag{1.21}$$

Eq. (1.20) becomes

$$A_x = \frac{g}{r}\tan\frac{\theta}{2}\sin\varphi, \quad A_x = \frac{g}{r}\tan\frac{\theta}{2}\cos\varphi, \quad A_z = 0. \tag{1.22}$$

There is a nodal line that goes from $z = 0$ to $z = \infty$ for $\theta = \pi$, and the Dirac condition is easily found if we compute the curvilinear integral [Eq. (1.18)] around this line for $\theta = \pi - \varepsilon$ and and $\varepsilon \to 0$. We must have

$$\frac{e}{\hbar c} = \int_{\Lambda\to 0} \mathbf{A}\cdot d\mathbf{l} = \frac{eg}{\hbar c}\int_{\theta=\pi-\varepsilon,\ \varepsilon\to 0} \frac{1}{r}\tan\frac{\theta}{2} r\sin\theta d\varphi = 2\pi n. \tag{1.23}$$

Therefore,

$$\frac{eg}{\hbar c}\int_{\varepsilon\to 0} \frac{\sin\varepsilon}{\tan\frac{\varepsilon}{2}} d\varphi = \frac{eg}{\hbar c} 2\times 2\pi = 2\pi n. \tag{1.24}$$

Here, we see that the factor 2 comes from the factor $\varepsilon/2$ in the tangent, and we could conclude from that that it is related to the fact that the nodal line begins at $\mathbf{r} = 0$. But this is wrong because the solution Eq. (1.20) or Eq. (1.22) chosen by Dirac depends on an arbitrary gauge; and in addition, his choice is not very good because *his potential has no definite parity*. Moreover, it must be stressed that with a *polar vector* **A**, the vector curl **A** is axial, so that Eq. (1.12) would be admissible only with a *pseudo-scalar constant g*, against which we have already objected. In the following discussion, we shall give the wave equation of a monopole in an electromagnetic field; our potential will not be **A**, but the pseudo-potential **B**, which will be a solution of the following equation (where e is the scalar electric charge):

$$curl\ \mathbf{B} = e\frac{\mathbf{r}}{r}. \tag{1.25}$$

B must be an axial vector, which is evident in Eq. (1.25), because *curl* **B** must be polar like **r**. *Mutatis mutandis*, Dirac's reasoning presented here will be true if we choose an axial solution of Eq. (1.25):

$$B_x = \frac{e}{r}\frac{yz}{x^2+y^2},\quad B_y = \frac{e}{r}\frac{-xz}{x^2+y^2},\quad B_z = 0,\quad r = \sqrt{x^2+y^2+z^2}. \tag{1.26}$$

This solution differs from the Dirac-like solution, which would be

$$B'_x = \frac{g}{r}\frac{-y}{r+z},\quad B'_y = \frac{g}{r}\frac{x}{r+z},\quad B'_z = 0. \tag{1.27}$$

In this, $\mathbf{B}'$ differs from **B** only by a gauge:

$$B - \mathbf{B}' = \nabla \arctan\frac{y}{x}. \tag{1.28}$$

In polar coordinates, Eq. (1.26) becomes

$$B_x = \frac{e}{r}\frac{\sin\varphi}{\tan\theta},\quad B_y = \frac{e}{r}\frac{-\cos\varphi}{\tan\theta},\quad B_z = 0. \tag{1.29}$$

Using Eq. (1.26) or (1.29) in Dirac's proof of the relation [Eq. (1.19)], the singular line goes from $-\infty$ to $+\infty$ instead of from 0 to $+\infty$ and the equality (1.24) becomes

$$2 \times \frac{eg}{\hbar c}\int_{\varepsilon\to 0}\frac{\sin\varepsilon}{\tan\varepsilon}\,d\varphi = 2\times\frac{eg}{\hbar c}2\pi = 2\pi n. \tag{1.30}$$

This result gives Eq. (1.19) again, but now the factor 2 is no longer due to $\tan\ \varepsilon/2$, but due to the fact that the singular line pierces the sphere in *two points*. Therefore, the factor $n/2$ in the Dirac formula [Eq. (1.19)] was not at all related to the fact that the singular line began in $r = 0$. Nevertheless, this answer is not good either, and we shall prove further that *the factor n/2 is actually a consequence of the double connexity of the rotation group*.

According to Eq. (1.19), if we choose the charge e of the electron as a unit electric charge, the magnetic charge is quantized. For $n = 1$, we obtain the unit magnetic charge as a function of the electron charge and of the fine structure constant:

$$g_0 = \frac{\hbar c}{2e^2} e = \frac{e}{2\alpha} = \frac{137}{2} e = 68.5e. \tag{1.31}$$

This is a large charge, which is of the same order as the electric charge of a nucleus in the region of lantanides, beyond the middle of Dmitri Mendeleev's classification (137e is even beyond the classification). Nevertheless, this does not mean that such a monopole interacts with atoms as strongly as an electric charge of the same order. On the contrary, it must be stressed that all the experiments on monopoles are performed directly in the atmosphere of the laboratory, often at distances of several meters that cannot be crossed, for instance, by electrons. It can be undertood from the formula [Eq. (1.8)] of Poincaré, which shows that the total Lagrange moment increases with the Poincaré constant λ (proportional to the magnetic charge, as will be confirmed in quantum mechanics), the vertex angle Θ' of the cone decreases with the charge because it varies as λ^{-1}. Finally, it is the angle Θ' that gives the deviation of monopoles by an electric charge.

It is noteworthy that Dirac's condition [Eq. (1.19)] is based on general assumptions of quantum mechanics and electromagnetism, which is confirmed (despite some differences) by our equation (1.30). Nevertheless, we cannot forget that it was not systematically proved—and indeed, it has even been contradicted by many authors. For instance, we have already quoted the systematic, but contradictory, experiments of Mikhailov (Mikhailov, 1985, 1987, 1993). A paper of Price and colleagues (Price *et al.*, 1975) also identifies a track as being either one of a heavy nucleus, or of a monopole with a Dirac charge. And we remember the well-known measure of Blas Cabrera that gave the Dirac charge (Cabrera, 1982), but it was an "irreproducible result."

1.3 Pierre Curie

Among the symmetry laws stated by Pierre Curie, there is at least one that is well known and applied even by many who don't know that he was the first who stated it, at the beginning of his memoir, (Curie, 1894a,b)[1]:

[1] In Lochak (1997a, b), part of the Curie paper is given in a modern form, with consequences for the charges, electromagnetic potentials, and quantum mechanics that will be given later in the book.

> When some causes produce some effects, the elements of symmetry of causes must be found in the produced effects

Reciprocally, it is evident that

> If some effects reveal some asymmetry, this dissymmetry must be found in the causes that gave rise to these effects.

These laws are only two introductive lines of Curie's great memoir, which plays an essential role in what has followed it because it is essentially devoted to electromagnetism. But, as it was said in the Foreword, we shall follow this memoir only for a few pages, to give a foundation to some definitions. Then, we shall use more modern language and introduce some extensions.

The Spatial Symmetry of an Electric Field

Consider an electric field generated by two parallel coaxial circular plates of different metals. It has the symmetry of the cause: a revolution field around the axis, and every plane passing it will be a plane of symmetry. This is the symmetry of a *truncated cone,* but not yet of a *cone,* because the symmetry could be greater (cylindrical or spherical).

To find the exact symmetry, Curie takes a conductive, electrically charged sphere in a uniform electric field: "A force will act on the sphere in the direction of the field." The asymmetry of the effect must be found in the cause: the force exerted on the sphere has no symmetry axis normal to its direction, so the system sphere-field (the cause) no longer has such an axis. On the other hand, the sphere has infinite axes of symmetry, such that the cause of assymmetry is not in the sphere but in the field itself. Conclusion: the electric field cannot have a cylindrical or a spherical symmetry and it has the symmetry of a cone and the field may be represented by a polar vector (in $\mathbb{R}^3$). The same is true for a current or an electric polarization.

The Spatial Symmetry of a Magnetic Field

Consider the magnetic field generated at the center of a circular wire carrying a permanent current. The axis of the wire is an axis of *isotropy* and the plane of the circle is a *plane of symmetry*. Therefore, a magnetic field has a plane of symmetry normal to the direction of the field[2].

[2] This paradoxical symmetry is curiously represented on a painting of René Magritte: *La reproduction interdite*, which shows a man before a mirror who turns his back to the viewer. His image in the mirror turns his back too—just like a magnetic field!

On the other hand, the field has no binary normal axis, for the following reason. Take a rectilinear *conductive* bar moving normally along its length. This moving bar has a binary axis parallel to its velocity. Now, let us introduce a magnetic field normal to the bar and to the velocity: an electromotive force is generated in the bar, normal to it, and the binary axis disappears. Therefore, this axis must be absent from the cause, which means that a magnetic field has no orthogonal binary axis: it has the symmetry of a rotating cylinder. It may be represented by an axial vector (in $\mathbb{R}^3$). The same is true for a magnetic current or a magnetic polarization. Maxwell already knew that (Maxwell, 1873), without speaking of symmetry.

Now, from the reasoning of Pierre Curie, we can easily deduce the symmetry of charges, which is not given in his papers. Let us take the preceding circular electrically charged plates. A symmetry with respect to a parallel and equidistant plane will exchange between themselves the plates and the charges. Are the latter modified or not? We don't know it a priori, but we know that the electric field between the plates will be reversed. Thus, the electric charges are not changed: electric charges e are P-invariant. The conclusion would be the opposite for magnetic charges because in a similar experiment, we see that the reflected magnetic field is not changed. Therefore, magnetic charges g are P-reversed:

$$P: \mathbf{E}\rightarrow -\mathbf{E};\quad \mathbf{H}\rightarrow\mathbf{H};\quad e\rightarrow e;\quad g\rightarrow -g. \tag{1.32}$$

We shall see later in this chapter that these conclusions are confirmed in quantum mechanics for $\mathbf{E}$, $\mathbf{H}$, and e, but not for g, at least in this formulation. Such a change of the sign of a physical constant, like g, would be astonishing because it would signify that the constant g is a pseudo-scalar: a unique case in physics, *while all the other constants are true scalars* (see the Foreword). We shall see that this is not the case in quantum mechanics, but in the meanwhile, we shall keep the classical variance in another form.

Time Symmetry of Electromagnetic Fields

Curie didn't speak of time symmetry, which was not considered in his time. We shall start from the Lorentz force exerted by a field $\mathbf{E}$, $\mathbf{H}$ on an electric or a magnetic charge:[3]

$$\mathbf{F}_{elec} = e(\mathbf{E} + (1/c)\,\mathbf{v}\times\mathbf{H});\quad \mathbf{F}_{magn} = g(\mathbf{H} - (1/c)\,\mathbf{v}\times\mathbf{E}) \tag{1.33}$$

[3] The formula for $\mathbf{F}_{magn}$ is easily found by applying the Lorentz transformation to the law $\mathbf{F} = g\mathbf{H}$ in the proper system.

These formulas cannot be contradicted by quantum mechanics because they must be found again at the geometrical optic limit. This is not enough to define variances, but it must be implicitly connected with them.

Now **F** is T-invariant (because $\mathbf{F} = m\gamma$), and **v** changes its sign with *t*, so we have, from Eq. (1.33):

$$T : e\mathbf{E} \rightarrow e\mathbf{E};\ \ e\mathbf{H} \rightarrow -e\mathbf{H};\ \ g\mathbf{H} \rightarrow g\mathbf{H};\ \ g\mathbf{E} \rightarrow -g\mathbf{E}. \tag{1.34}$$

Thus we have two possible variances:

$$\begin{array}{l} T_I : \mathbf{E} \rightarrow \mathbf{E},\ \ \mathbf{H} \rightarrow -\mathbf{H};\ \ e \rightarrow e, g \rightarrow -g \\ T_{II} : \mathbf{E} \rightarrow -\mathbf{E}, \mathbf{H} \rightarrow \mathbf{H};\ \ e \rightarrow -e, g \rightarrow g \end{array} \tag{1.35}$$

Such a case often happens: the electrodynamical phenomena only give a choice because they are able to define a link between the variances of several physical quantities, but not the variance of each quantity. It does not allow any possibility of an arbitrary choice[4]. Actually, in order to find the precise variances, we need some other phenomena, purely electric or purely magnetic (Curie, 1894a,b).

In this case, one can verify that to choose between the two possible laws [Eq. (1.35)], it is enough to find the variance of only one of the quantities $\mathbf{E}$, $\mathbf{H}$, e, g. We choose an *electrochemical phenomenon:* cathions heading to the anode with a current density: $\mathbf{J} = \rho\mathbf{v}$ (ρ is the density of cathions and **v** their velocity). Let us reverse the sign of time *t;* we do not know if the sign of charges is reversed, but in every case, the sign of ions and of the electrode remain opposite. Now, the sign of the velocity **v** is reversed; therefore, to conserve the density of current **J**, the sign of the electric charge must be reversed. Therefore, Law T_{II} is good and must be chosen.

Charge Conjugation and P, T, C Variances

In the forces [Eq. (1.33)], the fields **E** and **H** are exterior. Thus, they are independent of the charges *e* and *g* to which these fields are applied. But if a charge is reversed, the force is reversed, and thus we get

$$C : \mathbf{E} \rightarrow \mathbf{E},\ \ \mathbf{H} \rightarrow \mathbf{H},\ \ e \rightarrow -e,\ \ g \rightarrow -g. \tag{1.36}$$

Now, we can gather Eqs. (1.32), (1.35), and (1.36) into the **P**, **T**, **C** variances of fields and charges. As a result, we get the following table:

[4] For instance, such a choice is suggested in Jackson (1975, p. 249): "It is natural, convenient, and permissible to assume that charge is also a scalar under spatial inversion and even under time reversal." Of course, this is not an argument—and even if it were, it is wrong!

$$(I)\begin{bmatrix} P: \mathbf{E}\to -\mathbf{E}, & \mathbf{H}\to\mathbf{H}, & e\to e, & g\to -g \\ T: \mathbf{E}\to -\mathbf{E}, & \mathbf{H}\to\mathbf{H}, & e\to -e, & g\to g \\ C: \mathbf{E}\to\mathbf{E}, & \mathbf{H}\to\mathbf{H}, & e\to -e, & g\to -g \end{bmatrix}. \tag{1.37}$$

It must be emphasized that these P, T, C variances are directly deduced from experimental facts and from the laws of force [Eq. (1.33)], which are direct consequences of electromagnetism and relativity (and both are experimentally verified).

Symmetries of Electromagnetic Potentials

These symmetries are deduced from the definition of the electromagnetic fields **E** and **H**, which are related to the Lorentz potentials V and **A** or to the pseudopotentials W and **B**, which we cover later in this chapter[5]. W and **B** are the potentials "seen" by a magnetic pole, just as V and **A** are seen by an electric pole. Thus, we have two possible notations, for the electric case and for the magnetic case, respectively:

$$\mathbf{E} = -\nabla V - \frac{1}{c}\frac{\partial \mathbf{A}}{\partial t}, \quad \mathbf{H} = curl\,\mathbf{A}, \quad or: \mathbf{E} = curl\,\mathbf{B}, \quad \mathbf{H} = \nabla W + \frac{1}{c}\frac{\partial \mathbf{B}}{\partial t}. \tag{1.38}$$

From Eq. (1.37), we find the **P**, **T**, **C** variances of the potentials:

$$(II)\begin{bmatrix} P: \mathbf{A}\to -\mathbf{A}, & V\to V, & \mathbf{B}\to\mathbf{B}, & W\to -W, & e\to e, & g\to -g \\ T: \mathbf{A}\to\mathbf{A}, & V\to -V, & \mathbf{B}\to -\mathbf{B}, & W\to W, & e\to -e, & g\to g \\ C: \mathbf{A}\to\mathbf{A}, & V\to V, & \mathbf{B}\to\mathbf{B}, & W\to W, & e\to -e, & g\to -g \end{bmatrix}. \tag{1.39}$$

Let us make some remarks about these laws at this point:

a. The Lorentz transformation gathers the vector and scalar potentials $(\mathbf{A}, V)$ and $(\mathbf{B}, W)$, defined in $\mathbb{R}^3$, into two space-time quadrivectors:

$$A_\mu = (\mathbf{A}, iV); \quad iB_\mu = (\mathbf{B}, iW) \tag{1.40}$$

It is easy to introduce Eq. (1.39) into these expressions and to prove that A_μ and B_μ are polar and axial vectors in space and time, respectively (this is why there is an i before B_μ and not before A_μ).

b. The laws [Eq. (1.39)] give good $(P \;\; or \;\; T)$ variances $\mathbf{P}\to -\mathbf{P}, \;\; E\to E$ for the Lagrange momenta:

[5] Here, we retain the notation **B**

$$\mathbf{P} = \mathbf{p} + \frac{e}{c}\mathbf{A},\quad E = mc^2 + eV \text{ and } \mathbf{P} = \mathbf{p} + \frac{g}{c}\mathbf{B},\quad E = mc^2 + gW \tag{1.41}$$

c₁. One can verify that the laws [Eq. (1.37)] ensures the invariance of the Maxwell equations:

$$\begin{aligned} &-\frac{1}{c}\frac{\partial \mathbf{H}}{\partial t} = curl\,\mathbf{E},\quad \frac{1}{c}\frac{\partial \mathbf{E}}{\partial t} = curl\,\mathbf{H},\quad div\,\mathbf{H} = 0,\quad div\,\mathbf{E} = 0 \\ &\mathbf{H} = curl\,\mathbf{A},\quad \mathbf{E} = -\mathbf{grad}\,V - \frac{1}{c}\frac{\partial \mathbf{A}}{\partial t},\quad \frac{1}{c}\frac{\partial V}{\partial t} + div\,\mathbf{A} = 0. \end{aligned} \tag{1.42}$$

c₂. The laws [Eqs. (1.37) and (1.39)] ensure the invariance of the de Broglie equations of light, including the potentials (de Broglie, 1934):

$$\begin{aligned} &-\frac{1}{c}\frac{\partial \mathbf{H}}{\partial t} = curl\,\mathbf{E},\quad \frac{1}{c}\frac{\partial \mathbf{E}}{\partial t} = curl\,\mathbf{H} + k_0^2\mathbf{A} \\ &\quad div\,\mathbf{H} = 0,\quad div\,\mathbf{E} = -k_0^2 V \\ &\mathbf{H} = curl\,\mathbf{A},\quad \mathbf{E} = -\mathbf{grad}\,V - \frac{1}{c}\frac{\partial \mathbf{A}}{\partial t},\quad \frac{1}{c}\frac{\partial V}{\partial t} + div\,\mathbf{A} = 0 \end{aligned} \tag{1.43}$$

c₃. Finally, the same laws [Eqs. (1.37) and (1.39)] ensure the invariance of *the equations of the magnetic photon* to which we already alluded. We shall return to it later, more precisely (Lochak, 1995a,b, 2003). The role of the potentials is played by the pseudopotentials as follows:

$$\begin{aligned} &-\frac{1}{c}\frac{\partial \mathbf{H}}{\partial t} = curl\,\mathbf{E} + k_0^2\mathbf{B},\quad \frac{1}{c}\frac{\partial \mathbf{E}}{\partial t} = curl\,\mathbf{H} \\ &\quad div\,\mathbf{H} = k_0^2 W,\quad div\,\mathbf{E} = 0 \\ &\mathbf{H} = grad W + \frac{1}{c}\frac{\partial \mathbf{B}}{\partial t},\quad \mathbf{E} = curl\,\mathbf{B},\quad \frac{1}{c}\frac{\partial W}{\partial t} + div\,\mathbf{B} = 0. \end{aligned} \tag{1.44}$$

The Curie symmetries, in quantum mechanics, will be given later, and that discussion will provide a stronger basis to the CPT symmetries, where differences with some accepted principles appear. An important result of Tables (I) and (II) above (which is absent from Curie's results, but which was deduced owing to his methods) is that the electric charge e is *P-invariant*, but *T-reversed,* and that the inverse is true for the magnetic charge g. And this will be true in quantum mechanics.

Let us make a conclusive remark concerning Maxwell and Curie. It is well known that, in all domains, important ideas may be lost for a long time. In the domain of symmetry, we face a phenomenon of this kind. Despite the fact that modern physics is dominated by symmetry, such great pioneers as Maxwell and Curie knew some results in electromagnetism that now have been more or less forgotten.

CHAPTER 2

A Wave Equation for a Leptonic Monopole, Dirac Representation

As was stated in the Foreword, our theory is not based on the Dirac works on monopoles, but on his famous theory of the electron. Our theory is based on two main points:

- The massless Dirac equation has a second gauge invariance, which defines a second electromagnetic interaction that obeys the laws of a magnetic monopole and the symmetry laws predicted by Pierre Curie. The monopole and the anti-monopole are chiral particles that are mirror images, as are the neutrino and the antineutrino, but here, it is true for magnetically charged particles, as it was predicted more than a century ago by Curie (Curie, 1894a, b, 1994).
- Contrary to other theories, our theory predicts that such a monopole is associated not with strong interactions, but with weak ones. And contrary to these other theories, the prediction is confirmed by experimentation.

There are naturally two paths for the theory, following either Dirac or Weyl. This chapter is devoted to the *Dirac representation*, while the next one will be devoted to the *Weyl representation*.

2.1 THE TWO GAUGE INVARIANCES OF DIRAC'S EQUATION

Consider the Dirac equation without an external field:

$$\gamma_\mu \partial_\mu \psi + \frac{m_0 c}{\hbar} \psi = 0, \tag{2.1}$$

where: $x_\mu = \{x_k; ict\}$ are the relativistic coordinates and the matrices γ_μ are expressed through the following Pauli s_k matrices:

$$\gamma_k = i\begin{pmatrix} 0 & s_k \\ -s_k & 0 \end{pmatrix};\ k = 1,2,3;\ \gamma_4 = \begin{pmatrix} I & 0 \\ 0 & -I \end{pmatrix};$$
$$\gamma_5 = \gamma_1\gamma_2\gamma_3\gamma_4 = \begin{pmatrix} 0 & I \\ I & 0 \end{pmatrix}. \tag{2.2}$$

Now let us define a general form of gauge transformation, where Γ is a constant Hermitian matrix and θ a constant phase:

$$\psi \rightarrow e^{i\Gamma\theta}\psi \tag{2.3}$$

At this point, introduce the gauge [Eq, (2.3)] into Eq. (2.1):

$$\left(\gamma_\mu e^{i\Gamma\theta}\gamma_\mu\right)\gamma_\mu\partial_\mu\psi + \frac{m_0 c}{\hbar}e^{i\Gamma\theta}\psi = 0. \tag{2.4}$$

Now, develop Γ on Clifford algebra as follows:

$$\Gamma = \Sigma_{N=1}^{16} a_N\Gamma_N;\quad \Gamma_N = I, \gamma_\mu, \gamma_{[\mu}\gamma_{\nu]}, \gamma_{[\lambda}\gamma_\mu\gamma_{\nu]}, \gamma_5, \tag{2.5}$$

and remember the relation $\gamma_\mu\Gamma_N\gamma_\mu = \pm\Gamma_N$ (Pauli, 1936), where $(\pm)$ depends on μ and N. We have

$$\gamma_\mu e^{i\Gamma\theta}\gamma_\mu = e^{i\theta\ \Sigma_{N=1}^{16} a_N\gamma_\mu\Gamma_N\gamma_\mu} = e^{i\theta\ \Sigma_{N=1}^{16} \pm a_N\Gamma_N}. \tag{2.6}$$

Eq. (2.1) remains invariant under the transformation [Eq, (2.4)] if Γ commute or anticommute with all the γ_μ; thus, we must have, in the last term of Eq. (2.6), either + or − before Γ_N for all γ_μ. We find $\Gamma = I$, for the plus sign, and $\Gamma = \gamma_5$ for the minus sign, and no other possibility. So

$$\Gamma = I \Rightarrow \psi \rightarrow e^{i\theta}\psi \ \ or \ \ \Gamma = \gamma_5 \Rightarrow \psi \rightarrow e^{i\gamma_5\theta}\psi. \tag{2.7}$$

The great difference is as follows:

- In the first case, $\Gamma = I$ commutes with the γ_μ: Eq. (2.4) is identical to Eq. (2.1), which is invariant under the transformation [Eq. (2.3)]. And we have defined the phase invariance, and $\psi \rightarrow e^{i\theta}\psi$ for any value of m_0, and we know that this ensures the conservation of charge.
- In the second case, $\Gamma = \gamma_5$ anticommutes with the γ_μ, so that the differential term in Eq. (2.1) has a minus sign in the exponential, while a plus sign remains in the exponential of the mass term. Therefore, the transformation $\psi \rightarrow e^{i\gamma_5\theta}\psi$ defines a gauge invariance only for a massless particle, at least for a linear equation (we shall see later in this chapter that things become different for nonlinear equations).

But, even with the nonlinear case, the symmetry has not broken. It has just become another symmetry: a *chiral symmetry,* which knows the difference between left and right, as was the case for magnetism (Maxwell, 1873; Curie, 1894a). We went from an electric particle, like an electron, to a magnetic monopole (Curie, 1894b, Lochak, 1985, 1995a,b, 2006).

It may be asserted that a monopole is not necessarily a super-heavy scalar: it can be a massless pseudoscalar. Indeed, we shall prove that the chiral invariance entails the conservation of magnetism, but with some important differences with respect to the conservation of electricity:

1. The conservation of magnetism is weaker than the conservation of electricity because its conservation is broken by the introduction of a linear mass term in the equation. Despite some analogies, the equations for an electron and a monopole are very different because of their different gauge laws.
2. The second difference is that, in (2.7): θ is a scalar phase for an electron and a pseudoscalar for a magnetic monopole. This is because γ_5 is a pseudoscalar operator, which implies two different mathematical worlds.

2.2 THE EQUATION OF THE ELECTRON

The Dirac equation of the electron ensues from the first transformation [Eq. (2.7)] generalized by a local gauge, in which the abstract angle θ is replaced by a physical angle φ with physical coefficients:

$$\psi \rightarrow e^{i\frac{e}{\hbar c}\varphi}\psi \tag{2.8}$$

So, introducing Eq. (2.8) in the differential term of Eq. (2.1), we find (up to the exponential factor)

$$\partial_\mu \psi \rightarrow e^{i\frac{e}{\hbar c}\varphi}\left(\partial_\mu \Psi + i\frac{e}{\hbar c}\partial_\mu \varphi\, \psi\right). \tag{2.9}$$

Now, we can generalize Eq. (2.8) by the adjunction of a potential:

$$\psi \rightarrow e^{i\frac{e}{\hbar c}\varphi}\psi; \quad \mathrm{A}_\mu \rightarrow \mathrm{A}_\mu + \partial_\mu \varphi. \tag{2.10}$$

Owing to Eqs. (2.9) and (2.10), Eq, (2.1) may be replaced by the following equation:

$$\gamma_\mu\left(\partial_\mu - i\frac{e}{\hbar c}\mathrm{A}_\mu\right)\psi + \frac{m_0 c}{\hbar}\psi = 0, \tag{2.11}$$

which is the Dirac equation of the electron in the presence of an electromagnetic field deriving from a Lorentz potential A_μ and which is invariant under the local gauge transformation [Eq. (2.10)]. The gauge transformation is local because it depends on space and time through an external electromagnetic field deriving from the potential A_μ.

Eq. (2.11) implicitly defines a *minimal coupling* and a *covariant derivative:*

$$\nabla_\mu = \partial_\mu - i\frac{e}{\hbar c}A_\mu \tag{2.12}$$

In the gauge transformation [Eq, (2.10)], the Lorentz potential A_μ is a polar vector and φ a scalar angle.

2.3 THE SECOND GAUGE, THE SECOND COVARIANT DERIVATIVE, AND THE EQUATION FOR A MAGNETIC MONOPOLE

Now, consider the Dirac equation [Eq. (2.1)] with $m_0 = 0$:

$$\gamma_\mu \partial_\mu \psi = 0. \tag{2.13}$$

This equation is invariant under both gauges [Eq. (2.7)]. We shall now examine the second one in the local case; i.e., with a pseudoscalar phase φ depending on the coordinates:

$$\psi \rightarrow e^{i\frac{g}{\hbar c}\gamma_5 \varphi}\psi. \tag{2.14}$$

Introducing the transformation [Eq. (2.14)] in Eq. (2.13), we find $\gamma_\mu\left(\partial_\mu + i\frac{g}{\hbar c}\gamma_5\partial_\mu\varphi\right)\psi = 0$, which suggests a new minimal electromagnetic coupling by substituting the gradient of the pseudophase ϕ by the only possible potential, which is the pseudopotential defined in Eq. (1.40): $iB_\mu = (B, iW)$, from which we get a new covariant derivative:

$$\nabla_\mu = \partial_\mu - \frac{g}{\hbar c}\gamma_5 B_\mu. \tag{2.15}$$

In Eq. (2.15), i disppears because of the pseudoscalar character of γ_5. Finally, we find an new equation, which is the equation of a magnetic monopole (Lochak, 1983, 1984, 1985):

$$\gamma_\mu\left(\partial_\mu - \frac{g}{\hbar c}\gamma_5 B_\mu\right)\psi = 0. \tag{2.16}$$

This equation is relativistically invariant and gauge invariant, under the pseudoscalar transformation (with the same comment about i):

$$\psi \rightarrow e^{i\frac{g}{\hbar c}\gamma_5 \varphi}\psi;\quad B_\mu \rightarrow B_\mu + i\partial_\mu \varphi. \tag{2.17}$$

It will be proved later that the magnetic charge g is a scalar and not a pseudoscalar, which does not contradict Curie's laws because the pseudoscalar character of magnetism is not related to the number g but to the pseudoscalar magnetic charge operator $C = g\gamma_5$; i.e., to the pseudoscalar matrix γ_5. This matrix lies at the origin of the difference between classical and the quantum theories of magnetic monopoles.

2.4 THE DIRAC TENSORS AND THE "MAGIC ANGLE" *A* OF YVON-TAKABAYASI (FOR THE ELECTRIC AND THE MAGNETIC CASE)

It is known that in the Clifford basis [Eq. (2.5)], the Dirac spinor defines 16 bilinear tensorial quantities: a scalar, a polar vector, an antisymmetric tensor of rank 2, an antisymmetric tensor of rank 3 (an axial vector), and an antisymmetric tensor of rank 4 (a pseudoscalar):

$$\omega_1 = \overline{\psi}\psi; \;\; J_\mu = i\overline{\psi}\gamma_\mu\psi; \;\; M_{\mu\nu} = i\overline{\psi}\gamma_\mu\gamma_\nu\psi; \;\; \Sigma_\mu = i\overline{\psi}\gamma_\mu\gamma_5\psi; \;\; \omega_2 = -i\overline{\psi}\gamma_5\psi$$
$$\left(\overline{\psi} = \psi^+\gamma_4; \;\; \psi^+ = \psi \;\; h.c.\right). \tag{2.18}$$

If ω_1 and ω_2 do not vanish simultaneously, the Dirac spinor may be written as follows (Jacobi & Lochak, 1956a,b):

$$\psi = \rho\, e^{i\gamma_5\mathbf{A}} U\psi_O. \tag{2.19}$$

where ρ = amplitude, U = general Lorentz transformation, ψ_O = constant spinor, and A = the *pseudoscalar angle* of Yvon-Takabayasi:

$$\rho = \sqrt{\omega_1^2 + \omega_2^2}; \;\; A = \arctan\frac{\omega_2}{\omega_1}. \tag{2.20}$$

In Eq. (2.19), U is a product of six factors $e^{i\Gamma\,\vartheta}$, with three real Euler angles (rotations in R^3) and three imaginary angles (velocities in R^3). So we have seven angles in ψ: (1) three Euler angles, including the proper rotation angle φ, which gives a *half-scalar phase* $\varphi/2$ in the spinor Ψ, is conjugated by a *Poisson bracket* to the component J_4 of the *polar vector* J_μ; (2) the "imaginary three velocities," $i\frac{v_k}{c}$; (3) the *half-pseudoscalar* angle A conjugated to the Σ_4 component of the *axial vector* Σ_μ.

Both vectors J_μ and Σ_μ are defined in Eq, (2.18).

Angle A plays an important role in the Dirac theory of the electron because it appears in the tensor representation based on Eq, (2.18) (Takabayasi, 1957; Jacobi & Lochak, 1956a, b). Without A, the Dirac equation would be an equation of a classical relativistic spin fluid: the quantum properties are concentrated in the magic angle A, which appears in several tensorial equations deduced from the Dirac equation. The role played by the angle A in the theory of the magnetic monopole is even more fundamental.

For a discussion of all these questions, see Jakobi & Lochak (1956a, b), which give the classical-field Poisson brackets already noted in the Foreword and which are at the origin of the present theory of magnetic monopoles:

$$\left[\frac{\varphi}{2}, J_4\right] = \delta(r - r'); \quad \left[\frac{A}{2}, \Sigma_4\right] = \delta(r - r'). \tag{2.21}$$

In the electric case (Dirac theory of the electron), eJ_4 is a density of *electricity* and of *probability*, associated with the *phase invariance;* and the spatial part $e\mathbf{J}$ of eJ_μ is the current density of *electricity* or *probability*. As a result of the gauge invariance defined in Eq. (2.8), the Dirac equation [Eq. (2.11)] of the electron entails the conservation of electricity owing to the conservation of the polar vector eJ_μ:

$$\partial_\mu (eJ_\mu) = \left(\partial_\mu ie\overline{\psi}\gamma_\mu\psi\right) = 0. \tag{2.22}$$

In the magnetic case (equation of the monopole), the *polar electric current density* eJ_μ is replaced by *the axial magnetic current density* $K_\mu = g\Sigma_\mu$. The time and space components (K_4 and $\mathbf{K}$), of K_μ will be the *densities of magnetic charge and of magnetic current,* respectively. As a consequence of the *gauge invariance* [Eq. (2.17)], the equation of the monopole [Eq. (2.16)] entails the conservation of magnetism through the conservation of the axial vector density K_μ:

$$\partial_\mu K_\mu = 0; \quad \left\{K_\mu = g\Sigma_\mu = g\left(i\overline{\psi}\gamma_\mu\gamma_5\psi\right)\right\}. \tag{2.23}$$

Now, it must be noticed that owing to the expressions of J_μ and Σ_μ in terms of Ψ [Eq. (2.18)], one can prove the following:

1. J_μ is *polar,* Σ_μ is an *axial vector* or a *pseudovector:* the definition [Eq, (2.23)] shows that Σ_μ is the dual of a completely antisymmetric tensor of the third rank: $\{i\gamma_2\gamma_3\gamma_4, \ i\gamma_3\gamma_1\gamma_4, i\gamma_1\gamma_2\gamma_4, \gamma_1\gamma_2\gamma_3\}$.
2. J_μ is *timelike* and Σ_μ is *spacelike,* by virtue of the Darwin–de Broglie equalities:

$$-J_\mu J_\mu = \Sigma_\mu\Sigma_\mu = \omega_1^2 + \omega_2^2; \ J_\mu\Sigma_\mu = 0. \tag{2.24}$$

The expression Σ_μ for the magnetic current was already suggested by Salam (1966) for symmetry. But in this case, this is not an a priori definition–rather, it is a consequence of the wave equation and of the second gauge condition [Eq. (2.17)].

Here, it must be stressed that Dirac's theory defines only two vectors, without derivatives: J_μ and Σ_μ. Because J_μ is polar and *timelike,* it may be interpreted as a current density of electricity and probability. Because Σ_μ is axial, it may be interpreted as a current density of magnetism: this double coincidence is a remarkable example of harmony between physics and mathematics. We shall see a little later, in the discussion of the Weyl representation, that the spacelike character of Σ_μ is by no means an objection against its interpretation as a current: it will still reinforce this mathematical harmony[6].

2.5 *P, T, C* SYMMETRIES. PROPERTIES OF THE ANGLE *A* (NOT TO BE CONFUSED WITH THE LORENTZ POTENTIAL A)

Even though we shall be discussing the transformation of the wave function later in this chapter, it is interesting to say here that according to our theory, the *P, T, C* invariances are in perfect accordance with Curie's laws.

In the electric case, the correct transformations given by the P, T, C invariances of the Dirac equation [Eq. (2.11)] are the following, where A_k and A_4 are the Lorentz potentials (Lochak, 1997a, b):

$$\begin{aligned}
&P: e \to e;\ \ x_k \to -x_k;\ \ x_4 \to x_4;\ \ \psi \to \gamma_4\psi \\
&\ \ A_k \to -A_k;\ \ A_4 \to A_4 \\
&T: e \to -e;\ \ x_k \to x_k;\ \ x_4 \to -x_4;\ \ \psi \to -i\gamma_3\gamma_1\psi^* \\
&\ \ A_k \to A_k;\ \ A_4 \to -A_4 \\
&C: e \to -e;\ \ \psi \to \gamma_2\psi^*
\end{aligned} \tag{2.25}$$

where P and C are the Racah transformations (Racah, 1937), but T is not, because, as we have seen in Chapter 1, the electric charge e is reversed by the T transformation, which leads to the antilinear wave transformation $\Psi \to -i\gamma_3\gamma_1\Psi^*$, often known as *weak time reversal*[7].

We shall now adopt this law as the true time reversal. This is always true, including in the case of a magnetic charge, because one can easily prove that the P, T, C invariances of the monopole equation [Eq, (2.16)] are given by

[6] For a long time, Σ_μ was considered the spin vector, because its space components appeared in the Dirac expression of total angular momentum: $-i(x_j\partial_k - x_k\partial_j) + \sigma_k$ $(i, j, k = \textit{circular permutation},$ $\sigma_k = \textit{“spin matrices”})$.

[7] The Racah T transformation, $\psi \to \gamma_1\gamma_2\gamma_3\psi$, contradicts the transformation $e \to -e$.

$$\begin{aligned}
&P: g \to g;\ \ x_k \to -x_k;\ \ x_4 \to x_4;\ \ \psi \to \gamma_4 \psi \\
&\ \ B_k \to B_k;\ \ B_4 \to -B_4 \\
&T: g \to g;\ \ x_k \to x_k;\ \ x_4 \to -x_4;\ \ \psi \to -i\gamma_3\gamma_1\psi^* \\
&\ \ B_k \to B_k;\ \ B_4 \to -B_4 \\
&C: g \to g;\ \ \psi \to \gamma_2 \psi^*
\end{aligned} \tag{2.26}$$

In Eq. (2.25), contrary to Eqs. (1.37) and (1.39), the magnetic charge g is invariant in the three transformations P, T, C.

The pseudoscalar character of magnetism is not given by the constant g, but by the *charge-operator* $g\gamma_5$ which lies at the origin of all the differences between the classical and quantum theories of magnetic monopoles. Now it may be shown that $\omega_1 = \overline{\psi}\psi$ is really a scalar, and $\omega_2 = -i\overline{\psi}\gamma_5\psi$ a pseudoscalar, as a consequence of the P, T, C transformations of the spinor ψ given in Eqs. (2.25) and (2.26) and applied to the formulas of these quantities given in the list [Eq. (2.18)]. An elementary calculation gives

$$\begin{aligned}
&P: \omega_1 \to \omega_1;\ \ \omega_2 \to -\omega_2;\ \ T: \omega_1 \to \omega_1;\ \ \omega_2 \to -\omega_2; \\
&C: \omega_1 \to -\omega_1;\ \ \omega_2 \to -\omega_2.
\end{aligned} \tag{2.27}$$

Therefore, ω_1 represents P and T invariants, and ω_2 represents P and T pseudoinvariants. And they are both reversed by C so that they are not PTC invariants. On the contrary, it is easy to prove that they are both relativistic invariants.

The definition [Eq. (2.19)] of the angle **A** shows, owing to Eq. (2.26), that

1. The angle A is a relativistic invariant.
2. The sign of A is reversed by P **and by** T so that A is a relativistic pseudoscalar (in $\mathbb{R}^4$).
3. The angle A is C invariant. Therefore, A is PTC invariant.

Now a geometrical interpretation of the chiral gauge may be given. We shall first define a chiral plane, in which we consider a vector (ω_1, ω_2): actually, (ω_2) is reversed when x or t is reversed. By virtue of Eq. (2.20), the angle A is a pseudoangle, so that the vector with coordinates (ω_1, ω_2) may be defined by

$$\omega_1 = \rho \cos A;\ \ \omega_2 = \rho \sin A. \tag{2.28}$$

Now, consider a rotation θ in the plane (ω_1, ω_2), defined by a rotation $\theta/2$ of a spinor:

$$\psi' \to e^{i\gamma_5/2}\psi. \tag{2.29}$$

Using the definition [Eq. (2.18)] of $\omega_1 = \overline{\psi}\psi$ and $\omega_2 = -i\overline{\psi}\gamma_5\psi$, we find from Eq. (2.28) the rotation of the vector (ω_1, ω_2):

$$\begin{pmatrix} \omega_1' \\ \omega_2' \end{pmatrix} = \begin{pmatrix} \cos\theta & -\sin\theta \\ \sin\theta & \cos\theta \end{pmatrix} \begin{pmatrix} \omega_1 \\ \omega_2 \end{pmatrix} \Rightarrow A' = A + \theta. \tag{2.30}$$

Therefore, the second gauge invariance [Eq. (2.29)] is a rotation, just like the first one, but it is a rotation in the chiral plane, not in the physical space.

Now the quantity ρ will be called the *principal chiral invariant*. The rotation angle $\theta/2$ of the spinor is equal to half the rotation angle θ of a vector in the chiral plane, in accordance with the spinor geometry.

Finally, as we have seen, according to Eq. (2.26), the charge conjugation does not change the sign of the magnetic constant of charge g, which means that two monopoles with opposite constants g are not charge-conjugated: we shall see that a change of g to $-g$ signifies a change of the vertex angle of the Poincaré cone. In the next chapter, we shall see what charge conjugation means in the magnetic case, but it may be stated here that two conjugated monopoles have the same charge constant g.

We cannot create or annihilate pairs of monopoles with charges g *and* $-g$, as was the case for electric charges e *and* $-e$. As a result, there is no danger of an infinite polarization of the vacuum with such zero mass monopoles. Moreover, one has not to invoke great masses to explain the rarity of monopoles or the difficulty of observing them. There are other reasons for this, which will be explored later in this book.

CHAPTER 3

The Wave Equation in the Weyl Representation. The Interaction Between a Monopole and an Electric Coulombian Pole. Dirac Formula. Geometrical Optics. Back to Poincaré

This chapter will explore the same monopole equation as Chapter 2, but for the *Weyl representation*.

3.1 THE WEYL REPRESENTATION

We shall define the Weyl representation by the following transformation (Lochak, 1983, 2006), which divides the wave function ψ into the two-component spinors ξ and η:

$$\Psi \rightarrow U\Psi = \begin{pmatrix} \xi \\ \eta \end{pmatrix}; \quad U = U^{-1} = \frac{1}{\sqrt{2}}(\gamma_4 + \gamma_5); \quad \gamma_4 = \begin{pmatrix} 1 & 0 \\ 0 & -1 \end{pmatrix};$$

$$\gamma_5 = \begin{pmatrix} 0 & 1 \\ 1 & 0 \end{pmatrix}. \tag{3.1}$$

The matrix γ_5 and the magnetic charge operator C are diagonalized:

$$UBU^{-1} = Ug\gamma_5U^{-1} = g\gamma_4 = \begin{pmatrix} g & 0 \\ 0 & -g \end{pmatrix}. \tag{3.2}$$

Eqs. (3.1) and (3.2) show that ξ and η are eigenstates of B, with eigenvalues g and $-g$:

$$UBU^{-1}\begin{pmatrix} \xi \\ 0 \end{pmatrix} = g\begin{pmatrix} \xi \\ 0 \end{pmatrix}; \quad UBU^{-1}\begin{pmatrix} 0 \\ \eta \end{pmatrix} = -g\begin{pmatrix} 0 \\ \eta \end{pmatrix}. \tag{3.3}$$

Owing to Eqs. (3.1) and (1.40), Eq. (2.16) splits into a pair of uncoupled two-component equations in ξ and η, corresponding to the opposite eigenvalues of B:

$$\begin{aligned} &\left[\frac{1}{c}\frac{\partial}{\partial t} - \mathbf{s}.\nabla - i\frac{g}{\hbar c}(W + \mathbf{s}.\mathbf{B})\right]\xi = 0 \\ &\left[\frac{1}{c}\frac{\partial}{\partial t} + \mathbf{s}.\nabla + i\frac{g}{\hbar c}(W - \mathbf{s}.\mathbf{B})\right]\eta = 0. \end{aligned} \tag{3.4}$$

The P, T, C symmetries [Eq. (2.25)] take the form:

$$\begin{aligned} &P: g\rightarrow g;\ x_k\rightarrow -x_k;\ t\rightarrow t;\ B_k\rightarrow B_k;\ W\rightarrow -W;\ \xi\leftrightarrow\eta \\ &T: g\rightarrow g;\ x_k\rightarrow x_k;\ t\rightarrow -t;\ B_k\rightarrow -B_k;\ W\rightarrow W;\ \xi\rightarrow s_2\xi^*;\ \eta\rightarrow s_2\eta^* \\ &C: g\rightarrow g;\ \xi\rightarrow -is_2\eta^*;\ \eta\rightarrow is_2\xi^*. \end{aligned} \tag{3.5}$$

P and T exchange Eq. (3.4) between themselves.

Thus, we have a pair of *charge conjugated* particles—a *monopole* and an *antimonopole*—with the same charge constant g and *opposite helicities*. They are defined by the operator C, which shows that our monopole is a magnetically *excited neutrino because* Eq. (3.4) reduces to a pair of two-component neutrino equations if $g = 0$.

Eq. (3.4) is invariant under the following gauge transformation (with opposite signs of the phase of ξ and η, which is nothing but the Weyl representation of the gauge transformation [Eq. (2.16)]:

$$\xi \to \exp\left(i\frac{g}{\hbar c}\phi\right)\xi;\ \eta \to \exp\left(-i\frac{g}{\hbar c}\phi\right)\eta;\ W \to W + \frac{1}{c}\frac{\partial\phi}{\partial t};\ \mathbf{B} \to \mathbf{B} - \nabla\phi. \tag{3.6}$$

3.2 CHIRAL CURRENTS

The gauge [Eq. (3.6)] entails, for Eq. (3.4), the following *conservation laws*:

$$\frac{1}{c}\frac{\partial(\xi^+\xi)}{\partial t} - \nabla\xi^+\mathbf{s}\xi = 0;\ \frac{1}{c}\frac{\partial(\eta^+\eta)}{\partial t} + \nabla\eta^+\mathbf{s}\eta = 0. \tag{3.7}$$

Thus, we have *two currents* with several important properties. They are *isotropic* and *chiral,* and they *exchange between themselves by parity*:

$$\begin{aligned} &X_\mu = (\xi^+\xi, -\xi^+\mathbf{s}\xi);\ Y_\mu = (\eta^+\eta, \eta^+\mathbf{s}\eta); X_\mu X_\mu = 0; \\ &Y_\mu Y_\mu = 0;\ P \Rightarrow X_\mu \leftrightarrow Y_\mu. \end{aligned} \tag{3.8}$$

Owing to Eq. (3.1), we find a decomposition of the polar and axial vectors, as defined in Eq. (2.17):

$$J_\mu = X_\mu + Y_\mu;\ \Sigma_\mu = X_\mu - Y_\mu. \tag{3.9}$$

The chiral currents X_μ and Y_μ may be considered even more fundamental than electric and magnetic currents. We already know the relations [Eq. (2.23)], and it is easy to prove, by using Eqs. (2.17) and (3.1), that

$$\begin{aligned} &\omega_1 = \xi^+\eta + \eta^+\xi;\ \omega_2 = i(\xi^+\eta - \eta^+\xi); \\ &\rho^2 = \omega_1^2 + \omega_2^2 = 4(\xi^+\eta)(\eta^+\xi). \end{aligned} \tag{3.10}$$

It was noted in Chapter 2 that a consequence of Eq. (2.23) is that J_μ is timelike and Σ_μ is spacelike. Owing to Eq. (3.9), we can add that the fact that one of the vectors (J_μ, Σ_μ) is timelike and the other spacelike is a trivial property of the addition and subtraction of isotropic vectors. And if J_μ is *precisely* spacelike and Σ_μ spacelike, this is due to the + sign of $(\omega_1^2 + \omega_2^2)$ in Eq. (2.23).

Therefore, our magnetic current, $K_\mu = g\,\Sigma_\mu$, may be spacelike because the true magnetic currents are the isotropic currents $g\,X_\mu$ and $-g\,Y_\mu$, corresponding to the spinor states ξ and η. The pseudovector K_μ **is only their difference,** so it has no reason to be spacelike or timelike. Therefore, the relativistic type of the magnetic current K_μ has no importance; on the contrary, the fact that J_μ is timelike is very important because owing to this property, J_μ may be interpreted as a current density of probability or electricity. Moreover, J_μ is a polar vector, which is necessary for a current of probability or electricity, while Σ_μ is a pseudovector, as a magnetic current must be (Curie, 1894a,b). We have already noted this beautiful example of harmony between physics and mathematics.

3.3 A REMARK ABOUT THE DIRAC THEORY OF THE ELECTRON

The equations of current continuity [Eq. (3.7)] were deduced from the Weyl representation [Eq. (3.4)] of the equation of the magnetic monopole [Eq. (2.15)]. It is interesting to compare that result with the Weyl representation of the equation of the electron, applying the transformation [Eq. (3.1)] to Eq. (2.10) instead of (2.15).

Taking into account the equality $A_\mu = (\mathbf{A}, iV)$, we find a system that is the analog of Eq. (3.4) but equivalent to the Dirac equation:

$$\begin{aligned} &\left[\frac{1}{c}\frac{\partial}{\partial t} - \mathbf{s}.\nabla + i\frac{e}{\hbar c}(V + \mathbf{s}.\mathbf{A})\right]\xi + i\frac{m_0}{\hbar c}\eta = 0 \\ &\left[\frac{1}{c}\frac{\partial}{\partial t} + \mathbf{s}.\nabla + i\frac{g}{\hbar c}(V - \mathbf{s}.\mathbf{A})\right]\eta + i\frac{m_0}{\hbar c}\xi = 0. \end{aligned} \tag{3.11}$$

Let us notice some points here:

1. Eqs. (3.4) and (3.11) has the same differential part.
2. Thus, in the massless case, we find in both systems two separate equations for the chiral components ξ, η (i.e., for opposite helicities). It is known that in Eq. (2.15) or (3.4), the condition $m_0 = O$ is a consequence of the chiral gauge invariance $\Psi \rightarrow \exp\left(i\frac{g}{\hbar c}\gamma_5\phi\right)\Psi$. Nevertheless, the fact that chiral components obey separate equations [Eqs. (3.4) and (3.11)] depends only on the zero mass, whatever the reason for this zero mass may be and whatever the charge of the particle is.

3. Now, from the Dirac system [Eq. (3.11)], with $m_0 \neq O$ in the case of an electric interaction, it is easy to deduce the evolution law of isotropic currents:

$$\begin{aligned} &\frac{1}{c}\frac{\partial(\xi^+\xi)}{\partial t} - \nabla\xi^+\mathbf{s}\xi + i\frac{m_0 c}{\hbar}\left(\xi^+\eta - \eta^+\xi\right) = 0 \\ &\frac{1}{c}\frac{\partial(\eta^+\eta)}{\partial t} + \nabla\eta^+\mathbf{s}\eta - i\frac{m_0 c}{\hbar}\left(\xi^+\eta - \eta^+\xi\right) = 0. \end{aligned} \tag{3.12}$$

We see here that the law does not depend explicitly on electromagnetic interaction. The difference between electricity and magnetism appears in the presence of a mass term only in the case of the electron, a term that is excluded by the chiral gauge in the case of a magnetic monopole. The chiral gauge invariance is the true difference between the two theories because it introduces, in Eq. (3.4), the magnetic interaction that is responsible for new forces.

Taking Eqs. (3.8) and (3.10) into account, we find the following laws, which mean that the Dirac pseudoinvariant ω_2 is the source of chiral isotropic currents:

$$\partial_\mu X_\mu + i\frac{m_0 c}{\hbar}\omega_2 = 0; \quad \partial_\mu Y_\mu - i\frac{m_0 c}{\hbar}\omega_2 = 0. \tag{3.13}$$

Adding and subtracting these equalities, we find two well-known laws:

$$\partial_\mu J_\mu = 0; \quad \partial_\mu \Sigma_\mu + 2i\frac{m_0 c}{\hbar}\omega_2 = 0. \tag{3.14}$$

Eq. (3.13) expresses the conservation of electricity and probability, by the Dirac equation. Eq. (3.14) is called, in Dirac's theory, the *Uhlenbeck and Laporte equality*. Starting from our theory of the leptonic monopole, we see that Eq. (3.13) governs the evolution of the left and right isotropic currents generated by the Dirac pseudoinvariant, which implies that Eq. (3.14) of *Uhlenbeck and Laporte* governs their difference, Σ_μ.

At this point, it is important to notice a fundamental difference between electricity and magnetism; in the Dirac equation, there is conservation of neither isotropic currents X_μ and Y_μ, nor of their difference $\Sigma_\mu = X_\mu - Y_\mu$. As a result, there is no conservation of magnetism; on the contrary, the sum $J_\mu = X_\mu + Y_\mu$ is conserved, and this is the conservation of electricity. The latter is related only to the presence of a mass term, but the following must be underlined:

1. We cannot add to Eq. (3.11) a magnetic interaction because it would be contrary to the presence of the mass term.

2. We cannot introduce into Eq. (3.4) an electric interaction because there is no Dirac massless electron, which would not admit quantized states and would provoke difficulties with the creation and annihilation of pairs.

Therefore, "leptonic dyons" carrying both electric and magnetic charges cannot exist.

3.4 THE INTERACTION BETWEEN A MONOPOLE AND AN ELECTRIC COULOMBIAN POLE (ANGULAR FUNCTIONS)

To solve the problem of a central field, we must introduce W = 0 and either Eq. (1.26) or (1.29) of **B** in the chiral equations [Eq. (3.4)]. The Poincaré integral [Eq. (1.4)] takes, in the quantum case, the expressions given next, in Eq. (3.15), for the left and right monopole. For the time being, we shall admit that result without proof, which will be given in the next chapter, in a more general case:

$$\mathbf{J}_\xi = \hbar\left[\mathbf{r}\times(-i\nabla + D\,\mathbf{B}) + D\,\mathbf{r} + \frac{1}{2}\mathbf{s}\right]$$
$$\mathbf{J}_\eta = \hbar\left[\mathbf{r}\times(-i\nabla - D\,\mathbf{B}) - D\,\mathbf{r} + \frac{1}{2}\mathbf{s}\right]. \tag{3.15}$$

$\mathbf{J}_\xi$ and $\mathbf{J}_\eta$ differ only by the sign of D; i.e., by the sign of the eigenvalues of the charge operator $C = g\gamma_5$, defined in Eq. (2.16). The notations are

$$D = \frac{eg}{\hbar c},\quad \mathbf{B} = eB,\quad \widehat{\mathbf{r}} = \frac{\mathbf{r}}{r}, \tag{3.16}$$

where D is the *Dirac number,* which we already know from the *Dirac condition* [Eq. (1.19)]—the last will be found below a new form; and B is the pseudopotential [Eq. (1.26) or (1.29)].

As was said previously, the proof that $\mathbf{J}_\xi$ and $\mathbf{J}_\eta$ are first integrals of Eq. (3.4) will be given in the next chapter. But for now, it is easy to show that the components of $\mathbf{J}$ obey the relations:

$$[\mathbf{J}_2,\mathbf{J}_3] = i\hbar\,\mathbf{J}_1,\ [\mathbf{J}_3,\mathbf{J}_1] = i\hbar\,\mathbf{J}_2,\ [\mathbf{J}_1,\mathbf{J}_2] = i\hbar\,\mathbf{J}_3 \tag{3.17}$$

Here, we shall only find their proper states, restricting our demonstration to the plus sign of D; i.e., to the left monopole—the first expression in Eq. (3.4)—dropping the index ξ.

Now, let us write $\mathbf{J}$ as

$$\mathbf{J} = \hbar\left[\Lambda + \frac{1}{2}\mathbf{s}\right], \quad \Lambda = \mathbf{r} \times (-i\nabla + D\mathbf{B}) + D\hat{\mathbf{r}}. \tag{3.18}$$

One can see that $\hbar\Lambda$ is *the quantum form of the Poincaré first integral* [Eq. (1.4)] (Poincaré, 1896). $\mathbf{J}$ is the sum of the quantum form $\hbar\Lambda$ of the first integral and of the spin operator $\hbar\mathbf{s}$: $\mathbf{J}$ is the *total quantum angular momentum* of the monopole in an electric coulombian field, the generalization of the classical quantity. Of course, the components of $\hbar\Lambda$ obey the same relations [Eq. (3.16)] as the components of $\mathbf{J}$ because Λ commutes with $\mathbf{s}$.

In polar angles, from the definition [Eq. (3.18)] of Λ and from the polar form [Eq. (1.29)] of B, we find the following:

$$\begin{aligned}
\Lambda^+ &= \Lambda_1 + i\Lambda_2 = e^{i\varphi}\left(i\cot\theta + \frac{\partial}{\partial\varphi} + \frac{\partial}{\partial\theta} + \frac{D}{\sin\theta}\right)\\
\Lambda^- &= \Lambda_1 - i\Lambda_2 = e^{-i\varphi}\left(i\cot\theta + \frac{\partial}{\partial\varphi} - \frac{\partial}{\partial\theta} + \frac{D}{\sin\theta}\right)\\
\Lambda_3 &= -i\frac{\partial}{\partial\varphi}.
\end{aligned} \tag{3.19}$$

Let us note that, owing to our choice [Eq. (1.26)] for the electromagnetic gauge, there is no additional term in Λ_3, contrary to the findings of Wu and Yang (1975, 1976). Now, we need the eigenstates $Z(\theta, \varphi)$ of Λ^2 and Λ_3. By virtue of Eq. (3.16), the eigenvalue equations of Λ must be

$$\begin{aligned}
&\Lambda^2 Z = j(j+1)Z, \quad \Lambda_3 Z = mZ, \quad j = 0, \frac{1}{2}, 1, \frac{3}{2}, 3, \ldots,\\
&m = -j, -j+1, \ldots, j-1.
\end{aligned} \tag{3.20}$$

To simplify the calculation of $Z(\theta, \varphi)$, we shall introduce a new angle χ, the meaning of which will be given shortly. We write

$$\mathcal{D}(\theta, \varphi, \chi) = e^{iD\chi} Z(\theta, \varphi), \tag{3.21}$$

where the functions $\mathcal{D}(\theta, \varphi, \chi)$ are the eigenstates of operators R_k, which are easily derived from Eq. (3.19):

$$\begin{aligned}
R^+ &= R_1 + iR_2 = e^{i\varphi}\left(i\cot\theta + \frac{\partial}{\partial\varphi} + \frac{\partial}{\partial\theta} - \frac{i}{\sin\theta}\frac{\partial}{\partial\chi}\right)\\
R^- &= R_1 - iR_2 = e^{-i\varphi}\left(i\cot\theta + \frac{\partial}{\partial\varphi} - \frac{\partial}{\partial\theta} - \frac{i}{\sin\theta}\frac{\partial}{\partial\chi}\right)\\
R_3 &= -i\frac{\partial}{\partial\varphi}
\end{aligned} \tag{3.22}$$

Obviously, the eigenvalues are the same as those of Z:

$$R^2Z = j(j+1)Z, \quad R_3Z = mZ, \quad j = 0, \frac{1}{2}, 1, \frac{3}{2}, 3, \ldots,$$
$$m = -j, -j+1, \ldots, j-1. \tag{3.23}$$

The operators R_k are well known: they are the *infinitesimal operators of the rotation group written in* the *fixed referential.* The angles θ, φ, χ are the Euler angles of nutation, precession, and proper rotation. The role of the rotation group is not surprising because of the *spherical symmetry* of the system constituted by a monopole in a central electric field.

Our eigenfunction problem is thus trivially solved: instead of the cumbersome calculations of monopole harmonics that do not exist, we see, under the simple assumption of continuity of the wave functions with respect to the rotation group, that the angular functions are the *generalized spherical functions;* i.e., the matrix elements of the irreducible unitary representations of the rotation group (Gelfand, Minlos, & Shapiro, 1963; Lochak, 1959). These functions are also the eigenfunctions of the symmetrical top. This coincidence was noticed by Tamm in 1931 without explanation, but here the explanation is evident because we already know the analogy between a symmetrical top and a monopole in a central field. The eigenstates of R^2 and R_3 are

$$\mathcal{D}_j^{m',m}(\theta, \varphi, \chi) = e^{i(m\varphi + m'\chi)} d_j^{m',m}(\theta)$$

$$d_j^{m',m}(\theta) = N(1-u)^{\frac{-(m-m')}{2}}(1+u)^{\frac{-(m+m')}{2}}\left(\frac{d}{du}\right)^{j-m}\left[(1-u)^{j-m'}(1+u)^{j+m'}\right]$$

$$u = \cos\theta, \quad N = \frac{(-1)^{j-m} i^{m-m'}}{2^j}\left(\frac{(j+m)!}{(j-m)!(j-m')!(j+m')!}\right)^{1/2}$$

$$j = \frac{1}{2}, 1, \frac{3}{2}, 2, \ldots, \quad m, m' = -j, -j+1, \ldots, j-1, j. \tag{3.24}$$

The normalization factor N is so defined that rows and columns of the unitary (2j + 1) matrix of the representation $\mathcal{D}_j$ are normed to unity. To normalize the quantum states, we must take the factor Z in Eq. (3.21) in the form

$$Z_j^{m',m}(\theta, \varphi) = \sqrt{2j+1}\; \mathcal{D}_j^{m',m}(\theta, \varphi, 0). \tag{3.25}$$

The proper rotation angle χ does not appear in $Z_j^{m',m}(\theta,\varphi)$: it appears only in the phase $e^{iD\chi}$ (D = Dirac number) because the monopole was implicitly supposed to be a point contrary to the symmetric top that has a spatial extension. Nevertheless, there is a projection (different from zero), of the orbital angular momentum on the symmetry axis, due to the chirality of the magnetic charge. The eigenvalue associated to the projection is the quantum number m'. The crucial point is that, if we compare Eqs. (3.21) and (3.24), we see that the eigenvalue $\hbar m'$ of the projection must be equal to the Dirac number D.

The quantization of the Dirac number D, thus is a consequence of the continuity of the wave function, on the rotation group.

Taking into account Eq. (3.23), we find

$$D = \frac{eg}{\hbar c} = m' = -j, -j+1, \ldots, j-1, j; \; j = \frac{n}{2}. \tag{3.26}$$

Taking into account the definition [Eq. (3.16)], we see that the equality [Eq. (3.26)] is a new and more precise form of the Dirac condition [Eq. (1.19)]. In this new formula, the integer (or half-integer) m' is not an arbitrtay number as it was in the Dirac formula. Rather, m' is now defined by the projection of the angular momentum of the whole physical system on the symmetry axis passing through the two charges.

The condition [Eq. (3.26)], which implies the Dirac condition [Eq. (1.19)], appears as a consequence of the spherical symmetry of the system and of the continuity of the wave functions with respect to the rotation group. It is justified by a dynamical argument, not only formally derived.

As was already stated, the factor of one-half has nothing to do with the strings beginning at the origin: it is a consequence of the double connexity of the rotation group that appears in the presence of half-integers in the representations of the group, and thus in the corresponding values of j and m'. Let us draw attention to, concerning these questions, an important work of T. W. Barrett in which the role of the rotation group in electromagnetic field theories is extensively developed (Barrett, 1989).

Now, owing to Eqs. (3.16) and (3.26), we can define the values of the magnetic charges as functions of the charge of the electron, the Planck constant, and the velocity of light because the value g_0 of the fundamental magnetic charge is given for $n = 1$ by Eq. (3.26), and the others are multiples of this value:

$$g_0 = \frac{\hbar c}{2e^2} = \frac{1}{2\alpha} e = \frac{137}{2} e = 68,5 \; e, \quad g = n g_0. \tag{3.27}$$

In conclusion, it is useful to emphasize that the functions [Eq. (3.24)] are defined for all the values of the Euler angles (namely, $0 \leq \theta \leq 2\pi$, $0 \leq \varphi \leq 2\pi$, $0 \leq \chi \leq 2\pi$). These intervals are good for all angles, including θ, which is a so-called normal rotation angle, just as φ or χ is. Thus, the north pole is $\theta = 0$ and the south pole $\theta = 2\pi$. This fact seems shocking, but it is the reason for which the interval $0 \leq \theta \leq \pi$ is generally introduced, in order to obtain the univocity of Euler angles. But actually, it is better to describe the rotation group not in the physical space $\mathbb{R}^3$, but in $\mathbb{R}^4$, which is the SU^2 space and the space of the Euler-Olinde-Rodrigues parameters (Cartan, 1938; Lochak, 1959):

$$\begin{aligned} \xi_1 &= \sin\frac{\theta}{2}\cos\frac{\varphi-\chi}{2}, \quad \xi_2 = \sin\frac{\theta}{2}\sin\frac{\varphi-\chi}{2} \\ \xi_3 &= \cos\frac{\theta}{2}\sin\frac{\varphi+\chi}{2}, \quad \xi_4 = \cos\frac{\theta}{2}\cos\frac{\varphi+\chi}{2}. \end{aligned} \tag{3.28}$$

All is uniform in $\mathbb{R}^4$, including these parameters, the group representations, and the Euler angles. Now we must introduce the monopole harmonics with spin, obtained by the Clebsch-Gordan procedure (Lochak, 1985a,b, 1995):

$$\begin{aligned} \Omega_j^{m',m}(+) = \Omega_j^+ &= \begin{pmatrix} \left(\dfrac{j+m}{2j+1}\right)^{1/2} Z_j^{m',m-1} \\[2ex] \left(\dfrac{j-m+1}{2j+1}\right)^{1/2} Z_j^{m',m} \end{pmatrix}, \\ \Omega_j^{m',m}(-) = \Omega_j^- &= \begin{pmatrix} \left(\dfrac{j-m+1}{2j+1}\right)^{1/2} Z_j^{m',m-1} \\[2ex] -\left(\dfrac{j+m}{2j+1}\right)^{1/2} Z_j^{m',m} \end{pmatrix}. \end{aligned} \tag{3.29}$$

These harmonics correspond to the eigenvalues $k = j \pm 1/2$ of the total angular momentum $\mathbf{J}$. In the following discussion, we shall use the abbreviation Ω_j^-, Ω_j^-, as well as several relations, the first of which is directly deduced from Eq. (3.29):

$$\mathbf{J}^2\Omega_{j-1}^+ = \hbar^2 k(k+1)\,\Omega_{\mathrm{j}-1}^+, \quad \mathbf{J}^2\Omega_j^- = \hbar^2 k(k+1)\,\Omega_j^-. \tag{3.30}$$

The others are deduced from recurrence relations between generalized spherical functions (Gelfand et al., 1963):

$$\begin{aligned}
&\mathbf{s}.\,\widehat{\mathbf{r}}\,\Omega_{j-1}^{+} = \cos\Theta'\,\Omega_{j-1}^{+} + \sin\Theta'\,\Omega_{j}^{-}\\
&\mathbf{s}.\,\widehat{\mathbf{r}}\,\Omega_{j}^{-} = \sin\Theta'\,\Omega_{j-1}^{+} - \cos\Theta'\,\Omega_{j}^{+}\\
&\cos\Theta' = \frac{m'}{j} = \frac{D}{j},\quad \widehat{\mathbf{r}} = \frac{\mathbf{r}}{r}.
\end{aligned} \tag{3.31}$$

The angle Θ' is the vertex half-angle of the Poincaré-cone (previously shown in Figure 1.2) because $\hbar m'$ is the projection of the total orbital momentum $\hbar j$ on the symmetry axis of the system, as defined by the monopole and the Coulombian center. We already knew that in the classical case (as discussed in Chapter 1), and we shall find it again at the geometrical limit of quantum theory.

3.5 THE INTERACTION BETWEEN A MONOPOLE AND AN ELECTRIC COULOMBIAN POLE (RADIAL FUNCTIONS)

The calculation of radial functions is based on the wave equations [Eq. (3.4)]. We consider the ξ-equation [Eq. (3.4)] with $W=0$, making $B=1/e\ \mathbf{B}$, where $\mathbf{B}$ is given by Eqs. (1.26) and (1.29), and looking for a solution with an angular momentum $k=j-1/2$, taking into account Eq. (3.25). The ξ-equation becomes

$$\frac{i}{c}\frac{\partial\xi}{\partial t} = \mathbf{s}.(i\nabla - m'\mathbf{B})\xi. \tag{3.32}$$

To apply a classical integration method of the hydrogen atom in Dirac's theory, we introduce in Eq. (3.30) the following expansion for ξ, where $F^{\pm}(r)$ are the radial functions that we want:

$$\xi = e^{-i\omega t}\left[F_{j-1}^{+}(r)\Omega_{j-1}^{+} + F_{j}^{-}(r)\Omega_{j}^{-}\right]. \tag{3.33}$$

We find, multiplying by $\mathbf{s}.\,\widehat{\mathbf{r}}$,

$$\frac{\omega}{c}(\mathbf{s}.\,\widehat{\mathbf{r}})\left(F_{j-1}^{+}\Omega_{j-1}^{+} + F_{j}^{-}\Omega_{j}^{-}\right) = (\mathbf{s}.\,\widehat{\mathbf{r}})\ \mathbf{s}.(i\nabla - m'B)\left(F_{j-1}^{+}\Omega_{j-1}^{+} + F_{j}^{-}\Omega_{j}^{-}\right). \tag{3.34}$$

Using the equalities [Eqs. (1.26) and (3.18)], and the algebraic relation, we get

$$(\mathbf{s}.\mathbf{A}_1)(\mathbf{s}.\mathbf{A}_2) = \mathbf{A}_1.\mathbf{A}_2 + i(\mathbf{A}_1 \times \mathbf{A}_2).\,\mathbf{s} \tag{3.35}$$

Consequently, Eq. (3.34) takes the form

$$\frac{dF^+_{j-1}}{dr}\Omega^+_{j-1} + \frac{dF^-_j}{dr}\Omega^-_j = \left[\frac{1}{r}(\Lambda.\mathbf{s}) - \left(\frac{m'}{r} + i\frac{\omega}{c}\right)(\mathbf{s}.\,\mathbf{r})\right]\left(F^+_{j-1}\Omega^+_{j-1} + F^-_j\Omega^-_j\right). \tag{3.36}$$

We know, from Eqs. (3.18) and (3.20), that

$$\Lambda^2\Omega^\pm = \hbar^2 j(j+1)\Omega^\pm,\ \ \mathbf{J}^2\Omega^\pm = \hbar^2 k(k+1)\Omega^\pm (k = j - 1/2), \tag{3.37}$$

and

$$(\Lambda.\,\mathbf{s})\Omega^+_{j-1} = (j-1)\Omega^+_{j-1}\,,\ \ (\Lambda.\,\mathbf{s})\Omega^-_j = -(j+1)\Omega^-_j\,. \tag{3.38}$$

Multiplying Eq. (3.36) on the left by Ω^+_{j-1} and Ω^+_{j-1} in succession, and integrating on the angles, we can eliminate $\Omega^\pm$. Using Eq. (3.31), we find

$$\begin{gathered}\left[\frac{d}{dr} + \frac{1}{r} - \frac{j}{r}s_3 + \left(\frac{m'}{r} + i\frac{\omega}{c}\right)e^{-is_2(\Theta'/2)}s_3 e^{+is_2(\Theta'/2)}\right]F = 0,\\ F(r) = \begin{pmatrix} F^+_{j-1}(r) \\ F^-_j(r)\end{pmatrix}.\end{gathered} \tag{3.39}$$

At this point, let us introduce functions $B^+_{j-1}(r)$, $B^-_j(r)$ such that

$$F = \frac{e^{is_2(\pi/4 - \Theta'/2)}}{r}B,\ \ B = \begin{pmatrix} B^+_{j-1}(r) \\ B^-_j(r)\end{pmatrix}. \tag{3.40}$$

Eq. (3.39) now becomes

$$\left(\frac{d}{dr} - \frac{l}{r}s_3 + i\frac{\omega}{c}s_1\right)B = 0,\ \ l = j\sin\Theta' = \sqrt{j^2 - m'}. \tag{3.41}$$

We see that l is the projection of the total orbital angular momentum (monopole + field) on the plane orthogonal to the axis of the Poincaré cone (Poincaré, 1896). Differentiating Eq. (3.41), we obtain the Bessel equations (Ince, 1956):

$$\frac{d^2B^+_{j-1}}{dr^2} + \left[\left(\frac{\omega}{c}\right)^2 - \frac{l(l-1)}{r^2}\right]B^+_{j-1} = 0,\ \ \frac{d^2B^-_j}{dr^2} + \left[\left(\frac{\omega}{c}\right)^2 - \frac{l(l+1)}{r^2}\right]B^-_j = 0. \tag{3.42}$$

Using Eq. (3.41) and the recurrence formula, we get

$$zJ'_{\lambda}(z) + \lambda J_{\lambda}(z) = zJ_{\lambda-1}(z). \tag{3.43}$$

Finally, we have

$$B = \left(r\frac{\omega}{c}\right)^{1/2} \begin{pmatrix} iJ_{l-1/2}\left(r\frac{\omega}{c}\right) \\ J_{l+1/2}\left(r\frac{\omega}{c}\right) \end{pmatrix}. \tag{3.44}$$

Inserting this result in Eq. (3.40) and then in Eq. (3.33), we obtain the ξ spinor. A similar calculation would give the η spinor.

3.6 SOME GENERAL REMARKS

Here are some general remarks about this discussion so far:

1. Eq. (3.4) gives the correct expressions [i.e., Eq. (3.15)] for the angular momentum of a monopole in a coulombian field.
2. The Dirac relation for the product of an electric and a magnetic charge is deduced from our equation in a more precise form [i.e., Eqs. (3.26) and (3.27)], and the radial functions are also deduced from the equation. They are the same as those found for an electric charge in the field of an infinitely heavy monopole (Kazama, Yang, & Golhaber, 1977).
3. The classical analogy will be explained further in another chapter of this book.
4. ω is not quantized: the monopole in a coulombian electric field is always in a ionizing state.This fact, predicted by Dirac, might be *a priori* guessed for two reasons: (1) It is suggested by the spiraling motion on the cone described in the classical case by Poincaré, and we know that our equation has the Poincaré equation as a classical limit. (2) The potential **B** given in Eq. (1.26) has an infinite string and as a result, the wave equation cannot have square integrable solutions.
5. The fundamental difference between other theories and ours lies in the fact that the present theory is the only one based on a pseudoscalar charge operator $C = g\gamma_5$ and in which the charge constant g is a scalar, because the pseudoscalar character is confined in the operator γ_5. This entails that g is separately P, T, C invariant. To test what this difference means, let us introduce a pseudoscalar constant g instead of the operator $C = g\gamma_5$, in Eq. (2.15), which becomes $\left(\partial_\mu - \frac{g}{\hbar c}B_\mu\right)\Psi = 0$ (which is without i

because G_μ is a pseudovector). From (Lochak 5), Eq. (3.4) becomes $\left[\frac{1}{c}\frac{\partial}{\partial t}-\mathbf{s}.\nabla-i\frac{g}{\hbar c}(W+\mathbf{s}.\mathbf{B})\right]\xi=0,\ \left[\frac{1}{c}\frac{\partial}{\partial t}+\mathbf{s}.\nabla-i\frac{g}{\hbar c}(W-\mathbf{s}.\mathbf{B})\right]\eta=0$ with a difference with respect to Eq. (3.4). Both equations now have the same sign before i. This difference seems small, but actually it is important because, whereas the ξ and η equations exchange between themselves under the P and T transformations, as in the above mentioned system [Eq. (3.4)], the charge conjugation is now $C:g\rightarrow -g$, $-is_2\xi^*\rightarrow\eta$, $is_2\eta^*\rightarrow\xi$. The monopole and the antimonopole are thus not only chiral conjugated, they have opposite charges. Therefore, they can constitute pairs of magnetic charges and, by their masslessness, their annihilation induces a giant polarization. These particules are not true monopoles; rather, they are massless electric particles, "disguised in magnetic monopoles" (as stated in the Foreword and Lochak, 1985).

3.7 THE GEOMETRICAL OPTICS APPROXIMATION. BACK TO THE POINCARÉ EQUATION

Now we must verify that we have found the correct Poincaré equation and the Birkeland effect. Let us introduce in Eq. (3.4) the following expression of the spinor ξ:

$$\xi = a\, e^{iS/\hbar}, \tag{3.45}$$

where a is a two-component spinor and S a phase. At zero order in $\hbar$, we have

$$\left[\frac{1}{c}\left(\frac{\partial S}{\partial t}-gW\right)-\left(\nabla S+\frac{g}{c}\mathbf{B}\right).\,\mathbf{s}\right]a=0, \tag{3.46}$$

which is a homogeneous system with respect to a. A necessary condition for a nontrivial solution is

$$\frac{1}{c^2}\left(\frac{\partial S}{\partial t}-gW\right)^2-\left(\nabla S+\frac{g}{c}\mathbf{B}\right)^2=0, \tag{3.47}$$

which is a relativistic Jacobi equation with zero mass, and we can define the kinetic energy, the impulse, and the linear Lagrange momentum as follows:

$$E=-\frac{\partial S}{\partial t}+gW,\ \ \mathbf{p}=\nabla S+\frac{g}{c}\mathbf{B},\ \ \mathbf{P}=\nabla S. \tag{3.48}$$

The Hamiltonian function will equal

$$H = c\sqrt{\left(\mathbf{P} + \frac{g}{c}\mathbf{B}\right)^2} - gW, \tag{3.49}$$

and a classical calculation gives as an equation of motion:

$$\frac{d\mathbf{p}}{dt} = g\left(\nabla W + \frac{\partial \mathbf{B}}{\partial t}\right) - \frac{g}{c}\mathbf{v} \times curl\,\mathbf{B}, \tag{3.50}$$

which gives the classical form

$$\frac{d\mathbf{p}}{dt} = g\left(\mathbf{H} - \frac{1}{c}\mathbf{v} \times \mathbf{E}\right). \tag{3.51}$$

But we must not forget that the mass of our particle equals zero, so $\mathbf{v}$ is the velocity of light and we cannot write $\mathbf{p} = m\mathbf{v}$. But the equality $\mathbf{p} = (\mathrm{E}/c^2)\,\mathbf{v}$ still holds when the energy E is a constant, which will be the case in a coulombian electric field. So we have

$$\frac{d^2\mathbf{p}}{dt^2} = -\lambda\frac{1}{r^3}\frac{1}{c}\frac{d\mathbf{r}}{dt} \times \mathbf{r}, \quad \lambda = \frac{egc}{E} \tag{3.52}$$

This is exactly the Poincaré equation [Eq. (1.2), given previously in Chapter 1], with a minus sign because we chose the left monopole. Now, starting from the right monopole [i.e., from the second equation in Eq. (3.4)] and with the same approximation:

$$\eta = b\,e^{iS/\hbar}, \tag{3.53}$$

we find the following equation for b:

$$\left[\frac{1}{c}\left(\frac{\partial S}{\partial t} + gW\right) - \left(\nabla S - \frac{g}{c}\mathbf{B}\right).\,\mathbf{s}\right] b = 0. \tag{3.54}$$

Of course, Eq. (3.54) gives the same Poincaré equation [Eq. (3.52)] with a plus sign before λ.

3.8 THE PROBLEM OF THE LINK BETWEEN A LEPTONIC MAGNETIC MONOPOLE, A NEUTRINO, AND WEAK INTERACTIONS

The problem being explored in this chapter may be summarized as follows:

1. The Weyl representation splits the massless Dirac equation into two independent, two-component equations, which are considered since 1956 (by Lee and Yang and Landau) as the "neutrino two-component theory": one equation describes the neutrino and the other one the anti-neutrino.
2. We have shown that the massless Dirac equation admits a second gauge invariance: the *chiral gauge*. In addition, we have shown that not only C-symmetry, but also P-symmetry has the important property of being able to exchange between themselves the two Weyl equations, left and right, respectively, which implies the chirality of the neutrino and the anti-neutrino[8].
3. Now, we have proved that *the chiral gauge invariance of the massless Dirac equation entails a new electromagnetic interaction which corresponds to a magnetic monopole.* And obviously there is no other possibility. The new pair of Weyl-like equations with these new interaction terms remain separated as with a free field. And it must be emphasized that our massless monopole is a consequence of the second gauge invariance of the Dirac equation. It is profoundly rooted in electronics and electromagnetism, and for this reason, it is very different from the monopoles with enormous masses predicted by other theories. And the principal difference is that our monopole leaves observed characteristic tracks, it is created in several laboratories, and it gives physical observable consequences.
4. Now, the neutrino appears in this theory as a magnetic monopole with a zero charge. Remember that this charge is equal to n times a unit charge (including $n = 0$). And the laws of symmetry are identical. For this reason, we have presented the hypothesis that these leptonic monopoles, thanks to the neutrino symmetry, manifest low-energy interactions. This hypothesis was largely developed by Harald Stumpf, and we put forward here a couple of experimental arguments:

[8] There is a curious anecdote concerning this property. When Hermann Weyl, at the end of the 1920s, found his representation of the Dirac equation, he noticed, *in the massless case*, the splitting into the two equations that we are discussing (the difference is that we introduce the interaction with an electromagnetic field). So, he said, each half of the split equations acquires an independent sense. Pauli objected to this because such an equation is not P-invariant. That is true, of course, but he neglected the fact that there were, actually, *two equations*, which together were P-invariant, and it was unknown at that time that they were left and right. The funny part of this story is that Pauli a little later predicted the existence of the neutrino—i.e., the clue to the problem that was finally untangled a quarter of a century later. The sad part of the story is, that if Pierre Curie—the discoverer, if not the solver, of these problems—had been alive, perhaps all would have been evident to him from the very beginning.

a. A beta radioactive sample (normally emitting neutrinos), submitted to a magnetic field, emits leptonic monopoles (Ivoilov, 2006).
b. The lifetime of a beta radioactive sample is reduced when it is irradiated by leptonic monopoles (Ivoilov, 2006).
c. A great quantity of neutrinos is emitted by the sun (because of the great number of low-energy reactions). I have suggested the hypothesis that some of them could be excited as leptonic monopoles by strong solar magnetic fields. If this is the case, most of these monopoles would have to be trapped on the sun by the same magnetic fields, which could be a new hypothesis that could explain the lack of solar neutrinos received by the Earth. Nevertheless, some of these monopoles could escape and then follow trajectories directed toward the Earth. In such a case, they must follow the lines of the magnetic field directed to the Earth's magnetic poles. So, when the explorer Jean-Louis Etienne embarked on an expedition to the North Pole, we gave him some X-ray films, currently used in laboratories to register leptonic monopoles. And we have found on these films exactly the same characteristic lines of monopoles (Bardout et al., 2007).

3.9 SOME QUESTIONS ABOUT THE DIRAC FORMULA AND OUR FORMULA

At this point, let us recall this equation:

$$D = \frac{eg}{\hbar c} = \frac{n}{2}, \ (1.19), \text{ and : } D = \frac{eg}{\hbar c} = m' = -j, -j+1, \ldots, j-1, j; \ \left(j = \frac{n}{2}\right), \tag{3.26}$$

Dirac's conclusions and ours are different. Dirac was looking for the reason why all the electric charges that appear in the physical world are equal to either the electron charge or to a mutiple, and he was happy to find that, by virtue of his formula, an arbitrary electric charge e must equal $e = n\frac{\hbar c}{2g}$. Therefore, if there is even only one monopole in the world, all the elecric charges will be multiples of a unit charge that depends on the charge of this monopole.

Our position is different. We have a theory concerning a magnetic monopole, and we ask the question: What happens if that monopole interacts with an electric charge? We are, in principle, able to answer this question because we have a wave equation [namely, Eq. (2.15)]. But *the answer depends on the value of these charges*, contrary to what happens with two electric charges, which can always interact without any condition: the reason is that in our case, one charge is a scalar and the other is a pseudoscalar, which was shown in the Dirac case because Eqs. (1.12) and (1.25) present the same difficulty.

Thus, we find that Eq. (3.26) has an evident affinity with the Dirac condition, with the difference that in our case, the *electric charge—not the magnetic one—is given*, so that it seems that we must write Eq. (3.26) in inverse order (with α being the fine structure constant):

$$g = m'\frac{\hbar c}{e} = em'\frac{\hbar c}{e^2} = e\frac{m'}{\alpha} = 137\ \ em'$$
$$\text{with}: \ m' = -j, -j+1, \ldots, j-1, j \ \left(j = \frac{n}{2}\right). \tag{3.54}$$

Therefore, if we consider a magnetic charge g striking a particle with an electric charge e, the collision will be possible only when the charge g of the magnetic particle obeys the condition [Eq. (3.54)], depending on the electric charge and on the angular momentum (more precisely, on its *projection* on the symmetry axis): so that not only the charge but even the momentum must be good. And the problem is still more complicated because there are electric particles with greater charges: for instance, atomic nuclei with charges Ne, so that Eq. (3.54) becomes

$$g = m'\frac{\hbar c}{Ne} = em'\frac{\hbar c}{Ne^2} = e\frac{m'}{N\alpha} = e\frac{137}{N}m'. \tag{3.55}$$

So, we must conclude that, on account of the relation [Eq. (3.26)], it is impossible to conclude that the electric and the magnetic charges are both conservative quantities because the conservation laws deduced from their respective wave equations are not verified in the collisions.

But we have strong theoretical and experimental arguments in favor of an absolute conservation of the electric charge, at least in the frame of the actually recognized electromagnetic laws. Thus, it seems that we are obliged to admit that, despite the fact that Eqs. (2.16) and (3.4) of the leptonic magnetic monopoles seem to be correct for symmetry laws, for the link to weak interactions and for electromagnetic interactions with continuous fields, something is missing in the description of the interaction between magnetic and electric charges.

It is highly improbable that this problem results from a defect in Eq. (2.16) or (3.4) because the preceding arguments could be developed in the Dirac case, starting from Eq. (1.19), as in ours, starting from Eq. (3.26). And these relations are mutually reinforced not only by their analogy, but also because they are confirmed by different arguments.

It seems evident that the difficulties are in the facts, not in the method. Manifestly, these equations need to be generalized by the presence of operators that can describe quantum transitions between the different states, defined by the preceding conditions, which is not presently the case.

CHAPTER 4

Nonlinear Equations. Torsion and Magnetism

Until now, we have seen only linear equations of a magnetic monopole: Eqs. (2.16) and (3.4). This is quite natural, because our theory concerns the magnetic slope of the Dirac theory of the electron, which is itself linear. Actually, the strangeness of this theory is not the linearity, which is normal in quantum mechanics, but rather the fact that the monopole so described is *massless* for algebraic reasons, which plays a basic role in the theory and cannot be easily dismissed. It must be emphasized that I personally profoundly dislike it because of the strangeness of the fact in itself, and, I must confess, because I am a member of the de Broglie school, which always hated masslessness, even applied to the photon, from which this peculiarity was eliminated.

Nevertheless, at first glance it seems difficult to avoid this peculiarity in the case of our magnetic monopole because it is a consequence of the *chiral gauge invariance*, which itself lies at the origin of all the results of the theory, as follows:

- The conservation of magnetism and the correct electromagnetic interaction of a monopole
- The classical limit, which gives the Poincaré equation and the analogy with a symmetric top
- The accordance with the symmetry laws predicted by Pierre Curie (1894a,b), which are experimentally verified (first of all, the chiral symmetry)
- A more precise form of the Dirac relation between electric and magnetic charges
- The analogy between neutrinos and leptonic monopoles, the latter being considered as magnetically excited neutrinos and obeying the same laws of symmetry
- The influence of monopoles on the lifetime of β radioactivity

Thus, we have reason to believe that the γ_5 gauge is unavoidable. Therefore, if we wish to define a mass term, we must look for a new way to do so without abandoning chiral invariance. We have found such a way: namely, *nonlinearity,* because we have already found a *nonlinear chiral invariant,* which was given as Eq. (2.20) in Chapter 2 and Eq. (3.10) in Chapter 3 of this book.

As a result, we can introduce in the Lagrangian a function $F(\rho)$ of the chiral invariant as a mass term. Of course, we could just as easily introduce a function on the norm of the electric or magnetic currents: $J_\mu J_\mu$ or $\Sigma_\mu \Sigma_\mu$ (as cited in Chapter 2), as Heisenberg did in his nonlinear theory (Heisenberg, 1953, 1954, 1966; Dürr *et al.*, 1959; Borne, Lochak, & Stumpf, 2001; Lochak., 1985), but we know that these norms are, including the sign, equal to ρ^2 [Eq. (2.24)].

4.1 A NONLINEAR MASSIVE MONOPOLE

First, let us write the following Lagrangian (Lochak, 1985) in the Dirac representation[9]:

$$L = \overline{\psi}\gamma_\mu[\partial_\mu]\psi - \frac{g}{\hbar c}\overline{\psi}\gamma_\mu\gamma_5 B_\mu \psi + i\frac{\mathrm{M}(\rho)c}{\hbar}, \tag{4.1}$$

where ρ is given by Eq. (2.20) and $\mathrm{M}(\rho)$ is a scalar function of ρ with the dimension of a mass.

The corresponding equation is

$$\gamma_\mu\left(\partial_\mu - \frac{g}{\hbar c}\gamma_5 B_\mu\right)\Psi + i\frac{\mu(\rho)c}{\hbar}\frac{\omega_1 - i\gamma_5\omega_2}{\sqrt{\omega_1^2+\omega_2^2}}\Psi = 0, \quad \mu(\rho) = \frac{d\mathrm{M}(\rho)}{d\rho}. \tag{4.2}$$

In the Weyl representation, we get

$$\begin{aligned} L = {} & \xi^+\left(\frac{1}{c}[\partial_t] - \frac{g}{\hbar c}W\right)\xi - \xi^+\mathbf{s}.\left([\nabla] + \frac{g}{\hbar c}\mathbf{B}\right)\xi + \\ & +\eta^+\left(\frac{1}{c}[\partial_t] + \frac{g}{\hbar c}W\right)\eta + \eta^+\mathbf{s}.\left([\nabla] - \frac{g}{\hbar c}\mathbf{B}\right)\eta + i\frac{\mathrm{M}(\rho)c}{\hbar}, \end{aligned} \tag{4.3}$$

[9] Here, we use the Costa de Beauregard convention $[\partial] = \overrightarrow{\partial} - \overleftarrow{\partial}$.

which gives the following equations:

$$\frac{1}{c}\partial_t\xi - \mathbf{s}.\nabla\xi - i\frac{g}{\hbar c}(W + \mathbf{s}.\,\mathbf{B})\xi + i\frac{\mu(\rho)c}{\hbar}\sqrt{\frac{\eta^+\xi}{\xi^+\eta}}\,\eta = 0$$

$$\left(\mu(\rho) = \frac{dM(\rho)}{d\rho}\right)$$

$$\frac{1}{c}\partial_t\eta + \mathbf{s}.\nabla\eta + i\frac{g}{\hbar c}(W - \mathbf{s}.\,\mathbf{B})\eta + i\frac{\mu(\rho)c}{\hbar}\sqrt{\frac{\xi^+\eta}{\eta^+\xi}}\,\xi = 0 \qquad (4.4)$$

These equations are chiral-invariant, like the linear equation. The magnetic current [Eq. (2.23)] is conserved, and, owing to Eq. (2.24), the equations are *PTC*-invariant (Lochak, 1997a, b) and the isotropic chiral currents [Eq. (3.8)] are separately conserved. In spite of that, Eq. (4.4) is generally coupled, as opposed to Eq. (3.4). But this coupling is not strong: If the degree of $\mu(\rho)$ is greater than 1, the nonlinear term vanishes when $\rho = 2\sqrt{(\xi^+\eta)(\eta^+\xi)} = 0$, which happens either for $\xi = 0$ or $\eta = 0$, or on the light cone (the Majorana case), which will be examined later in this chapter.

Now one can see that, in Eq. (4.4), ξ and η are phase-independent. This is why we can consider plane waves with different frequencies ω and ω', as well as wave numbers k and k', for ξ and η:

$$\xi = a\, e^{i(\omega\, t - \mathbf{k}.\mathbf{r})};\ \eta = b\, e^{i(\omega' t - \mathbf{k}'.\mathbf{r})}. \qquad (4.5)$$

Introducing these expressions in Eq. (4.4) *without external field*, we find (Lochak, 1985, 1995a,b)

$$\left(\frac{\omega}{c} + \mathbf{s}.\mathbf{k}\right)a + \frac{\mu(\rho)c}{\hbar}\sqrt{\frac{b^+a}{a^+b}}\,b = 0$$

$$\left(\frac{\omega'}{c} - \mathbf{s}.\mathbf{k}'\right)b + \frac{\mu(\rho)c}{\hbar}\sqrt{\frac{a^+b}{b^+a}}\,a = 0. \qquad (4.6)$$

If we multiply the first equation by $\left(\frac{\omega'}{c} - \mathbf{s}.\mathbf{k}'\right)$, with the following definitions:

$$\left(\frac{\omega'}{c} - \mathbf{s}.\mathbf{k}'\right)\ \left(\frac{\omega}{c} + \mathbf{s}.\mathbf{k}\right) = \Omega + \mathbf{s}.\mathbf{K};\ \ \Omega = \frac{\omega\omega'}{c^2} - \mathbf{k}.\mathbf{k}',$$

$$\mathbf{K} = \frac{1}{c}(\omega'\mathbf{k} - \omega\mathbf{k}') + i\mathbf{k}\times\mathbf{k}', \qquad (4.7)$$

we have

$$(\Omega+\mathbf{s.K})a+\frac{\mu(\rho)c}{\hbar}\sqrt{\frac{b^{+}a}{a^{+}b}}\left(\frac{\omega'}{c}-\mathbf{s.k}'\right)b=0. \tag{4.8}$$

Then, owing to Eq. (4.6), we find

$$\left[\Omega+\mathbf{s.K}-\left(\frac{\mu(\rho)c}{\hbar}\right)^{2}\right]a=0, \tag{4.9}$$

and finally, we must make the determinant of this equation zero to find a nontrivial solution, which gives the dispersion relation

$$\left(\Omega-\left(\frac{\mu(\rho)c}{\hbar}\right)^{2}\right)^{2}-\mathbf{K}^{2}=0. \tag{4.10}$$

We shall find a more explicit expression going back to Eq. (4.7), from which

$$\mathbf{K}^{2}=\frac{1}{c^{2}}(\omega'\mathbf{k}-\omega\mathbf{k}')^{2}-(\mathbf{k}\times\mathbf{k}')^{2}, \tag{4.11}$$

and hence it is easy to deduce the following:

$$\Omega^{2}-\mathbf{K}^{2}=\left(\frac{\omega^{2}}{c^{2}}-k^{2}\right)\left(\frac{\omega'^{2}}{c'^{2}}-k'^{2}\right). \tag{4.12}$$

Thus, we get the dispersion relation

$$\left(\frac{\omega^{2}}{c^{2}}-\mathbf{k}^{2}\right)\left(\frac{\omega'^{2}}{c^{2}}-\mathbf{k}'^{2}\right)-2\left(\frac{\omega\,\omega'}{c^{2}}-\mathbf{k.k}'\right)\left(\frac{\mu(\rho)c}{\hbar}\right)^{2}+\left(\frac{\mu(\rho)c}{\hbar}\right)^{4}=0\,. \tag{4.13}$$

Now, let us take the case of a homogeneous equation in ξ and η:

$$\mathrm{M}(\rho)=\mu_{0}\,\rho;\;\;\mu(\rho)=\mu_{0}=\mathit{Const.} \tag{4.14}$$

Owing to Eq. (4.13), we easily find two interesting kinds of waves:

1. $\omega=\omega',\;\;\mathbf{k}=\mathbf{k}'$: Both monopoles have the same phase, and the dispersion relation reduces to

$$\frac{\omega^{2}}{c^{2}}=k^{2}+\mu_{0}^{2}\;\left(k=\sqrt{\mathbf{k}^{2}}\right). \tag{4.15}$$

This is the ordinary dispersion relation of a massive particle, known as a *bradyon*.

2. On the other hand, if we have $\omega = -\omega'$, $\mathbf{k} = -\mathbf{k}'$, the phases have opposite signs, and the dispersion relation becomes

$$\frac{\omega^2}{c^2} = k^2 - \mu_0^2 \tag{4.16}$$

This is the dispersion relation of a supraluminal particle, known as a *tachyon*.

The wave equations [Eqs. (4.2) and (4.4)] seem to be the first in which tachyons appear without any ad hoc condition, but only as a particular solution among others. These nonlinear equations can be considered in different ways, which were described in Lochak (2003). Let us state once more that *the chiral components of the nonlinear equations [Eq. (4.2)] of a monopole in a coulombian electric field cannot be separated,* as they were in the linear case [Eq. (3.4)].

4.2 THE NONLINEAR MONOPOLE IN A COULOMBIAN ELECTRICAL FIELD

We shall see shortly that, in a coulombian electrical field with the pseudopotential [Eq. (1.26)], not only the linear equations [Eq. (3.4)], but also the nonlinear equations [Eq. (4.4)] admit the same angular operator [Eq. (3.14)], as an integral of motion.

For technical reasons, we shall collect the operators [Eq. (3.15)] into a unique operator in the Dirac representation, and we shall introduce the following classical vectorial notation:

$$\mathbf{J} = \hbar\left[\mathbf{r} \times (-i\nabla + \gamma_4 D\mathbf{B}) + \gamma_4 D\frac{\mathbf{r}}{r} + \frac{1}{2}\mathbf{S}\right]; \quad \mathbf{S} = \begin{pmatrix} \mathbf{s} & 0 \\ 0 & \mathbf{s} \end{pmatrix};$$
$$D = \frac{eg}{\hbar c}; \ \mathbf{B} \to eB. \tag{4.17}$$

Of course, the commutation rules [Eq. (3.17)] are satisfied by the components of Eq. (4.17), and we shall prove that $\mathbf{J}$ is an integral of motion, but we must be more careful than in the linear case described in Chapter 3 because of the presence of a nonlinear term in the Hamiltonian. So we return to the definition of a first integral, which is not a commutation rule but rather the definition: *the mean value of the operator* $\mathbf{J}$ *is a constant in virtue of the wave equations [Eq. (4.2) or (4.4)]*. To do so, we introduce the Dirac form of quantum equations in a vectorial formulation:

$$\frac{1}{c}\frac{\partial\psi}{\partial t} = H\psi; \quad H = \alpha.\nabla + i\,D\,\mathbf{S}.\,\mathbf{B} + i\,\mu(\rho)\;(\omega_1\alpha_4 + \omega_2\alpha_5) \qquad (4.18)$$

$$\alpha = \begin{pmatrix} 0 & \mathbf{s} \\ \mathbf{s} & 0 \end{pmatrix}, \; \alpha_4 = \begin{pmatrix} I & 0 \\ 0 & -I \end{pmatrix} = \gamma_4; \; \mathbf{S} = \begin{pmatrix} \mathbf{s} & 0 \\ 0 & \mathbf{s} \end{pmatrix}, \; \sigma_4 = \begin{pmatrix} 0 & I \\ I & 0 \end{pmatrix} = \gamma_5$$

$$\sigma = \sigma_4\alpha, \quad -i\alpha_1\alpha_2\alpha_3 = \sigma_4, \quad \alpha_1\alpha_2\alpha_3\alpha_4 = \alpha_5, \quad \mathbf{s} \; = \; \text{Pauli matrices.} \qquad (4.19)$$

Now we must prove that

$$\frac{d}{dt}\int \Psi^+\mathbf{J}\,\Psi\;dxdydz = \int\left(\frac{\partial\Psi^+}{\partial t}\mathbf{J}\,\Psi + \Psi^+\mathbf{J}\,\frac{\partial\Psi}{\partial t}\right)\;dxdydz$$
$$= i\int \Psi^+[H\mathbf{J} - \mathbf{J}H\,]\Psi = 0. \qquad (4.20)$$

The classical $[H\,\mathbf{J} - \mathbf{J}\,H\,]$ commutator appears, but we must examine in more detail the following three terms:

$$\frac{1}{c}\frac{d}{dt}\int \psi^+\mathbf{J}\psi d\mathrm{v} = \mathbf{P} + \mathbf{Q} + \mathbf{R} \qquad (4.21)$$

which correspond to the x, y, z components of $\mathbf{J}$. For instance, we have

$$\mathbf{P}_x = \int \psi^+(\mathbf{J}_x\;\boldsymbol{\alpha}.\nabla - \boldsymbol{\alpha}.\nabla\,\mathbf{J}_x)\psi d\mathrm{v}$$
$$\mathbf{Q}_x = iD\int \psi^+\left(\mathbf{J}_x\;\frac{z(\sigma_1 y - \sigma_2 x)}{r(x^2+y^2)} - \frac{z(\sigma_1 y - \sigma_2 x)}{r(x^2+y^2)}\;\mathbf{J}_x\right)\psi d\mathrm{v} \qquad (4.22)$$
$$\mathbf{R}_x = \int \psi^+[\mathbf{J}_x\;(\omega_1\alpha_4 + \omega_2\alpha_5) - (\omega_1\alpha_4 + \omega_2\alpha_5)\,\mathbf{J}_x]\psi d\mathrm{v}.$$

Now, recall that in Chapter 1, we had

$$B_x = \frac{e}{r}\frac{yz}{x^2+y^2}, \quad B_y = \frac{e}{r}\frac{-xz}{x^2+y^2}, \quad B_z = 0, \quad r = \sqrt{x^2+y^2+z^2}, \qquad (1.26)$$

so

$$yB_z - B_y z = \frac{xr}{x^2+y^2} - \frac{x}{r}; \quad zB_x - xB_z = \frac{yr}{x^2+y^2} - \frac{y}{r}; \quad xB_y - yB_y = -\frac{z}{r}, \qquad (4.23)$$

and the operator [Eq. (4.17)] becomes

$$\hbar^{-1}\mathbf{J}_x = -i(\mathbf{r}\times\nabla)_x + D\frac{xr}{x^2+y^2}\sigma_4 + \frac{1}{2}\sigma_1$$
$$\hbar^{-1}\mathbf{J}_y = -i(\mathbf{r}\times\nabla)_y + D\frac{yr}{x^2+y^2}\sigma_4 + \frac{1}{2}\sigma_2 \tag{4.24}$$
$$\hbar^{-1}\mathbf{J}_z = -i(\mathbf{r}\times\nabla)_z + \frac{1}{2}\sigma_3.$$

So Eq. (4.22) is split into "triads," corresponding to $\mathbf{J}_x$, $\mathbf{J}_y$, $\mathbf{J}_z$. For instance, we have for x:

$$\mathbf{P}_x = \int \psi^+(\mathbf{J}_x\,\boldsymbol{\alpha}.\nabla - \boldsymbol{\alpha}.\nabla\,\mathbf{J}_x)\psi dv$$
$$\mathbf{Q}_x = iD\int \psi^+\left(\mathbf{J}_x\ \frac{z(\sigma_1 y - \sigma_2 x)}{r(x^2+y^2)} - \frac{z(\sigma_1 y - \sigma_2 x)}{r(x^2+y^2)}\ \mathbf{J}_x\right)\psi dv$$
$$\mathbf{R}_x = \int \psi^+[\mathbf{J}_x\mu(\rho)\ (\omega_1\alpha_4 + \omega_2\alpha_5) - \mu(\rho)(\omega_1\alpha_4 + \omega_2\alpha_5)\,\mathbf{J}_x]\psi dv. \tag{4.25}$$

Here, we shall consider only this x-case. Introducing Eq. (4.21), we have

$$\mathbf{P}_x = P_1 + P_2 + P_3;\ \ \mathbf{Q}_x = Q_1 + Q_2 + Q_3;\ \ \mathbf{R}_x = R_1 + R_2 + R_3. \tag{4.26}$$

If we gather Eq. (4.18) to Eq. (4.26), we find

$$P_1 = -i\int \psi^+[(\mathbf{r}\times\nabla)(\boldsymbol{\alpha}.\nabla) - (\mathbf{r}\times\nabla)(\boldsymbol{\alpha}.\nabla)]\psi dv$$
$$= \mathrm{i}\textstyle\int \psi^+(\alpha_2\partial_z - \alpha_3\partial_y)\psi dv$$
$$P_2 = -D\int \psi^+\left[(\boldsymbol{\alpha}.\nabla)\ \sigma_4\frac{xr}{x^2+y^2}\right]\psi dv \tag{4.27}$$
$$P_3 = -\mathrm{i}\int \psi^+(\alpha_2\partial_z - \alpha_3\partial_y)\psi dv = -P_1 \ \rightarrow\ P_1 + P_3 = 0.$$

We know that $\mathbf{P}_x = P_1 + P_2 + P_3$, and we find the following for $\mathbf{Q}_x$:

$$Q_1 = D\int \psi^+(y\partial_z - z\partial_y)\frac{z(\sigma_1 y - \sigma_2 x)}{r(x^2+y^2)}\psi dv$$
$$Q_2 = 0 \tag{4.28}$$
$$Q_3 = D\int \psi^+\frac{\sigma_3 xz}{r(x^2+y^2)}\psi dv \ \rightarrow\ Q_1 + P_2 = -Q_3.$$

Hence we see that $[H\mathbf{J} - \mathbf{J}H] = 0$ *for the first three linear terms of the Hamiltonian [Eq. (4.18)], which ensures* a fortiori *the conservation of* $\mathbf{J}$ *in the linear case presented in Chapter 3.*

Now, only the nonlinear part remains, which reduces Eq. (4.25) to the following condition:

$$\mathbf{R}_x = \int \psi^+ [\mathbf{J}_x \mu(\rho)\,(\omega_1\alpha_4 + \omega_2\alpha_5) - \mu(\rho)(\omega_1\alpha_4 + \omega_2\alpha_5)\,\mathbf{J}_x]\psi dv. \tag{4.29}$$

At first, we have

$$\begin{aligned} R_1 = -\,i \int \psi^+ [(\mathbf{r} \times \nabla)\mu(\rho)\,(\omega_1\alpha_4 + \omega_2\alpha_5) \\ - \mu(\rho)(\omega_1\alpha_4 + \omega_2\alpha_5)(\mathbf{r} \times \nabla)]\psi dv, \end{aligned} \tag{4.30}$$

which easily takes the following form:

$$R_1 = -i \int \psi^+ (\mathbf{r} \times \nabla) F(\rho)\psi dv = \mathrm{i} \int \psi^+ (\mathbf{r} \times \nabla) F\left(\omega_1^2 + \omega_2^2\right)\psi dv, \tag{4.31}$$

where $F(\omega_1^2 + \omega_2^2)$ is a function, the expression of which is not important because $(\omega_1^2 + \omega_2^2)$ does not depend on the angles, so that $R_1 = 0$. Finally, we find the following for the other components:

$$\begin{aligned} R_2 &= D \int \psi^+ \frac{\mathbf{r}}{x^2 + y^2} [\sigma_4\,(\omega_1\alpha_4 + \omega_2\alpha_5) - (\omega_1\alpha_4 + \omega_2\alpha_5)\sigma_4\,]\psi dv = 0 \\ R_3 &= \int \psi^+ \mu(\rho)[\,\sigma(\omega_1\alpha_4 + \omega_2\alpha_5) - (\omega_1\alpha_4 + \omega_2\alpha_5)\,\sigma]\psi dv = 0 \end{aligned} \tag{4.32}$$

because σ commutes and σ_4 anticommutes with α_4 and α_5.

So, the nonlinear equations [Eq. (4.2) or (4.3)] define the same angular momentum [Eq. (3.15)] as the linear equations. Therefore, the angular part must be the same in both cases; the difference is only in the radial factor.

4.3 CHIRAL GAUGE AND TWISTED SPACE. TORSION AND MAGNETISM

Let us take the particular case of Eq. (4.2) when $B_\mu = 0$; $\kappa(\rho) = \lambda\,\rho,\ \lambda = const$:

$$\gamma_\mu \partial_\mu \Psi + i\lambda(\omega_1 - i\gamma_5\omega_2)\Psi = 0. \tag{4.33}$$

Equivalent equations were considered by many researchers (Finkelstein, Lelevier, & Ruderman, 1951; Heisenberg, 1954; Dürr *et al.*, 1959; Weyl, 1950; Rodichev, 1961). Of these, Rodichev (1961) was the one to consider a space with an *affine connection,* and we shall briefly summarize this problem as follows:

1. No metric is introduced, and the theory is formulated only in terms of *connection coefficients* Γ^i_{rk}. One can define contravariant and covariant vectors T^i and T_i, and *covariant derivatives:*

$$\nabla_\mu T^i = \partial_\mu T^i + \Gamma^i_{r\mu} T^r; \quad \nabla_\mu T_i = \partial_\mu T_i - \Gamma^r_{i\mu} T_r. \tag{4.34}$$

2. Two important tensors are defined here[10], *curvature* and *torsion:*

$$-R^\mu_{\nu\sigma\lambda} = \partial_\sigma \Gamma^\mu_{\nu\lambda} - \partial_\lambda \Gamma^\mu_{\nu\sigma} + \Gamma^\mu_{\rho\sigma}\Gamma^\rho_{\nu\lambda} - \Gamma^\mu_{\rho\lambda}\Gamma^\rho_{\nu\sigma} \text{ and } S^\lambda_{[\mu\nu]} = \Gamma^\lambda_{\mu\nu} - \Gamma^\lambda_{\nu\mu}. \tag{4.35}$$

3. A *parallel transport* along a curve $x(t)$ is defined by $\nabla_\xi T = \xi^k \nabla_k T = 0$, $(\xi = x(t))$. A *geodesic line* is generated by the parallel transport of its tangent. Apart from a Euclidian space, a *geodesic rectangle is* broken by a gap in two terms: the first, in dt^2, depends on torsion, while the second, of the order of $o(dt^3)$, depends on curvature.
4. In a *twisted space* $(S^\lambda_{[\mu\nu]} \neq 0)$, a geodesic loop is an arc of helicoid, with a "thread" of the *second order:* the order of an area. Something similar happens in a *spin fluid:* the angular momentum of a droplet is of higher order than the spin (Costa de Beauregard, 1983; Weyl, 1950). Now, Rodichev considers the case of *a flat, twisted space:* with torsion $(\Gamma^\lambda_{[\mu\nu]} = S^\lambda_{[\mu\nu]} \neq 0)$ but straight geodesics $(\Gamma^\lambda_{(\mu\nu)} = 0)$, and with the following connection and covariant spinor derivative:

$$\Gamma_{\lambda[\mu\nu]} = S_{\lambda\mu\nu} = \Phi_{[\lambda\mu\nu]}; \quad \nabla_\mu \Psi = \partial_\mu \Psi - \frac{i}{4}\Phi_{[\lambda\mu\nu]}\gamma_\mu\gamma_\nu\Psi \tag{4.36}$$

and the following Lagrangian density:

$$L = \frac{1}{2}\{\overline{\psi}\,\gamma_\mu \nabla_\mu \psi - (\nabla_\mu \overline{\psi})\,\gamma_\mu \psi\}. \tag{4.37}$$

[10] When $R^i_{qkl} = S^\lambda_{[\mu\nu]} = 0$, the space is Euclidian.

Translating the last formula in our notation, it gives:

$$L = \frac{1}{2}\left\{\overline{\Psi}\,\gamma_\mu \partial_\mu \Psi - \left(\partial_\mu \overline{\Psi}\right)\,\gamma_\mu \Psi - \frac{i}{2}\Phi_{[\mu\nu\lambda]}\overline{\Psi}\gamma_\mu\gamma_\lambda\Psi\right\}. \tag{4.38}$$

Introducing the axial dual vector $\Phi_\mu = \frac{i}{3!}\varepsilon_{[\mu\nu\lambda\sigma]}\Phi_{[\nu\lambda\sigma]}$, the Lagrangian becomes

$$L = \frac{1}{2}\{\overline{\Psi}\,\gamma_\mu \partial_\mu \Psi - \left(\partial_\mu \overline{\Psi}\right)\gamma_\mu \Psi - \Phi_\mu \overline{\Psi}\gamma_\mu\gamma_5\Psi\}, \tag{4.39}$$

which gives the following equation:

$$\gamma_\mu\left(\partial_\mu - \frac{1}{2}\Phi_\mu\gamma_5\right)\Psi = 0. \tag{4.40}$$

With $\Phi_\mu = \frac{2g}{\hbar c}B_\mu$, this is *our equation* [Eq. (2.16)]. Let us note that Rodichev did not introduce Φ_μ as an external field: it was only a geometrical property. But in our case, we can say that a monopole plunged into an electromagnetic field induces a torsion in the surrounding space.

Rodichev ignored the monopole. He did not aim at the linear equation [Eq. (2.16)], but rather at a nonlinear equation, through the following Einstein-like action integral without an external field:

$$S = \int (L - bR)\, d^4x, \tag{4.41}$$

where L is given by Eq. (4.39), $b = Const$, R = total curvature, and

$$R = \Phi_{[\lambda\mu\nu]}\Phi_{[\lambda\mu\nu]} = -6\Phi_\mu\Phi_\mu. \tag{4.42}$$

Hence, Eq. (4.41) becomes

$$S = \int \frac{1}{2}\{\left[\overline{\Psi}\,\gamma_\mu\partial_\mu\Psi - \left(\partial_\mu\overline{\Psi}\right)\gamma_\mu\Psi - \Phi_\mu\overline{\Psi}\gamma_\mu\gamma_5\Psi\right] + 6b\Phi_\mu\Phi_\mu\}\, d^4x. \tag{4.43}$$

If we vary S with respect to Φ, we find

$$\begin{aligned} \Phi_\mu &= \frac{1}{24b}\overline{\Psi}\gamma_\mu\gamma_5\Psi \\ R &= -\frac{1}{4b}\left(\overline{\Psi}\gamma_\mu\gamma_5\Psi\right)\left(\overline{\Psi}\gamma_\mu\gamma_5\Psi\right). \end{aligned} \tag{4.44}$$

Now, the variation of S with respect to Ψ gives the following nonlinear equation, in which we recognize the Heisenberg equation (Borne, Lochak, & Stumpf, 2001), up to the coefficient

$$\gamma_\mu \partial_\mu \psi - \frac{1}{48b}\left(\overline{\psi}\gamma_\mu\gamma_5\psi\right)\gamma_\mu\gamma_5\psi = 0. \tag{4.45}$$

In so doing, *we come back once more to the monopole,* but now *in the nonlinear case* because Eq. (4.45) is a particular case of Eq. (4.2), by virtue of Eqs. (2.23), (2.24), and (4.44), which gives

$$R = \frac{1}{4b}\left(\omega_1^2 + \omega_1^2\right). \tag{4.46}$$

It means that the fundamental chiral invariant $\left(\omega_1^2 + \omega_1^2\right)$ that we defined, apart from a constant factor, is the curvature of the twisted space created by the self-action of the monopole, expressed in the equation by the identification of the torsion with the total curvature in Eq. (4.36). This confirms the link between our monopole and a torsion of the space.

CHAPTER 5

The Dirac Equation on the Light Cone. Majorana Electrons and Magnetic Monopoles

5.1 INTRODUCTION. HOW THE MAJORANA FIELD APPEARS IN THE THEORY OF A MAGNETIC MONOPOLE

In the first chapters of this book, we have developed the theory of a massless linear monopole, the quantized magnetic charge of which generalizes the Dirac formula. The neutrino appears as the fundamental zero state of the magnetic charge. The monopole is massless because the linear Dirac mass term would violate the *chiral gauge* invariance $\Psi \rightarrow \exp\left(i\frac{g}{\hbar c}\gamma_5\phi\right)\Psi$, which ensures the conservation of magnetism.

Nevertheless, in Chapter 4, we gave a generalization [Eq. (4.2)] of the linear equation [Eq. (2.16)], owing to the introduction of a *nonlinear mass term,* which is invariant with respect to the chiral gauge. There is an infinite family of such mass terms depending on an arbitrary function of a *chiral invariant* that is equal (up to a constant factor) to the space curvature.

Now we shall reexamine the problem of mass in another way. We shall consider *the Dirac equation on the relativistic light cone,* which gives a generalization of the *Majorana condition*. This result was achieved in Lochak (1987a,b, 1992, 2004). The main idea is that the *Majorana condition*, which reduces the Dirac equation to an abbreviated form, will be replaced by the condition that *the chiral invariant equals zero*, which is equivalent to writing the Dirac equation on the relativistic light cone if we define the light cone by the condition that the electric current (i.e., the velocity of the particle) is isotropic:

$$J_\mu J_\mu = 0, \tag{5.1}$$

However, by virtue of the algebraic relations [Eq. (2.24)]:

$$-J_\mu J_\mu = \Sigma_\mu \Sigma_\mu = \omega_1^2 + \omega_2^2 \ \left[= 4\left(\xi^+ \eta\right)\left(\eta^+ \xi\right) \text{ in the Weyl representation}\right].$$

Thus, Eq. (5.1) means that the chiral invariant equals zero on the light cone:

$$\rho = \sqrt{\omega_1^2 + \omega_2^2} = 0 \tag{5.2}$$

It must be noted that this definition is compatible with the conservation of electricity and magnetism because ρ is invariant under the ordinary gauge and the chiral gauge. Let us now consider the equations of the magnetic monopole (given in Chapter 4), with a nonlinear mass term.

So we have, in the Dirac representation:

$$\gamma_\mu \left(\partial_\mu - \frac{g}{\hbar c} \gamma_5 B_\mu \right) \Psi + \frac{1}{2} \frac{m(\rho)c}{\hbar} (\omega_1 - i\gamma_5 \omega_2) = 0. \tag{5.3}$$

And then, in the Weyl representation:

$$\begin{aligned} &\frac{1}{c}\partial_t \xi - \mathbf{s}.\nabla \xi - i\frac{g}{\hbar c}(W + \mathbf{s}.\,\mathbf{B})\xi + i\frac{m(2|\xi^+\eta|)c}{\hbar}\left(\eta^+ \xi\right)\ \eta = 0 \\ &\frac{1}{c}\partial_t \eta + \mathbf{s}.\nabla \eta + i\frac{g}{\hbar c}(W - \mathbf{s}.\,\mathbf{B})\eta + i\frac{m(2|\xi^+\eta|)c}{\hbar}\left(\xi^+ \eta\right)\ \xi = 0 \end{aligned} \quad \left\{B_\mu = (-i\mathbf{B}, W)\right\}. \tag{5.4}$$

These equations are invariant with respect to the chiral gauge transformation, and they represent a magnetic monopole. It was shown in Chapter 4 that the solutions of such equations as Eqs. (5.3) and (5.4) are divided into bradyon states (slower than light), tachyon states (faster than light), and luxon states (at the speed of light).

Just like the linear equations of the monopole, these nonlinear equations admit a nonlinear neutrino as a particular case for a zero charge $g = 0$, which means that such a nonlinear neutrino must have the same three states as the nonlinear monopole: bradyon, tachyon, and luxon. This hypothesis was previously formulated in another frame by Mignani and Recami (1975) and Recami and Mignani (1976).

Now the luxon state corresponds to the cancellation of the mass terms in Eqs. (5.3) and (5.4), which are thus reduced to the linear equations [Eqs. (2.16) and (3.4)]. But here, it does not mean a simple elimination of the mass term by the annihilation of a mass coefficient because m is not a simple coefficient, but a function. So that means a nonlinear condition on the wave functions:

$$\rho = 0 \;\Rightarrow\; \omega_1 = \omega_2 = 0 \;\Rightarrow\; \xi^+\eta = 0. \tag{5.5}$$

The cancellation of the nonlinear term under the condition [Eq. (5.5)] does not imply the cancellation of the wave. The condition [Eq. (5.5)] is not exactly equivalent to the Majorana condition (Majorana, 1937; McLennan, 1957), which reads as $\psi = \psi_c$ $(\psi_c = \psi_{charge\ \ conjugated})$. Rather, it gives a slightly more general condition (Lochak, 1985):

$$\psi = e^{2i\frac{e}{\hbar c}\theta}\gamma_2\psi^* = e^{2i\frac{e}{\hbar c}\theta}\psi_c \;\Rightarrow\; \xi = e^{2i\frac{e}{\hbar c}\theta}is_2\eta^*, \quad \eta = -e^{2i\frac{e}{\hbar c}\theta}is_2\xi^*, \tag{5.6}$$

where $\theta(x, t)$ is an arbitrary phase (the coefficient $2e/\hbar c$ will be useful later).

In other words, the ψ state defined by Eq. (5.6) is its own charge-conjugate, but up to an arbitrary phase: this is almost the Majorana condition, which gives not exactly the Majorana-abbreviated equation. Later, we shall consider an equation that will not be abbreviated from the linear Dirac equation of the electron, but from the nonlinear equation of the monopole.

The fact that such a condition arises from the monopole theory leads us to explore it more precisely. Since the abbreviated Majorana equation was already suggested as a possible equation for the neutrino, we can ask: why would this not be the case for a magnetic monopole?

Nevertheless, for now we shall consider not the magnetic case, but the electric one. And we want to issue an initial warning: Do not be disappointed that we will be looking at the electric case for a longer time than the magnetic one. The reason for this is that, the magnetic case is far much complicated than the electric one, and that the last is interesting in itself. And it is not so elementary—and not only that, it is illuminating for our subject.

5.2 THE ELECTRIC CASE: LAGRANGIAN REPRESENTATION AND GAUGE INVARIANCE OF THE MAJORANA FIELD

Several authors (e.g., McLennan, 1957; Case, 1957; Berestetsky, Lifschitz, & Pitaevsky, 1972) have written about the problem of a Lagrangian representation of the Majorana field, and they concluded that such a representation is impossible. We shall see that that is wrong, but it is interesting to see where the difficulty is.

Using Eq. (2.11) and the Majorana condition, $\psi = \psi_c$, for an electrically charged particle in the presence of an electomagnetic field, the Majorana equation may be written as

$$\gamma_\mu\left(\partial_\mu - \frac{ie}{\hbar c}A_\mu\right)\psi + \frac{m_0 c}{\hbar}\psi_c = 0. \tag{5.7}$$

If we try to find a Lagrangian for such an equation directly, it must contain a term like the following:

$$\overline{\psi}\psi_c = \psi^+\gamma_4\gamma_2\psi^*. \tag{5.8}$$

But we have, on the other hand:

$$\begin{aligned} \gamma_k &= i\alpha_4\alpha_k \ (k = 1,2,3), \ \ \gamma_4 = \alpha_4 \\ \alpha_k = \begin{bmatrix} 0 & s_k \\ s_k & 0 \end{bmatrix}&, \ \alpha_4 = \begin{bmatrix} I & 0 \\ 0 & -I \end{bmatrix}; \ (s_k = \text{Pauli matrices}). \end{aligned} \tag{5.9}$$

Introducing these expressions into Eq. (5.8), we have $\overline{\psi}\psi_c = 0$, and the corresponding term disappears from the Lagrangian, which is precisely the difficulty. But we shall proceed in another way: we consider the Majorana field as a constrained state of the Dirac field and express this constraint under the form of Eq. (5.5). Thus, we define the *Majorana Lagrangian* as a Dirac Lagrangian L_D, to which we add a constraint term with a Lagrange parameter λ:

$$L_M = L_D + \frac{\lambda}{2}\left(\omega_1^2 + \omega_2^2\right). \tag{5.10}$$

ω_1 and ω_2 are taken from Eq. (2.18), so that the variation of L_M, with respect to ψ, gives

$$\gamma_\mu\left(\partial_\mu - \frac{ie}{\hbar c}A_\mu\right)\psi + \frac{m_0 c}{\hbar}\psi + \lambda(\omega_1 - i\omega_2\gamma_5)\psi = 0. \tag{5.11}$$

This equation looks like our nonlinear equation [discussed in Chapter 4 and Lochak (1984, 1987a,b)], but here we have a mass term and an electric potential instead of the magnetic potential. In this form, the equation was found by Hermann Weyl (1950) and rediscovered later by other authors. The aim of Weyl (related to general relativity) was very different from ours.

Now, we vary the Lagrangian L_M [Eq. (5.10)] with respect to λ, which gives, using Eq. (5.6):

$$\gamma_\mu\left(\partial_\mu - \frac{ie}{\hbar c}A_\mu\right)\psi + \frac{m_0 c}{\hbar}e^{2i\frac{e}{\hbar c}\theta}\psi_c = 0 \tag{5.12}$$

It is the Majorana equation [Eq. (5.7)] with an *arbitrary* phase θ. We could write $\theta = 0$ in order to find Eq. (5.7), but that would be a bad idea because this phase is important: owing to this phase, Eq. (5.12) is *gauge-invariant,* while Eq. (5.7) is not. (By the way, nobody was worried about gauge-invariance)

In this case, the gauge invariance of Eq. (5.12) is a trivial consequence of the invariance of the Lagrangian [Eq. (5.10)]. But the invariance of Eq. (5.12) also can be directly demonstrated after the transformation:

$$\psi \to e^{i\frac{e}{\hbar c}\varphi}\psi, \quad A_\mu \to A_\mu - \partial_\mu\varphi, \quad \theta \to \theta + \varphi. \tag{5.13}$$

Now—and only now—the phase θ may be absorbed in the gauge and disappear. Therefore, we must first choose the gauge and only then cancel θ to find Eq. (5.7).

Therefore, the Majorana equation cannot be considered as independent: it is only the equation of a particular state (defined by a Lagrange multiplier) of the Dirac equation of the electron. And it is not gauge-invariant: only Eq. (5.11) is invariant. Nevertheless, we shall see that the Majorana equation may be considered itself, but this second interpretation is not equivalent to the preceding one.

5.3 TWO-COMPONENT ELECTRIC EQUATIONS. SYMMETRY AND CONSERVATION LAWS

Now, owing to Eq. (3.4), we find the Weyl representation of the class of the definite solutions of the Dirac equation:

$$\left(\pi_0 + \boldsymbol{\pi}.\mathbf{s}\right)\xi - im_0 c e^{\frac{2ie}{\hbar c}\theta} s_2 \xi^* = 0, \tag{5.14a}$$

$$\left(\pi_0 - \boldsymbol{\pi}.\mathbf{s}\right)\eta + im_0 c e^{2i\frac{e}{\hbar c}\theta} s_2 \eta^* = 0, \tag{5.14b}$$

$$\pi_0 = \frac{1}{c}\left(i\hbar\frac{\partial}{\partial t} + eV\right), \quad \boldsymbol{\pi} = \left(-i\hbar\nabla + \frac{e}{c}\mathbf{A}\right), \quad \{A_\mu = (\mathbf{A}, iV)\}. \qquad (5.15)$$

Eq. (5.14) is manifestly *C*-, *P*-, and *T*-invariant, but it is interesting to verify this property directly. Elementary calculations show, indeed, that the system [Eq. (5.14)] remains invariant by the following transformations using the Curie laws or those deduced from them (as covered in Chapter 4 and Poincaré, 1896):

$$\begin{aligned} &(C): \; i \to -i, \quad e \to -e, \quad \xi \to e^{2i\frac{e}{\hbar c}\theta} i s_2 \eta^*, \quad \eta \to -e^{2i\frac{e}{\hbar c}\theta} i s_2 \xi^* \\ &(P): \; \mathbf{x} \to -\mathbf{x}, \quad \mathbf{A} \to -\mathbf{A}, \quad \xi \to i\eta \quad \eta \to -i\xi \\ &(T): \; e \to -e, t \to -t, V \to -V, \eta \to s_2\xi^*, \xi \to -s_2\eta^*. \end{aligned} \qquad (5.16)$$

The *P* transformation can be written in another way:

$$(P): \; \mathbf{x} \to -\mathbf{x}, \quad \mathbf{A} \to -\mathbf{A}, \quad \xi \longleftrightarrow \eta, \quad \theta \to \theta + \frac{\pi}{2}\frac{\hbar c}{e}. \qquad (5.17)$$

And the gauge transformation takes the following form:

$$\xi \to e^{i\frac{e}{\hbar c}\varphi}\xi, \quad \eta \to e^{i\frac{e}{\hbar c}\varphi}\eta, \quad \mathbf{A} \to \mathbf{A} - \nabla\varphi, \quad V \to V + \frac{1}{c}\frac{\partial\varphi}{\partial t}, \quad \theta \to \theta + \varphi. \qquad (5.18)$$

It can be verified that the system [Eq. (5.14)] remains invariant under Eq. (5.18), which entails the conservation [Eq. (3.7)] of the chiral currents. It is important to note this conservation because it is true for a magnetic monopole (see Chapter 3), and here we see that it is also true for the solutions of the Dirac equation in the abbreviated case of an electron, restricted by the constraint $\rho = \sqrt{\omega_1^2 + \omega_2^2} = 0$ [Eq. (5.5)]. But this splitting into two equations is not applicable in the general case of the Dirac equation, which conserves only the electric current (the sum of the chiral currents), but not the magnetic current (see Chapter 3).

In the abbreviated electric case, the electric current is isotropic, the solutions of the Dirac equation are on the light cone, and the magnetic current disappears. Given that the expressions in Eq. (5.14) are split by the condition $\xi^+\eta = 0$, we can restrict ouselves, to only one of them—say the first one—and consider it in itself. This restricted equation is a chiral state of the electron. The second expression of Eq. (5.14) is the chiral conjugate of the first one, which means, owing to Eq. (5.16), that the image in a mirror is the time inverse of the first expression of Eq. (5.14).

5.4 THE CHIRAL STATE OF THE ELECTRON IN AN ELECTRIC COULOMB FIELD

Majorana considered that the equality $\psi = \psi_c$, introduced in the Dirac equation, gives something similar to a joint theory of the electron and the positron. But this is not the case because the preceeding equality is only a constraint imposed to the electron. Nevertheless, we have found a hybrid state: a kind of mixture of the electron and the positron. To show this, we shall solve the first expression of Eq. (5.14) in an electric coulomb field by introducing the following expressions:

$$eV = -\frac{e^2}{r}, \quad \mathbf{A} = 0, \quad \theta = \frac{\pi}{4}\frac{\hbar c}{e}. \tag{5.19}$$

These give the following equation:

$$\left[\frac{1}{c}\left(i\hbar\frac{\partial}{\partial t} - \frac{e^2}{r}\right) - i\hbar\mathbf{s}.\nabla\right]\xi + m_0 c s_2 \xi^* = 0. \tag{5.20}$$

The difficulty obviously lies in the complex conjugated ξ^*. So let us introduce the spherical functions with spin (Kramers, 1964; Bohm, 1960; Akhiezer & Berestetsky, 1965):

$$\Omega_\ell^m(+) = \begin{bmatrix} \left(\dfrac{\ell+m}{2\ell+1}\right)^{\frac{1}{2}} Y_\ell^{m-1} \\[2ex] \left(\dfrac{\ell-m+1}{2\ell+1}\right)^{\frac{1}{2}} Y_\ell^{m} \end{bmatrix}, \quad \Omega_\ell^m(-) = \begin{bmatrix} \left(\dfrac{\ell-m+1}{2\ell+1}\right)^{\frac{1}{2}} Y_\ell^{m-1} \\[2ex] -\left(\dfrac{\ell+m}{2\ell+1}\right)^{\frac{1}{2}} Y_\ell^{m} \end{bmatrix}, \tag{5.21}$$

in which Y_ℓ^m are the Laplace spherical functions ($\ell = 0, 1, 2, \ldots$; m $= -\ell, -\ell+1, \ldots, \ell-1, \ell$):

$$Y_l^m(\theta,\varphi) = \frac{(-1)^m}{2^l l!}\left(\frac{2l+1}{4\pi}\right)^{\frac{1}{2}}\left(\frac{(l+m)!}{(l-m)!}\right)^{\frac{1}{2}}\frac{e^{im\varphi}}{\sin^l\theta}\frac{d^{l-m}}{d\theta^{l-m}}sin^{2l}\theta. \tag{5.22}$$

Now, we have the following equalities (for more information, see Appendix A of this chapter):

$$\begin{aligned}
&\mathbf{s.n}\,\Omega^{m}_{l-1}(+) = \Omega^{m}_{l}(-);\ \ \mathbf{s.n}\,\Omega^{m}_{l}(-) = \Omega^{m}_{l-1}(+)\\
&\mathbf{s.n}\,s_2\,\Omega^{*m}_{l-1}(+) = i(-1)^{m+1}\Omega_{l}^{-m+1}(-)\\
&\mathbf{s.n}\,s_2\,\Omega^{*m}_{l}(-) = i(-1)^{m}\Omega_{l-1}^{-m+1}(+)
\end{aligned}\tag{5.23}$$

$$\begin{aligned}
&\mathbf{n} = \frac{\mathbf{r}}{r};\ \ x = r\cos\varphi\sin\theta,\ \ y = r\sin\varphi\sin\theta,\ \ z = r\cos\theta\overrightarrow{n} = \frac{\mathbf{r}}{r};\\
&x = r\cos\varphi\sin\theta,\ \ y = r\sin\varphi\sin\theta,\ \ z = r\cos\theta.
\end{aligned}\tag{5.24}$$

We look for a solution of Eq. (5.20) of the following form:

$$\xi = \sum_{m} F^{m}_{\ell-1}(t,r)\,\Omega^{m}_{\ell-1}(+) + \sum_{m'} G^{m'}_{\ell}(t,r)\,\Omega^{m'}_{\ell}(-\,). \tag{5.25}$$

But it is impossible to separate the variables t and r immediately. It is only possible to separate the angular variables φ and θ. Following a classical procedure in the Dirac theory (Kramers, 1964; Akhiezer & Berestetsky, 1965), we introduce Eq. (5.25) into Eq. (5.20), multiplying the left side by **s.n**. Owing to Eq. (5.23), we find

$$\begin{aligned}
&\frac{1}{c}\left(i\hbar\frac{\partial}{\partial t}-\frac{e^2}{r}\right)\left[F^{m}_{\ell-1}\,\Omega^{m}_{\ell}(-) + \sum_{m'} B^{m'}_{\ell}\,\Omega^{m'}_{\ell-1}(+)\right]\\
&= \ i\hbar\mathbf{s.n}\ \ \mathbf{s.}\nabla\left[\sum_{m} F^{m}_{\ell-1}\Omega^{m}_{\ell-1}(+) + \sum_{m'} B^{m'}_{\ell}\,\Omega^{m'}_{\ell}(-\,)\right]\\
&-im_0c\left[\sum_{m}(-1)^{m+1}F^{*m}_{\ell-1}\Omega_{\ell}^{-m+1}(-\,) + \sum_{m'}(-1)^{m'}B^{*m'}_{\ell}\Omega_{\ell}^{-m'+1}(+\,)\right].
\end{aligned}\tag{5.26}$$

The right-hand side is simplified owing to the classical relations:

$$\mathbf{s.n}\ \ \mathbf{s.}\nabla = \frac{\partial}{\partial r} - \frac{1}{r}\mathbf{s.}\Lambda, \tag{5.27}$$

where Λ is the orbital moment:

$$\Lambda = -i\mathbf{r}\times\nabla. \tag{5.28}$$

Now, we have other relations as follows (see Appendix B in this chapter):

$$\begin{aligned}
&\mathbf{s.}\Lambda\,\Omega^{m}_{\ell-1}(+) = (\ell-1)\,\Omega^{m}_{\ell-1}(+)\\
&\mathbf{s.}\Lambda\Omega^{m}_{\ell}(-) = -(\ell+1)\,\Omega^{m}_{\ell}(-\,),
\end{aligned}\tag{5.29}$$

so that, taking into account the fact that $\Omega_\ell^m(\pm)$ are orthonormal, we deduce from Eq. (5.26) the following system from which the angles are eliminated:

$$\begin{aligned}\left(\frac{1}{c}\frac{\partial}{\partial t}+i\frac{\alpha}{r}\right)F_{\ell-1}^m &= \left(\frac{\partial}{\partial r}+\frac{1+\ell}{r}\right)B_\ell^m+\chi(-1)^m F_{\ell-1}^{*-m+1}\\ \left(\frac{1}{c}\frac{\partial}{\partial t}+i\frac{\alpha}{r}\right)B_\ell^m &= \left(\frac{\partial}{\partial r}+\frac{1-l}{r}\right)F_{\ell-1}^m-\chi(-1)^m B_\ell^{*-m+1}\end{aligned} \tag{5.30}$$

$$m=-l,-l+1,\ldots,l-1,l,\quad \alpha=\frac{e^2}{\hbar c},\quad \chi=\frac{m_0 c}{\hbar}. \tag{5.31}$$

In a subsequent step, we take the complex conjugate form of Eq. (5.30), changing $m\to -m+1$:

$$\begin{aligned}\left(\frac{1}{c}\frac{\partial}{\partial t}-i\frac{\alpha}{r}\right)F_{l-1}^{*-m+1} &= \left(\frac{\partial}{\partial r}+\frac{1+l}{r}\right)B_l^{*-m+1}-\chi(-1)^m F_{l-1}^m\\ \left(\frac{1}{c}\frac{\partial}{\partial t}-i\frac{\alpha}{r}\right)B_l^{*-m+1} &= \left(\frac{\partial}{\partial r}+\frac{1-l}{r}\right)F_{l-1}^{*-m+1}+\chi(-1)^m B_l^m.\end{aligned} \tag{5.32}$$

We combine Eqs. (5.30) and (5.32), introducing the new functions:

$$\begin{aligned}\frac{P_{\ell-1}^m(r)}{r}e^{(-1)^m i\omega t} &= F_{\ell-1}^m+(-1)^m F_{\ell-1}^{*-m+1};\\ \frac{Q_{\ell-1}^m(r)}{r}e^{(-1)^m i\omega t} &= F_{\ell-1}^m-(-1)^m F_{\ell-1}^{*-m+1};\\ \frac{R_\ell^m(r)}{r}e^{(-1)^m i\omega t} &= B_\ell^m+(-1)^m B_\ell^{*-m+1};\\ \frac{S_\ell^m(r)}{r}e^{(-1)^m i\omega t} &= B_\ell^m-(-1)^m B_\ell^{*-m+1};\end{aligned} \tag{5.33}$$

with

$$Q_{l-1}^m=(-1)^{m+1}P_{l-1}^{*-m+1};\quad S_l^m=(-1)^{m+1}R_l^{*-m+1}. \tag{5.34}$$

With the definition in Eq. (5.34), the notations [Eq. (5.33)] are invariant under complex conjugation and $m\to -m+1$. Summing and subtracting Eqs. (5.30) and (5.32), we find a first-order system in r (see Ince, 1956):

$$r\frac{dX}{dr}=(M+Nr)X; \tag{5.35}$$

$$X=\begin{bmatrix} P^m_{\ell-1}(r) \\ Q^m_{\ell-1}(r) \\ R^m_{\ell}(r) \\ S^m_{\ell}(r) \end{bmatrix};\ M=\begin{bmatrix} \ell & 0 & 0 & i\alpha \\ 0 & \ell & i\alpha & 0 \\ 0 & i\alpha & -\ell & 0 \\ i\alpha & 0 & 0 & -\ell \end{bmatrix};$$

$$N=\begin{bmatrix} 0 & 0 & i\frac{\omega'}{c} & -\chi \\ 0 & 0 & \chi & i\frac{\omega'}{c} \\ i\frac{\omega'}{c} & \chi & 0 & 0 \\ -\chi & i\frac{\omega'}{c} & 0 & 0 \end{bmatrix};\ \omega'=(-1)^m\omega. \tag{5.36}$$

The matrix N is diagonalized by

$$S=\frac{1}{\sqrt{2}}\begin{bmatrix} 1 & 0 & \frac{\omega'}{\mu c} & i\frac{\chi}{\mu} \\ 0 & 1 & -i\frac{\chi}{\mu} & \frac{\omega'}{\mu c} \\ 1 & 0 & -\frac{\omega'}{\mu c} & -i\frac{\chi}{\mu} \\ 0 & 1 & i\frac{\chi}{\mu} & -\frac{\omega'}{\mu c} \end{bmatrix};\quad \mu=\sqrt{\frac{\omega^2}{c^2}-\chi^2}. \tag{5.37}$$

Introducing the new variable

$$Y=SX, \tag{5.38}$$

Eq. (5.35) takes the following form:

$$r\frac{dY}{dr}=\left\{i\mu r\begin{bmatrix} I & 0 \\ 0 & -I \end{bmatrix}+l\begin{bmatrix} 0 & I \\ I & 0 \end{bmatrix}+\frac{\alpha}{\mu}\begin{bmatrix} \frac{\omega'}{c}s_1+i\chi s_3 & 0 \\ 0 & -\frac{\omega'}{c}s_1-i\chi s_3 \end{bmatrix}\right\}Y. \tag{5.39}$$

Here, μ is defined in Eq. (5.37), I is the unit matrix of the second order, and s_1, s_3 are Pauli matrices. We shall now diagonalize Eq. (5.39), changing the functions once more:

$$Z = \begin{bmatrix} V & 0 \\ 0 & s_1 V \end{bmatrix} Y; \quad V = \left[\frac{\omega'}{2\mu c}\right]^{\frac{1}{2}} \begin{bmatrix} \left[\dfrac{\omega'/c}{\mu - i\chi}\right]^{\frac{1}{2}} & \left[\dfrac{\mu - i\chi}{\omega'/c}\right]^{\frac{1}{2}} \\ \left[\dfrac{\omega'/c}{\mu + i\chi}\right]^{\frac{1}{2}} & -\left[\dfrac{\mu + i\chi}{\omega'/c}\right]^{\frac{1}{2}} \end{bmatrix}. \tag{5.40}$$

V is chosen such that

$$V\left(\frac{\omega'}{c} s_1 + i\chi s_3\right) V^{-1} = \mu s_3. \tag{5.41}$$

The equation takes a new form:

$$r\frac{dZ}{dr} = \left\{ i\mu r \begin{bmatrix} I & 0 \\ 0 & -I \end{bmatrix} + \ell \begin{bmatrix} 0 & s_1 \\ s_1 & 0 \end{bmatrix} + i\alpha \begin{bmatrix} s_3 & 0 \\ 0 & s_3 \end{bmatrix} \right\} Z, \tag{5.42}$$

and by iteration, we find

$$\left[r\frac{d}{dr}\right]^2 Z = \left\{ -\mu^2 r^2 + \mu r \left(i \begin{bmatrix} I & 0 \\ 0 & -I \end{bmatrix} - 2\alpha \begin{bmatrix} s_3 & 0 \\ 0 & s_3 \end{bmatrix} \right) + \ell^2 - \alpha^2 \right\} Z. \tag{5.43}$$

All the matrices are diagonalized, and we find four independent equations for the components of Z:

$$\left[r\frac{d}{dr}\right]^2 Z_n = \left[-\mu^2 r^2 + \mu r (i\varepsilon - 2\alpha\varepsilon') + \ell^2 - \alpha^2 \right] Z_n \quad (n = 1, 2, 3, 4) \tag{5.44}$$

$$\varepsilon = 1, 1, -1, -1; \;\; \varepsilon' = 1, -1, -1, 1 \quad (\text{for} \quad n = 1, 2, 3, 4). \tag{5.45}$$

At this point, let us have the following:

$$r = \frac{i\rho}{2\mu}, \quad W_n = \rho^{\frac{1}{2}} Z_n. \tag{5.46}$$

Eq. (5.44) becomes, neglecting the suffix n,

$$\frac{d^2 W}{d\rho^2} + \left[-\frac{1}{4} + \frac{\frac{\varepsilon}{2} - i\alpha\varepsilon'}{\rho} + \frac{\frac{1}{4} + \alpha^2 - l^2}{\rho^2} \right] W = 0. \tag{5.47}$$

This is a Whittaker equation (Ince, 1956; Whittaker & Watson, 1958). The following coefficients are denoted here by k and m, keeping the classical notation for $W_{k,m}$. They are not to be confused with the previous other indices:

$$k = \frac{\varepsilon}{2} - i\alpha\varepsilon', \quad m = \sqrt{l^2 - \alpha^2}. \tag{5.48}$$

Thus, we can take the following Whittaker functions as radial functions, provided that they are square-integrable at the origin:

$$W_{k,m}(\rho) = W_{\frac{\varepsilon}{2}-i\alpha\varepsilon', \sqrt{l^2-\alpha^2}}\left(-2i\mu r\right). \tag{5.49}$$

However, in the vicinity of the origin, a regular solution of Eq. (5.47) may be written in the following form (Ince, 1956; Whittaker & Watson, 1958), taking into account Eqs. (5.46) and (5.48):

$$\left|W_{k,m}\right| = 2\mu r^{\frac{1}{2}+m}\Big(1 + O(r)\Big) \tag{5.50}$$

It must be noted that the same coefficient m appears in all the composants W_n, and thus in Z_n in Eq. (5.44); therefore, with the changes in Eqs. (5.46), (5.40), (5.38), (5.34), and (5.25), we can assert that

$$\xi^+\xi \approx r^2(m-1) \quad \text{(in the vicinity of } r = 0). \tag{5.51}$$

So the value of m [Eq. (5.48)] shows that $\xi^+\xi$ is always integrable at the origin because $l = 0, 1, 2...$ But even more interesting is the behavior at the infinity. From standard formulas, we have (Whittaker & Watson, 1958)

$$W_{k,m}(\rho) = e^{-\frac{1}{2}\rho}\rho^k\Big(1 + O(\rho^{-1})\Big) \quad \text{if } \left|\text{Arg}(-\rho)\right| < \pi. \tag{5.52}$$

The condition of validity is satisfied because $\rho = -2i\mu r$, by virtue of Eq. (5.46), so that, owing to Eq. (5.48):

$$W_{k,m}(\rho) = 2\mu r^{\frac{\varepsilon}{2}}(1 + O(r^{-1})\ \ [\varepsilon = \pm 1, \text{ as in Eq. (5.44)}]. \tag{5.53}$$

If we now consider the change of functions $\xi^+\xi$, we encounter some difficulty. In $r \approx 0$, we had the same exponents in Eq. (5.50) for all the components W and Z, but now the situation is different with the exponent $\frac{\varepsilon}{2}$ in Eq. (4.35). Using Eqs. (5.46), (5.40), (5.38), (5.34), and (5.25) once more, we find for $\xi^+\xi$ the following asymptotic form:

$$\xi^+\xi = \sum a_{nn'} r^{\varepsilon_n + \varepsilon_{n'} - 3} \quad (\text{for } r \to \infty), \tag{5.54}$$

where, according to Eq. (5.46), ε_n take the values $\varepsilon = \pm 1$ for the different components of Z, which leads to several conclusions, discussed next.

5.5 CONCLUSIONS FROM THE PHYSICAL BEHAVIOR OF A CHIRAL STATE OF A DIRAC ELECTRON (A MAJORANA ELECTRON), IN AN ELECTRIC COULOMBIAN FIELD

The asymptotic form [Eq. (5.54)] shows that $\xi^+\xi$ *would be integrable in the whole space only if*, in the sum of the second member of Eq. (5.54), ε_n is never equal to 1. The different values of ε_n and $\varepsilon_{n'}$ give terms with r^{-5} (for $\varepsilon_n + \varepsilon_{n'} = -2$); r^{-3} (for $\varepsilon_n + \varepsilon_{n'} = 0$); r^{-1} (for $\varepsilon_n + \varepsilon_{n'} = 2$).

Now, only the first type of term gives a convergent integral as $r \to \infty$. In order for the integral of $\xi^+\xi$ to converge, we must exclude the terms with $\varepsilon_n = 1$, which implies the annihilation of the components Z_1 and Z_2 in Eq. (5.42). But if we do this, we get $Z \equiv 0$ and the wave function disappears.

$\xi^+\xi$ is, thus, *never integrable* on the whole space. Therefore, the Majorana electron has no bound states in a Coulomb field: the spectrum is continuous and there are only ionized states. It must be noticed that the sign of α in Eq. (5.42) does not play just any role: the Majorana electron—more precisely, the Majorana state of the Dirac electron—has a diffusive behavior of the same type, independent of a positive or negative charge of the coulomb field.

It is easy to understand why this is so. In the state of ξ [Eq. (5.25)], which is associated with a value $\frac{l-1}{2}$ of the total kinetic momentum, the terms corresponding to the different values m have, according to Eq. (5.33), exponential factors $e^{(-1)^m \omega t}$, where ω is the energy such that ξ *is a superposition of states with positive and negative energies,* corresponding to the *electron or positron states.*

Thus, the Majorana theory is not a "simultaneous theory of the electron and of the positron". It is only a hybrid state of the Dirac electron, "which does not know" the sign of its electric charge. We understand why it cannot be in a bound state. But its diffusing states will be very different from the state of a fast, "normal" electron state because the wave functions are different from the wave functions of a Keplerian system in a ionized state.

To make this fact more understandable, we shall carry out the preceding calculation in the classical limit, and we shall see that all the trajectories are hyperbolic, as it might be guessed, but the hyperbolas are *not Keplerian.* And given that the classical limit does not know the quantum superposition, there are two kinds of hyperbolas corresponding respectively to the diffusion, in an attractive or a repulsive field.

5.6 THE GEOMETRICAL OPTICS APPROXIMATION OF THE STATES OF THE MAJORANA ELECTRON

Consider the general equation [the first expression of Eq. (5.14)] for ξ, with the definitions [Eq. (5.15)] and the electromagnetic gauge [Eq. (5.19)]. Now, in the first expression of Eq. (5.14), we introduce the following expression [$a(t,\mathbf{r})$ and $b(t,\mathbf{r})$ are new *spinors*]:

$$\xi = a(t,\mathbf{r})e^{-\frac{i}{\hbar}S(t,\mathbf{r})} + b(t,\mathbf{r})e^{+\frac{i}{\hbar}S(t,\mathbf{r})}. \tag{5.55}$$

Neglecting the $\hbar$ terms, we have the following equation:

$$\begin{aligned}&\left\{\left[\frac{1}{c}\left(\frac{\partial S}{\partial t}+eV\right)-\left(\nabla S-\frac{e}{c}\mathbf{A}\right).\mathbf{s}\right]a+m_0cs_2b^*\right\}e^{-\frac{i}{\hbar}S}\\&-\left\{\left[\frac{1}{c}\left(\frac{\partial S}{\partial t}-eV\right)-\left(\nabla S+\frac{e}{c}\mathbf{A}\right).\mathbf{s}\right]b-m_0cs_2a^*\right\}e^{+\frac{i}{\hbar}S}=0.\end{aligned} \tag{5.56}$$

For $\hbar \to 0$, the phases $\pm\frac{S}{\hbar}$ become infinitely fast, and, multiplying Eq. (5.56) by $e^{\frac{iS}{\hbar}}$ and $e^{\frac{-iS}{\hbar}}$, alternately, we find the geometrical optics approximation:

$$\begin{aligned}&\left[\frac{1}{c}\left(\frac{\partial S}{\partial t}+eV\right)-\left(\nabla S-\frac{e}{c}\mathbf{A}\right).\mathbf{s}\right]a+m_0cs_2b^*=0\\&\left[\frac{1}{c}\left(\frac{\partial S}{\partial t}-eV\right)+\left(\nabla S+\frac{e}{c}\mathbf{A}\right).\mathbf{s}\right]b-m_0cs_2a^*=0.\end{aligned} \tag{5.57}$$

Now we introduce a new spinor $\widehat{b}$:

$$\widehat{b} = s_2b^*. \tag{5.58}$$

Taking the complex conjugate of the second equation [Eq. (5.57)] multiplied on the left by s_2 (taking into account that s_2 is imaginary, which gives the plus sign in the second equation), one obtains for Eq. (5.57):

$$\begin{aligned}&\left[\frac{1}{c}\left(\frac{\partial S}{\partial t}+eV\right)-\left(\nabla S-\frac{e}{c}\mathbf{A}\right).\mathbf{s}\right]a+m_0c\widehat{b}=0\\&\left[\frac{1}{c}\left(\frac{\partial S}{\partial t}-eV\right)+\left(\nabla S+\frac{e}{c}\mathbf{A}\right).\mathbf{s}\right]\widehat{b}+m_0ca=0.\end{aligned} \tag{5.59}$$

Multiplying the first equation by the matrix before $\widehat{b}$ in the second equation, we get

$$\left\{\left[\frac{1}{c}\left(\frac{\partial S}{\partial t}-eV\right)+\left(\nabla S+\frac{e}{c}\mathbf{A}\right).\mathbf{s}\right]\left[\frac{1}{c}\left(\frac{\partial S}{\partial t}+eV\right)\right.\right.$$
$$\left.\left.-\left(\nabla S-\frac{e}{c}\mathbf{A}\right).\mathbf{s}\right]-m_0^2c^2\right\}a=0 \tag{5.60}$$

or

$$\left\{\begin{array}{l}\dfrac{1}{c}\left(\dfrac{\partial S}{\partial t}+eV\right)\left(\dfrac{\partial S}{\partial t}-eV\right)-\left(\nabla S+\dfrac{e}{c}\mathbf{A}\right)\left(\nabla S-\dfrac{e}{c}\mathbf{A}\right)-m_0^2c^2\\ +\ 2\dfrac{e}{c}\left[V\nabla S+\dfrac{1}{c}\dfrac{\partial S}{\partial t}\mathbf{A}+i\nabla S\times\mathbf{A}\right].\mathbf{s}\end{array}\right\}a=0. \tag{5.61}$$

For $a\neq 0$, we must set equal to zero the determinant of the matrix, which gives a Hamilton-Jacobi equation that reads, for $\mathbf{A}=0$:

$$\left[\frac{1}{c^2}\left(\frac{\partial S}{\partial t}\right)^2-(\nabla S)^2-\frac{e^2}{c^2}V^2-m_0^2c^2\right]^2-\frac{4e^2}{c^2}V^2(\nabla S)^2=0. \tag{5.62}$$

The factorization of the difference of two squares gives two equations that take the following form in the coulomb case:

$$\frac{1}{c}\left(\frac{\partial S}{\partial t}\right)^2-\left(\left|\nabla S\right|-\frac{\varepsilon e^2}{c}\frac{1}{r}\right)^2-m_0^2c^2=0\quad(\varepsilon=\pm 1). \tag{5.63}$$

We can see that the sign of the charge does not play any role because $\varepsilon=\pm 1$ not due to the charge, but to the factorization. And, still more important, these Hamilton-Jacobi equations are different from those that are found in the well-known problem of an electron in a coulomb field. In the latter case, we have the following equations with two signs $\varepsilon=\pm 1$ as well, but they are now due to the sign of the charge, and they correspond to two kinds of trajectories, ellipses, or hyperbolas:

$$\frac{1^2}{c}\left(\frac{\partial S}{\partial t}-\frac{\varepsilon e^2}{r}\right)^2-(\nabla S)^2-m_0^2c^2=0\quad(\varepsilon=\pm 1). \tag{5.64}$$

Now, if we introduce in Eq. (5.63) the decomposition

$$S=-Et+W, \tag{5.65}$$

we find

$$\frac{E^2}{c^2} - m_0^2c^2 = \left[\left|\nabla W\right| - \frac{\varepsilon e^2}{c}\frac{1}{r}\right]^2, \tag{5.66}$$

from which it follows immediately that

$$E \;\geq\; m_0c^2. \tag{5.67}$$

This means that there is not any bound state, and thus no closed trajectories. Actually, we have in Eq. (5.66) two equations:

$$(\nabla W)^2 = \frac{1}{c^2}\left(\sqrt{E^2 - m_0^2c^4} + \frac{\varepsilon e^2}{r}\right)^2 \quad (\varepsilon = \pm 1), \tag{5.68}$$

and thus, in polar coordinates:

$$(\nabla W)^2 = \left(\frac{\partial W}{\partial r}\right)^2 + \frac{1}{r^2}\left(\frac{\partial W}{\partial \varphi}\right)^2. \tag{5.69}$$

Now, if we write

$$W = J\varphi + f(r) \quad (J = \text{Const.}), \tag{5.70}$$

Eq. (5.68) is transformed into

$$f(r) = \int \left(A^2 + \frac{2B}{r} + \frac{C}{r^2}\right)^{\frac{1}{2}} dr, \tag{5.71}$$

with

$$A = \frac{1}{c}\sqrt{E^2 - m_0^2c^4}, \quad B = \frac{A\varepsilon e^2}{c}, \quad C = \frac{e^4}{c^2} - J^2. \tag{5.72}$$

The discriminant in Eq. (5.71) is positive:

$$\Delta' = B^2 - A^2C = A^2J^2 \geq 0 \tag{5.73}$$

As a result, the roots are real:

$$\frac{1}{r} = \frac{A\left(\frac{\varepsilon e^2}{c} \pm J\right)}{J^2 - \frac{e^4}{c^2}} \quad (\varepsilon = \pm 1). \tag{5.74}$$

Suppose now that $J \neq 0$. Given that we are in the limit of a quantum problem, we can write

$$J \approx \hbar = \frac{e^2}{\alpha c} = 137\frac{e^2}{c} \gg \frac{e^2}{c} \tag{5.75}$$

This approximation is not essential, but it is convenient for what follows because we can write the positive root [Eq. (5.74)] in a simplified form:

$$\frac{1}{r^*} = \frac{A}{J} \quad \text{with} : \quad r^* = \frac{Jc}{\sqrt{E^2 - m_0^2c^4}}. \tag{5.76}$$

So the trajectory defined by Eqs. (5.70) and (5.71) is now

$$\frac{\partial W}{\partial J} = \varphi_0 \rightarrow \varphi - \varphi_0 = J\int_{r^*}^{r} - \frac{1}{\left(A^2 + \frac{2B}{r} + \frac{C}{r^2}\right)^{\frac{1}{2}}} \frac{dr}{r^2}. \tag{5.77}$$

Taking $\varphi_0 = 0$, with the approximation [Eq. (5.76)], the equation of the trajectory becomes

$$\frac{1}{r} = \frac{e^2\sqrt{E^2 - m_0^2c^4}}{J^2c^2}\left(\varepsilon + \frac{Jc}{e^2}\cos\varphi\right) \quad (\varepsilon = \pm 1). \tag{5.78}$$

It is a *hyperbola* because by virtue of Eq. (5.75), its eccentricity is greater than 1:

$$\frac{Jc}{e^2} > 1. \tag{5.79}$$

It must be underscored that the hyperbolic character of the trajectory already had been determined by Eq. (5.67) and not only by the simplified form [i.e., Eq. (5.76)]. In conclusion, there are not any bound state as it had previously been noted, but do not forget that there are two possible types of trajectories because $\varepsilon = \pm 1$, the two signs corresponding to the two equations [Eq. (5.63)]. To wit:

- If $\varepsilon = +1$, the *concavity* of the trajectory is oriented to the central field and the motion is *attractive*.
- If $\varepsilon = -1$, the *convexity* of the trajectory is oriented to the central field and the motion is *repulsive*.

Therefore, in accordance with the quantum treatement, both cases are possible, whatever the charges and the central field might be.

It is interesting to compare these results with the classical case of a relativistic electron in a coulombian potential: we consider the classical equation

[Eq. (5.64)] again, introducing Eqs. (5.69) and (5.70), which gives an integral of the same form as Eq. (5.71):

$$f(r) = \int \left(A^2 + \frac{2B'}{r} + \frac{C}{r^2}\right)^{\frac{1}{2}} dr \tag{5.80}$$

$$A = \frac{1}{c}\sqrt{E^2 - m_0^2c^4}, \quad B' = \frac{E\varepsilon e^2}{c^2}, \quad C = \frac{e^4}{c^2} - J^2. \tag{5.81}$$

In the case of $E \geq m_0c^2$, in comparison with Eq. (5.72), one can see that the only coefficient B remains, while the factor A is substituted by E/c, which means the coincidence of these two cases for the limit $E \rightarrow m_0c^2$. But it must be noted that, in the preceding case, the condition $E \geq m_0c^2$ [Eq. (5.67)] was necessary, while here, in the classical case, it is only one of two possibilities because we could have $E < m_0c^2$, which would correspond to elliptic trajectories (i.e., bound states).

Taking the preceding calculation again with the constants [Eq. (5.81)], we find the trajectories as follows:

$$\frac{1}{r} = \frac{e^2E}{J^2c^2}\left(\varepsilon + c\frac{\sqrt{(E^2 - m_0^2c^4)J^2 + m_0^2c^2e^4}}{Ee^2}\cos\varphi\right). \tag{5.82}$$

This formula, which is good only for $E > m_0c^2$, differs from the classical formula only by the absence of the precession factor in the argument of the cosine, which we have neglected by virtue of Eq. (5.75), and the preceding approximation, which actually results in the replacement of C by $-J^2$. On the contrary, the approximation would not be valuable under the root sign in the expression of the eccentricity except if $E \gg m_0c^2$, which is the limit to which Eqs. (5.78) and (5.82) tend.

But the interesting case arises when $E - m_0c^2$ is small, because the eccentricity of the classical hyperbola depends on E and

$$c\frac{\sqrt{(E^2 - m_0^2c^4)J^2 + m_0^2c^4}}{Ee^2} \rightarrow 1 \quad \text{if } E \rightarrow m_0c^2. \tag{5.83}$$

Thus, the classical *parabolic* trajectory results when $E \rightarrow m_0c^2$.

On the contrary, the eccentricity of the hyperbola [Eq. (7.78)] is independent of energy and, consequently, from the angle between the asymptotes, while

$$\frac{1}{p} = \frac{e^2\sqrt{E^2 - m_0^2c^4}}{J^2c^2} \rightarrow 0 \quad \text{if} \ E \rightarrow m_0c^2. \tag{5.83'}$$

Therefore, the parameter approches infinity, while Eq. (5.82) shows that in the classical case, when $E \rightarrow m_0c^2$, the parameter tends toward a finite value. Consequently, for low energies, we find two different ways of behavior that could be experimentally distinguished, provided that one could create this strange, constrained state of the electron described by the Majorana field.

5.7 HOW COULD ONE OBSERVE A MAJORANA ELECTRON?

We have seen that in a coulomb field, at the geometrical optic approximation, the Majorana electron behaves either like a particle with a negative charge or like a particle with a positive charge, but it remains different from an electron or a positron because its motion is not Keplerian.

Nevertheless, this is only a problem of trajectories; that is, a problem of the rays of the wave given by the Jacobi equation. If we introduce the corresponding approximate expression of the action S in the expression of the wave function [Eq. (5.55)], we find an approximate solution of the equations [Eq. (5.57)].

Therefore, we shall find that, despite that trajectories seem to "choose" their charge (+ or −), the wave function evidently remains a superposition state of two waves with *opposite phases;* that is, waves with conjugated charges. Let us apply that concept to plane waves.

We write Eq. (5.55) with constant spinors a and b:

$$\xi = a \ e^{i(\omega t - \mathbf{k}.\mathbf{r})} + b \ e^{-i(\omega t - \mathbf{k}.\mathbf{r})}, \tag{5.84}$$

and we introduce Eq. (5.84) into the first expression in Eq. (5.14) with $V = A = 0$, and an angle θ, which is defined in Eq. (5.19). Analogous to the one of the § 5.6, a simple computation gives

$$\frac{\omega^2}{c^2} = k^2 + \frac{m_0^2c^2}{\hbar^2} \tag{5.85}$$

$$\xi = a \ e^{i(\omega t - \mathbf{k}.\mathbf{r})} - \frac{\hbar}{m_0c}\left(\frac{\omega}{c} - \mathbf{s}.\mathbf{k}\right)s_2a^* \ e^{-i(\omega t - \mathbf{k}.\mathbf{r})}. \tag{5.86}$$

This is a superposition of two waves with energies of opposite signs. But let us return to the Dirac equation; that is, the two expressions in Eq. (5.14) linked by Eq. (5.6), with the condition [Eq. (5.19)]. Therefore, it is not exactly the Majorana field but the Dirac field that is constrained by Eq. (5.5). In other words, it is the equation [Eq. (5.12)] with the value of Eq. (5.19) for the angle θ, and $A_\mu = 0$.

Now we must find the wave ψ, owing to Eq. (5.86) and

$$\eta = s_2 \xi^*, \quad \psi = \frac{1}{\sqrt{2}} \begin{bmatrix} \xi + \eta \\ \xi - \eta \end{bmatrix}. \tag{5.87}$$

We shall take $0z$ for the direction of propagation of the wave and

$$a = \begin{bmatrix} a_1 \\ a_2 \end{bmatrix} \quad \mathbf{k} = \{0, 0, k\}, \tag{5.88}$$

with a_1 and a_2 = components of a, in Eq. (5.86). So we find

$$\psi = \frac{1}{\sqrt{2}} (\psi_1 + \psi_2) \tag{5.89}$$

$$\psi_1 = a_1 \begin{bmatrix} 1 + \dfrac{\hbar}{m_0 c}\left(\dfrac{\omega}{c} + k\right) \\ 0 \\ 1 - \dfrac{\hbar}{m_0 c}\left(\dfrac{\omega}{c} + k\right) \\ 0 \end{bmatrix} e^{i(\omega t - kz)} - i a_1^* \begin{bmatrix} 0 \\ 1 + \dfrac{\hbar}{m_0 c}\left(\dfrac{\omega}{c} + k\right) \\ 0 \\ -\left[1 - \dfrac{\hbar}{m_0 c}\left(\dfrac{\omega}{c} + k\right)\right] \end{bmatrix} e^{-i(\omega t - kz)} \tag{5.90}$$

$$\psi_2 = a_2 \begin{bmatrix} 0 \\ 1 + \dfrac{\hbar}{m_0 c}\left(\dfrac{\omega}{c} - k\right) \\ 0 \\ 1 - \dfrac{\hbar}{m_0 c}\left(\dfrac{\omega}{c} - k\right) \end{bmatrix} e^{i(\omega t - kz)} + i a_2^* \begin{bmatrix} 1 + \dfrac{\hbar}{m_0 c}\left(\dfrac{\omega}{c} - k\right) \\ 0 \\ -\left[1 - \dfrac{\hbar}{m_0 c}\left(\dfrac{\omega}{c} - k\right)\right] \\ 0 \end{bmatrix} e^{-i(\omega t - kz)}. \tag{5.91}$$

Here, ψ is the superposition of two waves ψ_1 and ψ_2 with the constants a_1 and a_2. Each wave ψ_1 and ψ_2 depends on energy and helicity, which is easy to define because if Oz is the direction of propagation so that the spin is projected in the same direction, and

$$\sigma_3 = \begin{bmatrix} s_3 & 0 \\ 0 & s_3 \end{bmatrix} = \begin{bmatrix} 1 & 0 & 0 & 0 \\ 0 & -1 & 0 & 0 \\ 0 & 0 & 1 & 0 \\ 0 & 0 & 0 & -1 \end{bmatrix}, \tag{5.92}$$

then we see the following:

1. $\psi_1$is a superposition of two waves with the same sign of helicity and charge (+ and −, respectively, for each wave).
2. ψ_2 is a superposition of two waves with opposite helicities and charges.

The relative phase of the components of ψ_1 or ψ_2 (i.e., $a_{1,2}$ and $a^*_{1,2}$) has no physical meaning because the constant θ in Eq. (5.12) or (5.14) is arbitrary. Now, for low energies,

$$|k| \ll \frac{\omega}{c}, \quad \frac{\omega}{c} = \frac{m_0 c}{\hbar}, \tag{5.93}$$

we have, in a first approximation:

$$\psi_1 = \begin{bmatrix} a_1 e^{i(\omega t - kz)} \\ -i a_1^* e^{-i(\omega t - kz)} \\ 0 \\ 0 \end{bmatrix}, \quad \psi_2 = \begin{bmatrix} i\ a_2 e^{-i(\omega t - kz)} \\ a_2^* e^{i(\omega t - kz)} \\ 0 \\ 0 \end{bmatrix}. \tag{5.94}$$

In conclusion, if we could "keep alive" (i.e., keep from destroying) two parallel-beam of electrons and positrons with the same energy and this polarization for a sufficiently long time, the definite couples would have the behavior of a Majorana electron. In particular, in a coulomb field, an electron in such a state would exhibit the strange behavior just described instead of following the classical Kepler laws.

5.8 THE EQUATION IN THE MAGNETIC CASE

We have recalled in Eq. (5.3) the general nonlinear equation of a magnetic monopole, and we know that the chiral gauge invariance is broken and the magnetic charge is no more conserved if we add a linear mass term (it is the reason for which the Dirac equation does not conserve the magnetic charge). As we know (as discussed in Chapter 3), the Majorana condition ensures the conservation of chiral currents, and thus of magnetism.

Such an equation is not really chiral gauge-invariant, but in this case, it admits a *subset of gauge-invariant solutions.* Now, remember that the chiral invariance is an invariance with respect to the rotations in the chiral plane $\{\omega_1, \omega_2\}$ (i.e., with respect to the rotations of an angle A), which can be obtained in two ways:

1. The first way is to introduce in the Lagrangian a mass term that depends only on the norm of the vector $\{\omega_1, \omega_2\}$; this was done until now, and it results in Eq. (5.3).
2. The second way is to add to the Lagrangian of the linear monopole an arbitrary mass term that is not necessarily chiral-invariant (as was the norm of $\{\omega_1, \omega_2\}$), but which is such that the obtained equation has a subset of solutions that annihilates the chiral invariant:

$$\rho = \left(\omega_1^2 + \omega_2^2\right)^{1/2} = 0. \tag{5.95}$$

Such solutions thus obey the generalized Majorana condition [Eq. (5.6)], which we write here in a simpler form:

$$\Psi = e^{i\theta}\gamma_2\psi^* = e^{i\theta}\psi_c. \tag{5.96}$$

Actually, we can put $\theta = 0$, as discussed later in this chapter. A priori, we could start from an arbitrary term of mass, but for simplicity, we shall choose the linear mass term of the Dirac equation. So now we can introduce, in the equation of the massless monopole, the mass term of Eq. (2.1) under the condition Eq. (5.95) or (5.96), which will be expressed by means of a Lagrange multiplier. Thus, we have the Lagrangian:

$$L = \overline{\Psi}\gamma_\mu\left[\partial_\mu\right]\Psi - g\hbar c\overline{\Psi}\gamma_\mu\gamma_5 B_\mu\Psi - m_0 c\hbar\overline{\psi}\psi + \lambda\left(\omega_1^2 + \omega_2^2\right), \tag{5.97}$$

from which, varying $\overline{\psi}$, we deduce the following equation, which looks like our nonlinear equation from Chapter 4, but with a linear term:

$$\gamma_\mu\left(\partial_\mu - g\hbar c\gamma_5 B_\mu\right)\Psi - m_0 c\hbar\psi + \lambda(\omega_1 - i\omega_2\gamma_5)\Psi = 0. \tag{5.98}$$

The difference between this equation and the equation of our nonlinear monopole is the presence of a linear mass term and of the constant λ instead of $m(\rho^2)$. But the linear term will be transformed, and the nonlinear term itself will disappear because we must vary L with respect to the Lagrange multiplier λ, in order to find Eq. (5.95). Thus, we have

$$\omega_1 = \omega_2 = 0, \tag{5.99}$$

which gives Eq. (5.97) and annihilates the λ term in Eq. (5.98). The Lagrange multiplier thus remains undetermined, since it does not appear in the equation. If we introduce Eq. (5.96), **we find the Majorana equation up to a phase factor** $e^{i\theta}$, with a magnetic interaction instead of an electric one:

$$\gamma_\mu\left(\partial_\mu - g\hbar c\gamma_5 B_\mu\right)\Psi - m_0 c\hbar e^{i\theta}\gamma_2\psi^* = 0. \tag{5.100}$$

It is a new, nonlinear equation of a magnetic monopole, different from the one found earlier. In the Weyl representation (discussed in Chapter 3), Eq. (5.100) splits into two equations that are formally separated, but are actually linked to each other:

$$\begin{aligned} \left(\pi_0^+ + \pi^+ \cdot s\right)\xi - im_0 c e^{i\theta} s_2 \xi^* &= 0 \\ \left(\pi_0^- - \pi^- \cdot s\right)\eta + im_0 c e^{i\theta} s_2 \eta^* &= 0, \end{aligned} \tag{5.101}$$

with the following definitions:

$$\begin{aligned} \pi_0^+ &= \frac{1}{c}\left(i\hbar\frac{\partial}{\partial t} + gW\right), \quad \pi^+ = -i\hbar\frac{\partial}{\partial t} + \frac{g}{c}\,\mathbf{B} \\ \pi_0^- &= \frac{1}{c}\left(i\hbar\frac{\partial}{\partial t} - gW\right), \quad \pi^- = -i\hbar\frac{\partial}{\partial t} - \frac{g}{c}\,\mathbf{B}. \end{aligned} \tag{5.102}$$

We can remark that, *in the electric case, we had only one operator* $\{\pi_0, \pi\}$, while *in the magnetic case, we have two operators: right and left.* Before examining Eq. (5.101), we must take a moment to specify some points concerning Eq. (5.98).

First, this equation was found a long time ago by Weyl (1950), only *for a free wave (i.e., without interaction),* and with another aim. For Weyl, the nonlinear term was not a Lagrange condition. Rather, it was a change of the Dirac equation, owing to which the nonlinear Weyl equation (contrary to the Dirac linear equation) has the property of keeping the same form, in general relativity, if it were expressed in metric form with an affine connection $\Gamma_{\mu\lambda\nu}$, depending on $g_{\mu\nu}$; or with coefficients $\Gamma_{\mu\lambda\nu}$, independent of $g_{\mu\nu}$.

In various forms, the same equation without interactions was later found again by several authors and reexamined from different points of view. Two papers are particularly interesting with respect to this problem:

1. The first (Rodichev, 1961) already has been discussed in Chapter 4. Just recall that it is based on a particular case of Eq. (5.98), where λ is an ordinary constant:

$$\gamma_\mu\partial_\mu\Psi + \lambda(\Omega_1 - i\Omega_2\gamma_5)\Psi = 0. \tag{5.103}$$

It was shown (both in Chapter 4 and Lochak, 1985e) that the chiral invariant is equal, up to a constant factor, to the total curvature. But in this case, the space is flat and the curvature is reduced to the torsion, so that when we show that the Majorana condition [Eq. (5.96)] is equivalent to the condition [Eq. (5.95)] it actually signifies that the Majorana condition annihilates the torsion of the space.

2. Now we give results due to A. Bachelot (1988a,b), who solved the global Cauchy problem for Eq. (5.103) without electromagnetic interaction, but with initial conditions that are not supposed to be small: they are only so to the extent that the chiral invariant $\rho = \left(\omega_1^2 + \omega_2^2\right)^{1/2}$ is small. In other words, it remains in the vicinity of the condition [Eq. (5.95)], which, as already established, is close to the generalized Majorana condition.

To prove his theorem, Bachelot first proved the following lemma, which is of great interest in itself:

- Consider the Dirac equation without interaction, but with a mass term M, possibly depending on space and time:

$$\gamma_\mu \partial_\mu \Psi + M\psi = 0 \tag{5.104}$$

Bachelot proved that if the chiral invariant $\rho = \left(\omega_1^2 + \omega_2^2\right)^{1/2}$ vanishes at a given instant in the whole space, it remains equal to zero later. It is easy to generalize the lemma of Bachelot in the presence of a *magnetic interaction,* and we shall directly formulate and prove it in this more general case.

- Given the equation

$$\gamma_\mu\left(\partial_\mu \Psi - g\hbar c\gamma_5 B_\mu\right)\Psi - m_0 c\hbar\psi = 0, \tag{5.105}$$

if at a given instant, the chiral invariant $\rho = \left(\Omega_1^2 + \Omega_2^2\right)^{1/2}$ (and so, the torsion of the space) vanishes in the whole space, il remains equal to zero. Bachelot starts from two conservation laws:

$$\partial_\mu \overline{\psi}\gamma_\mu\psi = 0, \quad \partial_\mu \overline{\tilde{\psi}}\gamma_2\gamma_4\gamma_\mu\psi = 0 \quad \left(\tilde{\psi} = \text{transposed } \psi\right). \tag{5.106}$$

The first law is the conservation of the Dirac current (i.e., of electricity). It must be noted that the chiral currents are not separately conserved, contrary to Eq. (3.7), because of the presence of a linear mass term in Eq. (5.105); but their sum is conserved as in the Dirac equation, and this sum is precisely the Dirac electric current that appers in Eq. (5.106).

The second law is the conservation of the crossed current between charge-conjugated states. Bachelot deduced it from Eq. (5.104), but it is also true for Eq. (5.105), with a magnetic interaction. On the contrary, the second conservative law would be wrong in the case of an ordinary Dirac equation with an electric interaction. Indeed, we get in this case:

$$\partial_\mu \tilde{\psi} \gamma_2 \gamma_4 \gamma_\mu \psi + iA_\mu \tilde{\psi} \gamma_2 \gamma_4 \gamma_\mu \psi = 0. \tag{5.107}$$

Now, if these two laws [Eq. (5.106)] are true, Bachelot uses the conservation laws, as follows:

$$\int_{\mathbf{R}^3} |\psi|^2 dx = \text{Const}, \quad \int_{\mathbf{R}^3} \tilde{\psi} \gamma_2 \psi dx = \text{Const}, \tag{5.108}$$

provided that these integrals do exist. This reservation must be demanded because we know that there are no bound states between an electric and a magnetic charge (Lochak, 1983, 1984), so that this result is not general.

Under the preceding restriction, we find from Eq. (5.108):

$$\int_{\mathbf{R}^3} \left| \Psi - e^{i\theta} \gamma_2 \psi^* \right|^2 dx = 2 \int_{\mathbf{R}^3} \left\{ |\Psi|^2 - \Re e^{-i\theta} \tilde{\psi} \gamma_2 \psi^* \right\} dx = \text{Const}. \tag{5.109}$$

Therefore, if at a given instant, Eq. (5.95), or, equivalently, Eq. (5.96) is realized, it also will be realized in the future. This is known as the *lemma of Bachelot,* and we know that it is true not only for Eq. (5.104), but also for Eq. (5.105).

If the preceding formulas are true, the condition to which Eqs. (5.100) and (5.101) were submitted through the Lagrange multipliers will be strongly weakened because instead of a constraint imposed at every instant, we have only an initial condition. Therefore, the Majorana magnetic states are simply particular solutions of the Dirac equation that have a magnetic interaction.

More precisely, these states are *monopole states of the Dirac equation* because it will be shown that Eq. (5.100) or (5.101) is actually chiral-invariant, despite the fact that Eq. (5.105) is not chiral-invariant. They represent a couple of monopoles and, in order to make them appear, it is sufficient to satisfy an initial condition, at least in certain cases.

Let us emphasize, once more, that by virtue of Eq. (5.107), this conclusion, which is true in the magnetic case, is not true in the electric case. So that, if we are able to satisfy the conditions [Eq. (5.95)], we shall obtain monopoles, but not electrons.

5.10 ANOTHER POSSIBLE EQUATION: THE GAUGE INVARIANCE PROBLEM

At this point, let us introduce the transformation $\Psi \rightarrow e^{ig\hbar c\gamma_5\Phi}\Psi$ (discussed in Chapter 4) in Eq. (5.100). Here, we find

$$\gamma_\mu\left[\partial_\mu\Psi - g\hbar c\gamma_5\left(B_\mu + i\partial_\mu\Phi\right)\right] e^{ig\hbar c\gamma_5\Phi}\Psi - m_0c/\hbar e^{i\theta}\gamma_2 e^{-ig\hbar c\gamma_5\Phi}\psi^* = 0. \tag{5.110}$$

And then, taking into account the anticommutation rules of γ matrices:

$$\gamma_\mu\left[\left(\partial_\mu\Psi - g\hbar c\gamma_5\left(B_\mu + i\partial_\mu\Phi\right)\right]\Psi - m_0c/\hbar e^{i\theta}\gamma_2\psi^* = 0. \tag{5.111}$$

We find the correct interaction term with the B_μ potentials, but with a phase factor Φ, the origin of this factor is the angle A. The chiral gauge invariance is not obvious, as could be expected, because this invariance in its general form appears only in the equations in which the chiral angle A does not appear, whereas in the present case, we started from Eq. (5.105), which is not chiral gauge-invariant. We have just imposed one of the conditions [i.e., Eq. (5.95)], which does not make the angle A disappear, but it is undetermined. That is, this angle appears in the equation, but its value can be eliminated if it is a polar angle around a nil rotation vector.

Finally, the preceding phase factor is eliminated by a choice of angle A because the gauge invariance is lost, and the phase θ may be eliminated as well because it plays no dynamical role. Thus, we can write, as a consequence of Eq. (5.95):

$$\Psi = \gamma_2\psi^* = \psi_c\xi = is_2\eta^*\eta = -is_2\xi^*. \tag{5.112}$$

And instead of Eqs. (5.100) and (5.101), we have (without θ)

$$\gamma_\mu\left(\partial_\mu - g\hbar c\gamma_5 B_\mu\right)\Psi - m_0c\hbar\gamma_2\psi^* = 0 \tag{5.113}$$

and [see Eq. (5.102)]

$$\begin{aligned} \left(\pi_0^+ + \pi^+\cdot s\right)\xi - im_0cs_2\xi^* &= 0 \\ \left(\pi_0^- + \pi^-\cdot s\right)\eta + im_0cs_2\eta^* &= 0. \end{aligned} \tag{5.114}$$

5.11 GEOMETRICAL OPTIC APPROXIMATION

Until now, all seems well, but it would be desirable to test the qualities of the preceding equations for a well-known case, such as the interaction of a monopole with an electric charge, as it was done previously in the comparable case of the electron, replacing Eq. (5.14) with Eq. (5.114). This seems easy because of the apparent identity of both equations. Unfortunately, that is not so because the potentials hidden in these formulas are fundamentally different, so the magnetic case is far more complicated than the electric one. And this case is more difficult than the case of a linear, massless monopole (as discussed in Chapter 3), precisely because of the nonlinear mass term.

For these reasons, we shall be content with the classical approximation. Thus, we shall take Eq. (5.114) with π^+ and π^- defined in Eq. (5.102), with the following expressions:

$$\xi = a\exp(-iS/\hbar) + b\exp(iS/\hbar), \quad \eta = -is_2\xi^*. \tag{5.115}$$

A calculation analogous to the one in § 5.6 gives an equation of the Hamilton-Jacobi type:

$$\left[\left(\frac{1}{c}\frac{\partial S}{\partial t}\right)^2 - \left(\nabla S + \frac{g\mathbf{B}}{c}\right)^2 - m_0^2c^2\right]\left[\left(\frac{1}{c}\frac{\partial S}{\partial t}\right)^2 - \left(\nabla S - \frac{g\mathbf{B}}{c}\right)^2 - m_0^2c^2\right] = 4m_0^2g^2\mathbf{B}^2. \tag{5.116}$$

This is very different from Eq. (5.61) because of the difference in the potentials (see Eq. (1.26) in Chapter 1):

$$W = 0, \; B_x = \frac{e}{r}\frac{yz}{x^2+y^2}, \; B_y = \frac{e}{r}\frac{-xz}{x^2+y^2}, \; B_z = 0, \quad r = \sqrt{x^2+y^2+z^2}. \tag{5.117}$$

The electric coulomb field is written as

$$\mathbf{E} = \mathit{curl}\,\mathbf{B} = e\mathbf{r}/r^3. \tag{5.118}$$

Now, it must be recognized that Eq. (5.116) is not the equation of a classical magnetic monopole in the presence of an electric charge, which would be

$$\mathbf{E} = \mathit{curl}\,\mathbf{G} = e\mathbf{r}/r^3$$

and so

$$\mathit{either}: \left[(\partial S/\partial t)^2/c^2 - (\nabla S + g\mathbf{B}/c)^2 - m_0^2c^2\right] = 0 \tag{5.119}$$

$$\mathit{or}: \left[(\partial S/\partial t)^2/c^2 - (\nabla S - g\mathbf{B}/c)^2 - m_0^2c^2\right] = 0. \tag{5.120}$$

depending on the sign of the magnetic charge. The simultanuous presence of both brackets, $(\nabla S + g\,\mathbf{B}/c)$ and $(\nabla S - g\,\mathbf{B}/c)$, in Eq. (5.116) suggests that the equation contains a couple of monopoles of opposite signs, and one can observe that, far from the center of the electric charge, $B \rightarrow 0$; and that Eq. (5.116) splits into two components: Eqs. (5.119) and (5.120). Thus, we actually find, asymptotically, a couple of classical monopoles. This may be called the *zero approximation*. The first-order equations (nearer to the center) may be written as

$$\begin{aligned} \left[(\partial S/\partial t)^2/c^2 - \left(\vec{\nabla} S + g\,\mathbf{B}/c\right)^2 - m_0^2 c^2\right] &= 2m_0 g|\mathbf{G}| \\ \left[(\partial S/\partial t)^2/c^2 - \left(\vec{\nabla} S - g\,\mathbf{B}/c\right)^2 - m_0^2 c^2\right] &= -2m_0 g|\mathbf{G}|. \end{aligned} \tag{5.121}$$

These equations have additive terms with respect to Eqs. (5.119) and (5.120).

It is interesting to introduce, in one of these equations, these two quantities:

$$p = \nabla S = m\frac{d\mathbf{r}}{dt}, \quad \lambda = \frac{egc}{\varepsilon} \quad (\varepsilon = \text{energy}), \tag{5.122}$$

which results in the following equation:

$$d^2 r/dt^2 = -\lambda/r^3 \cdot dr/dt \times r - m_0 g|\mathbf{B}|. \tag{5.123}$$

Without the second term of the second member, this is the Poincaré equation [i.e., Eq. (1.2)] of the interaction between an electric and a magnetic charge in classical mechanics. Remember that we already obtained such an equation, at the geometrical optic limit of our equation of a massless monopole in Chapter 3.

Nevertheless, we cannot neglect this strange additional term that appears in Eq. (5.123), and which is the same term as in Eqs. (5.116), (5.121), and (5.122): a kind of reminder of the sin of introducing a linear mass term, which disappears far from the center of the charge but which calls for a certain caution concerning the "Majorana monopole."

It must be added that the importance of the Poincaré equation comes not only from the fame of its author, but from the fact that this equation is experimentally verified by the Birkeland effect (as discussed in Chapter 1). It is the equation of the motion of a beam of cathodic rays in the presence of a pole of a linear magnet—actually, a magnetic monopole. This is why we have attached great importance to the fact that the Poincaré equation is the classical limit of our equation of the magnetic monopole.

It is, thus, impossible to be indifferent to a violation of the Poincaré equation. Yet it is not a total invalidation of the Majorana monopole because the additive term is always the same, and it tends to zero in two cases:

1. If the proper mass tends to zero, *this monopole tends to our massless monopole.* But this case is not significant because *it is evident from* Eqs. (5.113) *and* (5.114) *that the nonlinear Majorana term tends to zero.*
2. *Far from the center of the charge,* which is due to the potentials and is not evident from Eqs. (5.113) and (5.114). This gives the Majorana monopole an asymptotic significance. However, it must be noted that, far from the center, the additive terms become negligible; but unfortunately, the potential terms become negligible too, so *the Majorana monopole tends to our massless monopole when it is no more a monopole.*

APPENDIX A

In this appendix, let us give a proof of Eq. (4.6). First, we know that by the very definition of $\Omega_l^m(-)$ *and* $\Omega_l^m(+)$, we get

$$\mathbf{J}^2\Omega_{l-1}^m(+) = j(j+1)\Omega_{l-1}^m(+), \quad j = l - \frac{1}{2} \text{ and} \tag{A.1}$$

$$\mathbf{J}_z\Omega_l^m(-) = u\Omega_l^m(-), \ \mathbf{J}_z\Omega_{l-1}^m(+) = u\Omega_{l-1}^m(+), \quad u = m - \frac{1}{2}, \tag{A.2}$$

where we have

$$\mathbf{J} = \mathbf{L} + \mathbf{s}, \quad \mathbf{L} = -i\ \mathbf{r} \times \nabla. \tag{A.3}$$

Now, we can easily verify that the operator

$$\mathbf{s.n} = \frac{1}{r}\mathbf{s.r} = \begin{bmatrix} \cos\theta & \sin\theta e^{-i\phi} \\ \sin\theta e^{i\phi} & -\cos\theta \end{bmatrix} \tag{A.4}$$

commutes with $\mathbf{J}$:

$$[\mathbf{J}, \mathbf{s.n}] = 0. \tag{A.5}$$

Therefore, $\mathbf{s.n}$ transforms a subspace Ω that belongs to the subspace of eigenvalues of J^2 and J_z in an element of the same subspace. For instance, we have

$$\mathbf{s.n}\ \Omega_l^m(+) = A\Omega_l^m(+) + B\Omega_{l+1}^m(-), \tag{A.6}$$

where the constants A and B do not depend on m. We shall compute them for particular values of m and of the polar angles as follows:

$$m = l + 1, \quad \theta = \frac{\pi}{2}, \quad \phi = 0. \tag{A.7}$$

From Eq. (5.21), we have

$$\Omega_l^{l+1}(+) = \begin{bmatrix} Y_l^l\left(\frac{\pi}{2}, 0\right) \\ 0 \end{bmatrix}, \quad \Omega_{l+1}^{l+1}(-) = \begin{bmatrix} \left(\frac{l}{2l+3}\right)^{12} Y_{l+1}^l\left(\frac{\pi}{2}, 0\right) \\ -\left(\frac{2l+2}{2l+3}\right)^{12} Y_{l+1}^{l+1}\left(\frac{\pi}{2}, 0\right) \end{bmatrix}. \tag{A.8}$$

Now, from Eq. (5.22),

$$Y_{l+1}^l\left(\frac{\pi}{2}, 0\right), \quad \left(\frac{2l+2}{2l+3}\right)^{12} Y_{l+1}^{l+1}\left(\frac{\pi}{2}, 0\right) = -Y_l^l\left(\frac{\pi}{2}, 0\right), \tag{A.9}$$

and Eq. (A.4) gives

$$\mathbf{s.n}\left(\frac{\pi}{2}, 0\right) = \begin{bmatrix} 0 & 1 \\ 1 & 0 \end{bmatrix}. \tag{A.10}$$

Finally, it is sufficient to introduce Eqs. (A.8), (A.9), and (A.10) into Eq. (A.7) to find

$$A = 0, \quad B = 1, \tag{A.11}$$

which proves the first relation [Eq. (5.23)]. The second relation is evident because

$$(\mathbf{s.n})^2 = I. \tag{A.12}$$

Thus, we have

$$Y_l^{m*}(\theta, \phi) = (-1)^m Y_l^{-m}. \tag{A.13}$$

APPENDIX B

To prove Eq. (5.29), remember that in Eq. (A.3), **L** and **s** commute, so from Eqs. (5.21) and (A.1), we have

$$\mathbf{J}^2\Omega_l^m(\pm) = j(j+1)\Omega_l^m(\pm), \quad \mathbf{L}^2\Omega_l^m(\pm) = l(l+1)\Omega_l^m(\pm)$$
$$\mathbf{S}^2\Omega_l^m(\pm) = s(s+1)\Omega_l^m(\pm) = \frac{3}{4}\Omega_l^m(\pm). \tag{B.1}$$

Thus, applying Eq. (A.3), we get

$$\begin{aligned}(\mathbf{L}+\mathbf{S})^2\Omega_l^m(\pm) &= \left(\mathbf{L}^2+\mathbf{S}^2+2\mathbf{L}.\mathbf{S}\right)\Omega_l^m(\pm)\\ &= \left(\mathbf{L}^2+\mathbf{S}^2+\mathbf{L}.\mathbf{S}\right)\Omega_l^m(\pm)\end{aligned} \tag{B.2}$$

so that Eq. (B.1) gives $j(j+1)\Omega_l^m(\pm) = \left[l(l+1)+\frac{3}{4}+\mathbf{L}.\mathbf{s}\right]\Omega_l^m(\pm)$ and Eq. (5.29).

CHAPTER 6

A New Electromagnetism with Four Fundamental Photons: Electric, Magnetic, with Spin 1 and Spin 0

6.1 THEORY OF LIGHT

6.1.1 Theory of Light and Wave Mechanics: A Historical Summary

This chapter presents an introduction to a new theory of light and gravitation (the last at a linear approximation) that generalizes, owing to the idea of the magnetic monopole, the de Broglie theory of light and gravitation based on his theory of spin particles. The idea of leptonic monopole—and its consequences—are the new concepts added to de Broglie's theory. On the contrary, other ideas that appear in the new theory, including the "magnetic photon," were implicitly present (in a hidden form) in the de Broglie theory of spin particles; but curiously, they remain unexploited (or even noticed) until recent years. This is the reason for the following short historical summary.

The de Broglie theory of spin particles started from his work on the theory of light that began as a dynamic theory of the Einstein photon. At that time (1922), the wave mechanics did not yet exist: it appeared a little later, precisely from this dynamic theory of Einstein's "light quanta" (de Broglie, 1922).

De Broglie initially tried a test of the photon hypothesis, going as far as possible with the radiation theory, in a purely corpuscular way, in the spirit of Newton, but introducing relativistic mechanics, kinetic theory, and thermodynamics; nevertheless, they did not use electromagnetism because de Broglie *aimed to find where and in what form the waves become necessary.*

He considered Einstein's "light quanta," which were not yet called *photons,* to be true particles (as he put it, "atoms of light") with a *small proper mass,* obeying the laws of relativistic mechanics. Starting from a purely corpuscular point of view, he got several results previously considered as the consequences of electromagnetism:

- For instance, if $E = mc^2 = 1/\sqrt{1 - v^2/c^2} =$ *total energy*, the relativistic form of the momentum B is $B = mc = E/c$, from which de Broglie obtained the correct relation $p = \rho/3$ between the pressure and energy density of black radiation (de Broglie, 1922), first proved by Boltzmann and later ascribed to Maxwell's theory.[11]
- Applying relativity, de Broglie gave the correct mean energy 3 kT for the photon instead of the half-value $(3/2)kT$ of the classical theory of gas. This energy was usually considered as the sum of electric and magnetic energies, whereas it is a simple consequence of relativistic kinematics.
- Finally, de Broglie obtained the formula of the Doppler effect from the relativistic addition of velocities and Planck's law of quanta.

After these results, de Broglie realized that his ideas were not restricted to light and photons, but rather could be said about every particle. Therefore, he attached a frequency to each material particle via the expression $mc^2 = h\nu$. This brought him, if not yet to the wave, at least to a frequency that he ascribed to an "internal clock" of the particle, which was not far from Newton's conceptions. But he rapidly understood that such an interpretation is not relativistically invariant because if ν is an internal frequency of a particle, it is submitted to the slowing of the clocks, while m will increase with the velocity. The de Broglie "illuminating idea" (in his own words) was that, on the contrary, the frequency of a wave would have the same variance as m so that the expression $mc^2 = h\nu$ becomes relativistically invariant and defines univocally ν from m. This was the start of wave mechanics.

[11] It is curious to note that Planck found twice this result, due to the omission of relativity (which is absolutely astonishing coming from Max Planck). So, he wrote $E = (1/2)mv^2 \Rightarrow B = mv = 2W/v \Rightarrow$ $p = 2\rho/3$, with an erroneous factor of 2, considered by the opponents to Einstein as an argument against the photon hypothesis (*de Broglie*, 1922) …

It must be stressed that de Broglie considered from the very beginning that the photon had a mass: namely, a mass far smaller than the one of an electron, but it was a "true" mass that includes the photon in a description of all the particles of the universe. Nevertheless, such a theory of light could not be developed with the Schrödinger or Klein-Gordon equation because the first is nonrelativistic and the waves of both equations are not polarized.

The situation became different with the appearance of the Dirac equation for which de Broglie was immediately enthusiastic because he saw in it a possible beginning for a theory of light (de Broglie, 1932a–c): the equation was relativistic, with a four-component wave function (and, therefore, a polarization) and a spin: the axial vector that he had predicted for light[12]; and a second-rank tensor, $M_{\mu\nu} = \overline{\psi}\gamma_\mu\gamma_\nu\psi$, which is antisymmetric as the electromagnetic tensor, despite the fact that it was not a wave.

Nevertheless, the elements of the Dirac equation could not be directly applied to a photon: the wave does not have the variance either of a vector or of an antisymmetric tensor [such a tensor $(\overline{\psi}\gamma_\mu\gamma_\nu\psi)$ is present in the theory, but it is not the wave]; the spin rotates twice as slow and the particle is a fermion, not a boson, as was already well known. Nevertheless, the way was not obstructed as it had been because the different elements did exist, but in a distorted form.

After some initial attempts (de Broglie, 1932a, b), de Broglie realized that a photon cannot be an elementary particle, but the fusion of a pair: perhaps of a spin-1/2 corpuscle and its "anticorpuscle" (this word appearing here for the first time), both obeying a Dirac equation (de Broglie, 1932b).

The creation and annihilation of pairs suggested that a photon could result from the "fusion" of an electron-positron pair linked by an electrostatic force. The smallness of the photon mass could be a consequence of a defect in relativistic mass. But the introduction of an electrostatic force is a source of confusion because a theory of photons is a theory of electromagnetism, so the electrostatic force must be a *consequence* of the theory, not an a priori hypothesis.

[12] De Broglie (1922) wrote: "A more complete theory of quanta of light must introduce a polarization in such a way that: to each atom of light would be linked an internal state of right or left polarization represented by an axial vector with the same direction as the propagation velocity." It was shown later that when the velocity of a particle tends to the velocity of light, the space components of the vector spin lies along the velocity.

So, recognizing that the choice of conjugated particles was impossible, de Broglie supposed that the photon is a neutrino-antineutrino pair or, more generally, the center of mass of a couple of Dirac particles. He published the equation in 1932 (de Broglie 1932b, c) and he developed the theory during many years.

6.1.2 De Broglie's Method of Fusion

First, let us take as an example a pair of identical, ordinary particles of mass m, obeying the Schrödinger equation, with respective coordinates (x_1, y_1, z_1) and (x_2, y_2, z_2). Their center of mass is

$$x = \frac{x_1 + x_2}{2}, \quad y = \frac{y_1 + y_2}{2}, \quad z = \frac{z_1 + z_2}{2}. \tag{6.1}$$

The Schrödinger equation of the center of mass is definite, using the coordinates in Eq. (6.1):

$$-i\hbar \frac{\partial \phi}{\partial t} = \frac{1}{2M} \Delta \phi \ (M = 2m). \tag{6.2}$$

But such a procedure cannot be extended to a pair of Dirac particles because there is no quantum (or even a classical) relativistic theory of systems of particles. Therefore, de Broglie suggested a formal way that is easier to generalize. He associated the particles with two different waves, ψ and φ, without making any distinction between their coordinates. So we have the following equations with the same coordinates x_k:

$$-i\hbar \frac{\partial \psi}{\partial t} = \frac{1}{2m} \Delta \psi; \quad -i\hbar \frac{\partial \varphi}{\partial t} = \frac{1}{2m} \Delta \varphi \tag{6.3}$$

Now, the fusion conditions, expressing the equality of moment and energy in the case of plane waves, are

$$\frac{\partial \psi}{\partial t} \varphi = \psi \frac{\partial \varphi}{\partial t} = \frac{1}{2} \frac{\partial (\psi \varphi)}{\partial t}; \quad \frac{\partial^2 \psi}{\partial x_k^2} \varphi = \frac{\partial \psi}{\partial x_k} \frac{\partial \varphi}{\partial x_k} = \psi \frac{\partial^2 \varphi}{\partial x_k^2} = \frac{1}{4} \frac{\partial^2 (\psi \varphi)}{\partial x_k^2}. \tag{6.4}$$

Multiplying the first equation in Eq. (6.3) by φ and the second by ψ, we find for $\phi = (\varphi \psi)$ Eq. (6.2) again. Then de Broglie applied the same conditions to all the waves without restriction to the plane waves, and he applied it to the relativistic case: it is called the *fusion postulate*.

6.1.3 De Broglie's Equations of Photons

Consider the Dirac equations of two particles of mass $\frac{\mu_0}{2}$:

$$\begin{aligned} \frac{1}{c}\frac{\partial \psi}{\partial t} &= \alpha_k \frac{\partial \psi}{\partial x_k} + i\frac{\mu_0 c}{2\hbar}\alpha_4 \psi \\ \frac{1}{c}\frac{\partial \varphi}{\partial t} &= \alpha_k \frac{\partial \varphi}{\partial x_k} + i\frac{\mu_0 c}{2\hbar}\alpha_4 \varphi\ , \end{aligned} \tag{6.5}$$

where $\{\alpha_k,\ \alpha_4\}$ are the Dirac matrices[13]:

$$\alpha_k = \begin{pmatrix} 0 & \sigma_k \\ \sigma_k & 0 \end{pmatrix};\ \alpha_k = \begin{pmatrix} I & 0 \\ 0 & -I \end{pmatrix};\ (\sigma_k = \textit{Pauli matrices}). \tag{6.6}$$

In analogy with Eq. (6.4), de Broglie put the *fusion conditions* on the Dirac wave-components as follows:

$$\frac{\partial \psi_n}{\partial t}\varphi_m = \psi_n \frac{\partial \varphi_m}{\partial t} = \frac{1}{2}\frac{\partial(\psi_n \varphi_m)}{\partial t};\ \frac{\partial \psi_n}{\partial x_k}\varphi_m = \psi_n \frac{\partial \varphi_m}{\partial x_k} = \frac{1}{2}\frac{\partial(\psi_n \varphi_m)}{\partial x_k}. \tag{6.7}$$

So he found for $\phi = \{\phi_{nm} = \psi_n \varphi_m\}$ a new equation, which he extended by postulate to all the ϕ functions even if their form is not $\phi_{nm} = \psi_n \varphi_m$:

$$\begin{aligned} \frac{1}{c}\frac{\partial \phi}{\partial t} &= a_k \frac{\partial \phi}{\partial x_k} + i\frac{\mu_0 c}{\hbar} a_4 \phi \\ \frac{1}{c}\frac{\partial \phi}{\partial t} &= b_k \frac{\partial \phi}{\partial x_k} + i\frac{\mu_0 c}{\hbar} b_4 \phi. \end{aligned} \tag{6.8}$$

The matrices a and b are defined as

$$\begin{aligned} a_r &= \alpha_r \times I,\ (a_r)_{ik,lm} = (\alpha_r)_{il}\delta_{km} \\ b_r &= I \times \alpha_r,\ (b_r)_{ik,lm} = (-1)^{r+1}(\alpha_r)_{km}\delta_{il}\quad (r = 1, 2, 3, 4). \end{aligned} \tag{6.9}$$

They separately verify the relations of the Dirac matrices, where a and b commute:

$$a_r a_s + a_s a_r = 2\delta_{rs};\ b_r b_s + b_s b_r = 2\delta_{rs};\ a_r b_s - b_s a_r = 0. \tag{6.10}$$

With this finding, it is easy to prove that the components of ϕ obey the Klein-Gordon equation.

Eq. (6.8) with the definitions (6.9) are the *de Broglie photon equations* and we shall see that they include the Maxwell equations.

[13] For the beginning of the theory, we keep the old notations that de Broglie used.

First, however, we must examine some other representations of the photon equations:

The Quasi-Maxwellian Form

First, it must be noted that there are too many equations in Eq. (6.8): 32 equations for only 16 components of the wave ϕ. There is a problem of compatibility. To solve the problem, de Broglie added and subtracted the two systems in Eq. (6.8):

$$(A)\ \frac{1}{c}\frac{\partial\phi}{\partial t} = \frac{a_k + b_k}{2}\frac{\partial\phi}{\partial x_k} + i\frac{\mu_0 c}{\hbar}\frac{a_4 + b_4}{2};\quad (B)\ 0 = \frac{a_k - b_k}{2}\frac{\partial\phi}{\partial x_k} + i\frac{\mu_0 c}{\hbar}\frac{a_4 - b_4}{2}\phi. \tag{6.11}$$

Furthermore, it will be shown that Eq. (6.8) exactly contains the Maxwell equations (up to mass terms), but Eq. (6.11) is already an outline of these equations because this system is divided into a group (*A*) of "evolution equations" that resembles the Maxwell equations in $\partial\mathbf{E}/\partial t$ and $\partial\mathbf{H}/\partial t$, and a group (*B*) of "condition equations," of the same kind as $div\mathbf{E} = 0$ and $div\mathbf{H} = 0$. In de Broglie (1934b), it gave only the group (*A*), but it is easy to prove, in analogy with the Maxwell equations, the following:

- Owing to Eq. (6.10), (*B*) is a consequence of (*A*).
- Actually, (*B*) is only satisfied by the solutions of (*A*) whose Fourier expansion does not contain a zero frequency. But the zero frequencies are automatically absent from the solutions of (*A*) if $\mu_0 \neq 0$.
- Therefore, *iff* $\mu_0 \neq 0$, the condition (*B*) is a consequence of the evolution equations (*A*).

Canonical Form

Eq. (6.8) can be transformed in another way:

$$\begin{aligned}(C)\quad & \frac{1}{c}\frac{a_4 + b_4}{2}\frac{\partial\phi}{\partial t} = \frac{b_4 a_k + a_4 b_k}{2}\frac{\partial\phi}{\partial x_k} + i\frac{\mu_0 c}{\hbar}a_4 b_4\phi \\ (D)\quad & \frac{1}{c}\frac{a_4 - b_4}{2}\frac{\partial\phi}{\partial t} = \frac{b_4 a_k - a_4 b_k}{2}\frac{\partial\phi}{\partial x_k}.\end{aligned} \tag{6.12}$$

This new system is at the basis of the Lagrangian derivation of the theory and of its tensorial form, and it was used by de Broglie to quantize the photon field and to describe the photon-electron interaction (de Broglie, 1940–1942). Just as in Eq. (6.11), (*D*) is a consequence of (*C*) *if* $\mu_0 \neq 0$, which is proved by applying to (*C*) the operator: $\frac{1}{c}\frac{a_4 - b_4}{2}\frac{\partial}{\partial t}$, taking into account Eq. (6.10). It is noteworthy that the strongest arguments in favor of a massive photon are not the answers to particular experimental objections, but the arguments imposed by the fusion theory, which are linked to the very structure of the theory.

6.1.4 Introduction of a Square-Matrix Wave Function

Now, let us return to the initial system [Eq. (6.8)], but in terms of relativistic coordinates $x_k = (x, y, z)$, $x_4 = ict$ with γ matrices $(\mu, \nu = 1, 2, 3, 4)$:

$$\begin{aligned} &\gamma_\mu \gamma_\nu + \gamma_\nu \gamma_\mu = 2\delta_{\mu\nu};\;\; \mu, \nu = 1, 2, 3, 4;\;\; \gamma_k = i\alpha_4 \alpha_k; \\ &\gamma_4 = \alpha_4;\;\; \gamma_5 = \gamma_1 \gamma_2 \gamma_3 \gamma_4. \end{aligned} \tag{6.13}$$

Multiplying Eq. (6.8) by $i\alpha_4$, we find that owing to Eq. (6.9), the following new system, which is not written in terms of a 16-line column wave function ϕ but in terms of a 4 ×4 square matrix wave function ψ:

$$\begin{aligned} &\partial_\mu \gamma_\mu \Psi - \frac{\mu_0 c}{\hbar} \Psi = 0 \\ &\partial_\mu \Psi \tilde{\gamma}_\mu - \frac{\mu_0 c}{\hbar} \Psi = 0 \end{aligned} \quad \left(\mu, \nu = 1, 2, 3, 4;\;\; \tilde{\gamma}_\mu = \gamma_\mu \;\; \textit{transp.}\right). \tag{6.14}$$

The transposed matrices $\tilde{\gamma}$ are easily eliminated because, if two sets of Dirac matrices γ_μ and $\tilde{\gamma}_\mu$, verify the relations [Eq. (6.13)], there are two (and only two) nonsingular matrices, Λ **and** Γ, such that

$$\tilde{\gamma}_\mu = \Lambda \gamma_\mu \Lambda^{-1};\;\; \tilde{\gamma}_\mu = -\Gamma \gamma_\mu \Gamma^{-1};\;\; \Lambda = \Gamma \gamma_5;\;\; \mu = 1, 2, 3, 4. \tag{6.15}$$

γ_5 is given in Eq. (6.13), and Eq. (6.15) is true for $\tilde{\gamma}_\mu$ transposed from γ_μ; A solution is as follows:

$$\Gamma = -i\gamma_2 \gamma_4;\;\; \Lambda = \Gamma \gamma_5 = -i\gamma_3 \gamma_1. \tag{6.16}$$

The Λ case in Eq. (6.15) was given in Pauli (1936), and the Γ case was given by de Broglie to eliminate $\tilde{\gamma}_\mu$ in Eq. (6.16). Indeed, introducing Γ into Eq. (6.14), we find the system given by Tonnelat, de Broglie, and Pétiau (Tonnelat, 1938; de Broglie, 1940–1942):

$$\begin{aligned} &\partial_\mu \gamma_\mu (\psi \Gamma) - \frac{\mu_0 c}{\hbar} (\psi \Gamma) = 0 \\ &\partial_\mu (\psi \Gamma) \gamma_\mu + \frac{\mu_0 c}{\hbar} (\psi \Gamma) = 0. \end{aligned} \tag{6.17}$$

The equations obtained by substituting Λ to Γ [Eq. (6.15)] were given recently (Lochak, 2014):

$$\begin{aligned} &\partial_\mu \gamma_\mu (\psi\Lambda) - \frac{\mu_0 c}{\hbar} (\psi\Lambda) = 0 \\ &\partial_\mu (\psi\Lambda)\gamma_\mu - \frac{\mu_0 c}{\hbar} (\psi\Lambda) = 0. \end{aligned} \tag{6.18}$$

The apparently small formal difference (a minus sign) between the two systems [Eqs. (6.17) and (6.18)] entails a great physical difference because the solutions of these equations exchange between themselves by a mutiplication by γ_5: they are dual in space-time, and we shall prove that it signifies the exchange between electric and magnetic charges.

So, the substitution of Λ to Γ in the representation by square matrices of the initial de Broglie's equations [Eq. (6.5)] gives two kinds of photons:

- Electric and magnetic photons
- The electromagnetic formulas of the photon equations

The fundamental electromagnetic formulas were given by de Broglie in his first papers, starting from Eq. (6.8) (de Broglie 1934a, b, 1936). For the sake of simplicity, we start from Eqs. (6.17) and (6.18), applying a procedure suggested by M.A. Tonnelat and then used by de Broglie (1934b).

Let us expand a 4 ×4 matrix Θ on the Clifford algebra in $\mathbb{R}^{+---}$:

$$\Lambda\Psi = \Theta = I\varphi_0 + \gamma_\mu\varphi_\mu + \gamma_{[\mu\nu]}\varphi_{[\mu\nu]} + \gamma_\mu\gamma_5\varphi_{\mu 5} + \gamma_5\varphi_5 \,, \tag{6.19}$$

where φ_0 is a scalar, φ_μ a polar vector, $\varphi_{[\mu\nu]}$ an antisymmetric tensor of rank 2, $\varphi_{\mu 5}$ an axial vector (the dual of an antisymmetric tensor of rank 3) and φ_5 a pseudoscalar (the dual of an antisymmetric tensor of rank 4). These expressions correspond in $\mathbb{R}^3$ to a scalar I_1; the Lorentz potentials **A**, V (linked to the electric charges); the electromagnetic fields **H**, **E**; the pseudo potentials **B**, W (linked to magnetic charges); and a pseudoscalar I_2[14]:

$$\begin{aligned} &\mathbf{H} = Kk_0\left(\varphi_{[23]}, \varphi_{[31]}, \varphi_{[12]}\right);\;\; \mathbf{E} = Kk_0\left(i\varphi_{[14]},\; i\varphi_{[24]},\; i\varphi_{[34]}\right) \\ &\mathbf{A} = K(\varphi_1,\, \varphi_2,\, \varphi_3);\;\; iV = K\varphi_4 \\ &-i\mathbf{B} = K\left(\varphi_{[15]},\, \varphi_{[25]},\, \varphi_{[35]}\right);\;\; W = K\varphi_{[45]} \\ &I_1 = K\varphi_0;\;\; iI_2 = K\varphi_5\;\; \left(k_0 = \frac{\mu_0 c}{\hbar};\;\; K = \frac{\hbar}{2\sqrt{\mu_0}}\right). \end{aligned} \tag{6.20}$$

[14] Remember that **B** is not an induction. The Lorentz polar quadripotential (V, **A**) remains linked to the electric Einstein photon, while the pseudoquadripotential (W,**B**) is linked to the magnetic photon.

Now, if we develop Eqs. (6.17) and (6.18) owing to Eqs. (6.19) and (6.20), we find two sets of equations, discussed in the next sections.

6.1.4 The Equations of the "Electric Photon" (Γ Matrix).

The expansion of the matrix wave-function $\Psi = \psi\,\Gamma$ according to Eq. (6.19) splits Eq. (6.17) into two systems (de Broglie 1940–1942, 1943), that we now refer to as the *electric photon* because the vector potential **A** appears in (6.21):

$$(M)\left(\begin{array}{l} -\frac{1}{c}\frac{\partial \mathbf{H}}{\partial t} = curl\mathbf{E};\quad \frac{1}{c}\frac{\partial \mathbf{E}}{\partial t} = curl\,\mathbf{H} + k_0^2\mathbf{A} \\ div\mathbf{H} = 0;\quad div\mathbf{E} = -k_0^2 V \\ \mathbf{H} = curl\,\mathbf{A};\quad \mathbf{E} = -gradV - \frac{1}{c}\frac{\partial \mathbf{A}}{\partial t};\quad \frac{1}{c}\frac{\partial V}{\partial t} + div\mathbf{A} = 0 \end{array}\right) \quad (6.21)$$

$$(NM)\left(\begin{array}{l} -\frac{1}{c}\frac{\partial I_2}{\partial t} = k_0 W;\quad grad\ I_2 = k_0\mathbf{B};\quad \frac{1}{c}\frac{\partial W}{\partial t} + div\,\mathbf{B} = k_0 I_2 \\ curl\,\mathbf{B} = 0;\quad gradW + \frac{1}{c}\frac{\partial \mathbf{B}}{\partial t} = 0;\quad \{(k_0 I_1 = 0;\ k_0 \neq 0) \Rightarrow I_1 = 0\} \end{array}\right). \quad (6.22)$$

Actually, de Broglie fixed his attention essentially on the first system of equations [Eq. (6.21)], which he denoted as (M) ("Maxwellian"), for obvious reasons, and he considered it as the equations of the photon (M: spin 1). This was the great victory of his theory: the deduction of Maxwell's equations from Dirac's equation.

Curiously, de Broglie was rather puzzled by the second system spin 0, that he named negatively: **NM** ("non-Maxwellian"), without giving any clear interpretation. He thought at first of a meson but then abandoned the idea. Here, we shall adopt the following very simple interpretation.

It is natural to find two systems of equations because the fundamental equations [Eq. (6.8)] are not the equations of a particle of spin 1, but of a particle of maximum spin 1: a combination of two particles of spin ½, as de Broglie underlined it. For this reason, just as for a diatomic molecule, we find two states described by two systems of equations: an orthostate of spin 1 = ½ + ½ (parallel spins) and a parastate of spin 0 = ½ − ½ (opposite spins). We shall adopt this interpretation.

Both states have equal rights with respect to the symmetry laws because both are linked by a symmetry of form and both have a physical sense, despite the fact that one of them (the orthostate: spin 1) is related to a far more celebrated case: the Maxwell equations, while the orthostate is related to the smaller Aharonov-Bohm effect, as will be shown later in this chapter.

Thus, we have two photons—more precisely, two spin states: 1 and 0, of a photon described by the systems [Eqs. (6.21) and (6.22)]. And it is not a general photon, but only an electric photon; this is because we shall find another one: a magnetic photon. For the moment, we have just the electric photon with two photon states: a spin 1 state (M), "Maxwellian," and a spin 0 state (NM), "non-Maxwellian."

The (M) equations are Maxwell's equations, but with two differences:

1. The first difference is the presence of the *mass terms,* which introduces a link between fields and potentials, the latter becoming physical quantities and losing their gauge invariance.
2. The second difference is the automatic definition of fields, through the Lorentz potentials, with the following Lorentz gauge condition:

$$\mathbf{H} = \mathit{curl}\,\mathbf{A};\ \ \mathbf{E} = -\mathbf{grad}\,V - \frac{1}{c}\frac{\partial \mathbf{A}}{\partial t};\ \ \frac{1}{c}\frac{\partial V}{\partial t} + \mathit{div}\mathbf{A} = 0. \tag{6.23}$$

These relations are not arbitrarily added to the field equations, as they were in the classical theory: they appear automatically and they are themselves field equations, as a consequence of the massive photon. Of course, they were already present in a hidden form in Eqs. (6.8), (6.11), (6.12), and (6.17).

A consequence of Eq. (6.21). is that the fields and potentials do not obey the d'Alembert wave equation, but rather the Klein-Gordon equation:

$$\Box F + k_0^2 F = 0;\ \ (F = \mathbf{E},\ \mathbf{H},\ \mathbf{A},\ V,\ \mathbf{B},\ W,\ I_1,\ I_2). \tag{6.24}$$

The electrostatic solution is not the Coulomb potential $\frac{1}{r}$, but the Yukawa potential $V = e^{-r/k_0}/r$, which remains a long-range potential because of the smallness of the Compton wave number $k_0 = \mu_0 c/\hbar$.

The (NM) equations were previously considered by de Broglie (as was already said) as describing an independent spin 0 meson with a far greater μ_0 mass than the mass of the photon. This fact is astonishing, considering that the equations (M) and (NM) came from the decomposition of the same system of equations, so that both rest masses are obliged to be equal. Of course, we shall abandon this idea, which was actually later abandoned by de Broglie himself. Our interpretation will be based, on the contrary, on the link between the two systems (M) and (NM), admitted as a fact.

The system (NM) describes a chiral particle because I_1 is a true invariant, but $I_1 = 0$; actually, the particle is definite by the second invariant I_2 which is a pseudoinvariant, dual of an antisymmetric tensor in $\mathbb{R}^{+---}$ (with $I_2 \neq 0$), and by the pseudo-quadrivector (**B**,W) in $\mathbb{R}^{+---}$.

It must be noted that de Broglie remarked (de Broglie, 1940–1942) that the situation could be interpreted in another way, defining a second electromagnetic field (which he called an "anti-field") which *equals zero* by virtue of Eq. (6.22), as follows:

$$\mathbf{H}' = -\frac{1}{c}\frac{\partial \mathbf{B}}{\partial t} + grad\,W; \quad \mathbf{E}' = curl\,\mathbf{B}. \tag{6.25}$$

We shall follow the second interpretation here, on the basis of a symmetry between electricity and magnetism developed in our papers concerning the photon (Lochak 20) and the magnetic monopole (Lochak, 1992, 2000)[15]. We consider the systems [Eqs. (6.21)–(6.22)] as simply describing, with equal weight the orthohydrogene state (spin 1) and the parastate (spin 0) of an **electric photon**, for the following reasons:

1. In the system (M), we have an electromagnetic field $(\mathbf{E}, \mathbf{H})$ and a *polar* 4-potential $(V, \mathbf{A})$, related to $(\mathbf{E}, \mathbf{H})$ by the Lorentz formulas [Eq. (6.23)]. These fields and potentials enter into the dynamics of an *electric* charge. Because $k_0 \neq 0$, we have in general $div\mathbf{E} \neq 0$, so that the electric field **E** is not transversal, contrary to the magnetic field **H**: and **E** has a small longitudinal component, of the order of k_0.
2. In the (NM) equations, we have a *pseudo*-invariant I_2 and an *axial* 4-potential $(\mathbf{B}, W)$, to which may be added the invariant I_1 and the anti-field $\{\mathbf{E}', \mathbf{H}'\}$, defined in Eq. (6.25), and which will be related to magnetism. But here, $I_1 = \mathbf{E}' = \mathbf{H}' = 0$, which confirms the electric character of the (NM) photon by the annihilation of magnetic quantities.

The difference between the de Broglie interpretation and mine is that now (NM) is no longer separated from the spin 1 (state M): it is the spin 0 state of the same photon. The electric photon is the whole system [Eqs. (6.21)–(6.22)] with two values of spin.

6.1.5 The Equations of the Magnetic Photon (Λ Matrix).

This second photon is given by Eq. (6.18) with $\Lambda = \Gamma\gamma_5$ [Eq. (6.15)] instead of Γ in [Eq. (6.17)]. The primed new field components are the

[15] The de Broglie definition [Eq. (6.25)] of $\mathbf{H}'$ and $\mathbf{E}'$, in terms of a *pseudo*-quadripotential $(\mathbf{B}, W)$, was later rediscovered by Cabibbo and Ferrari (1962).

dual of the preceding ones, which means that the matrix γ_5 exchanges electricity and magnetism (Lochak, 1992, 2000):

$$
(\mathrm{M})\quad \left(\begin{array}{l}
-\frac{1}{c}\frac{\partial \mathbf{H}'}{\partial t} = curl\,\mathbf{E}' + k_0^2\mathbf{B}'; \quad \frac{1}{c}\frac{\partial \mathbf{E}'}{\partial t} = curl\,\mathbf{H}' \\
div\mathbf{H}' = k_0^2 W'; \quad div\mathbf{E}' = 0 \\
\mathbf{H}' = grad\,W' + \frac{1}{c}\frac{\partial \mathbf{B}'}{\partial t}; \quad \mathbf{E}' = curl\,\mathbf{B}'; \quad \frac{1}{c}\frac{\partial W'}{\partial t} + div\mathbf{B}' = 0
\end{array}\right) \tag{6.26}
$$

$$
(\mathrm{NM})\quad \left(\begin{array}{l}
-\frac{1}{c}\frac{\partial I_1}{\partial t} = k_0 V'; \quad gradI_1 = k_0\mathbf{A}'; \quad \frac{1}{c}\frac{\partial V'}{\partial t} + div\mathbf{A}' = k_0 I_1 \\
curl\mathbf{A}' = 0; \quad grad\,V' + \frac{1}{c}\frac{\partial \mathbf{A}'}{\partial t} = 0; \quad \{(k_0 I_2 = 0;\ k_0 \neq 0) \Rightarrow I_2 = 0\}
\end{array}\right) \tag{6.27}
$$

The new photon is associated, as before, with a couple of fields. But the situation is inverted in the following ways:

1. The anti-field $(\mathbf{E}', \mathbf{H}')$ and the *axial* 4-potential $(W', \mathbf{B}')$ satisfy the Maxwell-type (M) system [Eq. (6.26)]. The definition [Eq. (6.25)] of the anti-fields now appears in Eq. (6.26) automatically (and not by an a priori definition), as one of the field equations. Now $(\mathbf{E}', \mathbf{H}')$ are not equal to zero. The fields $(\mathbf{E}', \mathbf{H}')$ are exactly those that enter into the dynamics of a magnetic charge: a monopole (Lochak 1985, 1995b and Chapters 2, 3 of this book).

Besides, symmetrically to the electric case, we now have $div\mathbf{H}' \neq 0$, so that, in a plane wave, the *magnetic* field $\mathbf{H}'$ (instead of the electric one $\mathbf{E}'$) has a small longitudinal component of the order of k_0, while $\mathbf{E}'$ is transversal. We have a *magnetic photon*.

2. Now, the *polar* potentials $(V', \mathbf{A}')$ dual from $(W', \mathbf{S}')$ appear in the (NM) system (i.e., in the *spin 0 state*). The invariant I'_1 and the pseudoinvariant I'_2 invert their roles: we have now $I'_1 \neq 0$ and $I'_2 = 0$. The electromagnetic field $(\mathbf{E}', \mathbf{H}')$ defined by the Lorentz formulas [Eq. (6.23)] gives $(\mathbf{E}' \neq 0,\ \mathbf{H}' \neq 0)$ in the Maxwellian formulas (M) and $(\mathbf{E}' = \mathbf{H}' = 0)$ in the non-Maxwellian formulas (NM), as opposed to what we had in the electric case.

It is a remarkable fact that de Broglie's *fusion* of two Dirac equations not only gives the classical Maxwell equations (as was proved by de Broglie), but also defines *two classes* of photons, corresponding respectively to electric or magnetic charges. The algebraic symmetry excludes any other possibility.

The symmetry between the two electromagnetic fields is all the more interesting in that such a symmetry already appears in the Dirac equation itself, in the form of *two minimal interactions* corresponding to electric and magnetic charges, associated with the two kinds of fields (Chapter 2 of this book and Lochak, 1995b). Symmetries of Dirac's and de Broglie's equations are thus mutually reinforced. Now we must address other issues, to wit:

- We have two kinds of photons: the electric and the magnetic photon.

But is their physical difference given by the difference between the two pairs of equations: Eqs. (2.10)−(2.11) and Eqs. (6.17)−(6.18) or the Dirac gauge and equation and the chiral gauge and equation? Yes, because it is the difference between the motion of an electron or a monopole in an electrodynamic field. For instance, in a linear electric field, the electron is linearly accelerated, while the monopole rotates around the field, and the reverse is true for a linear magnetic field.

- Actually, there are not only two but four kinds of photons because they can have a spin 1 or a spin 0.

The preceding answer is only related to spin 1. We must now answer a new question: are the spin 0 photons already known? The answer is yes and there is a wellknown example.

6.1.6 The Aharonov−Bohm Effect

Consider the equations of (NM) potentials: Eqs. (6.22) and (6.27):

Spin 0 electric photon: $-\frac{1}{c}\frac{\partial I_2}{\partial t} = k_0 W$; $grad\, I_2 = k_0\mathbf{B}$; $\frac{1}{c}\frac{\partial W}{\partial t} + div\,\mathbf{B} = k_0 I_2$ and the associated equations in Eq. (6.22).

Spin 0 magnetic photon: $-\frac{1}{c}\frac{\partial I_1}{\partial t} = k_0 V'$; $grad\, I_1 = k_0\mathbf{A}'$; $\frac{1}{c}\frac{\partial V'}{\partial t} + div\mathbf{A}' = k_0 I_1$ and the associated equations in Eq. (6.27).

We must remember that the spin 1 electric photon is associated with a magnetic spin 0 photon by the pseudo − invariant I_2, while the spin 1 magnetic photon is associated with an electric spin 0 photon by the true invariant I_1. The preceding relations immediately imply that the spin 0 potentials are the gradients of relativistic invariants, which verifiy the Klein-Gordon equation:

$$\partial_\mu I_1 = k_0 A_\mu;\quad \Box I_1 + k_0^2 I_1 = 0;\quad \partial_\mu I_2 = k_0 B_\mu;\quad \Box I_2 + k_0^2 I_2 = 0. \tag{6.28}$$

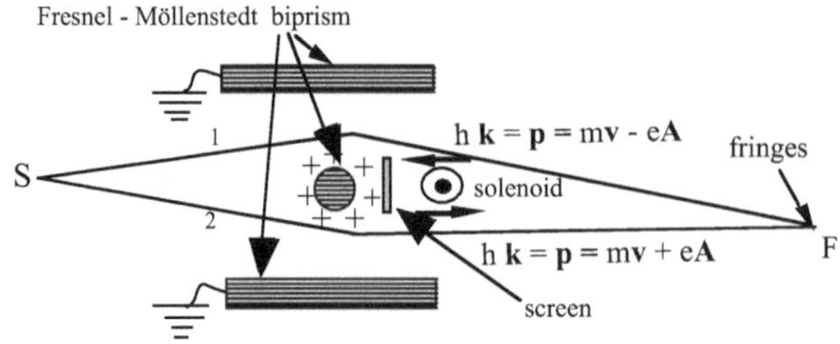

Figure 6.1 Aharonov-Bohm experiment.

We know that by virtue of Eqs. (6.22) and (6.27), or Eq. (6.28), the corresponding electromagnetic fields equal zero. The question is: how can the spin 0 photon be detected? More precisely, since these fieldless potentials are unable to generate a force, what could be observed? The answer is, of course, the phase, first characteristic of a wave. The Aharonov-Bohm effect was imagined at first by David Bohm[16] to answer this question, and to prove that contrary to a common idea, the electromagnetic potentials are not only mathematical intermediates (even if they can play this role): they are observable physical quantities. The effect had been predicted ten years earlier by Ehrenberg and Siday (1949).

6.1.7 The Effect

The idea suggested by Bohm (Aharonov and Bohm, 1959; Tonomura, 1998; Peshkin and Tonomura, 1989; Olariu and Popescu, 1985; Lochak, 1983) was to modify electron interference by a fieldless magnetic potential created by a magnetic string or by a thin solenoid orthogonal to the plan of interfering electron trajectories, as shown in Figure 6.1. The Young slits are obtained by means of a Fresnel–Möllenstedt biprism.

The solenoid must be infinitely long (in principle), so the magnetic field emanating from the extremities cannot disrupt the experiment: it is assumed in the calculations, but actually a few millimeters are sufficient because the transverse dimensions of the device are of the order of microns. This arrangement of the solenoid has led to the idea that the magnetic flux through the trajectories' quadrilateral plays an essential role. Many disagree with that idea (Lochak, 1983).

The problem of eliminating this hypothesis was elegantly solved by Tonomura (see: Peshkin and Tonomura, 1989) by substituting the rectilinear string by a microscopic torus (10 μm): one of the electron beams passes through the torus and the other outside, the magnetic lines being trapped in the torus.

[16] I know that because I was acquainted with Bohm who lived in Paris in that time.

Let us give an intuitive interpretation of the experiment. The principle is that the wave vector of an electron in a magnetic potential is given by the de Broglie wave (de Broglie, 1934c), which is a direct consequence of the identification of the principles of Fermat and of least action ($\mathbf{p}$ is the Lagrange momentum):

$$\frac{h}{\lambda}\mathbf{n} = \hbar\mathbf{k} = \mathbf{p} = m\mathbf{v} + e\mathbf{A}. \tag{6.29}$$

It is obvious from Eq. (6.29) that interference and diffraction phenomena are influenced by the presence of a magnetic potential independent of the presence of the field because the interferences depend only on the phase. It is well known in optics: an interference figure is shifted in a Michelson interferometer by introducing a plate of glass in one of the virtual beams, which causes a phase shift and thus a change of the optical path without any additive force.

These phenomena are manifestly gauge dependent: if we add something to $\mathbf{A}$, whether a gradient or not, in the de Broglie wave λ [Eq. (6.29)], the last is modified. This is evident even in the classical de Broglie formula: $\lambda = h/mv$ when $\mathbf{A} = 0$, which is gauge dependent too, a fact often emphasized by de Broglie himself, who said, "If gauge invariance were general in quantum mechanics, the electron interferences could not exist."

In the case of the Aharonov-Bohm experiment, there is an additive phase with both interfering waves in opposite directions, which doubles the shift of the interference fringes. Let us recall a proof of the effect, independent from the fact that a potential generates forces or not (Lochak, 1983).

6.1.8 The Magnetic Potential of an Infinitely Thin and Infinitely Long Solenoid

We consider the case corresponding to the Aharonov-Bohm experiment: electrons diffracted on Young slits and falling on a magnetic solenoid orthogonal to the plane electron trajectories, according to the Figure 6.1 and, further, to the schematic shown in Figure 6.2, the solenoid is along Oz. To simplify the

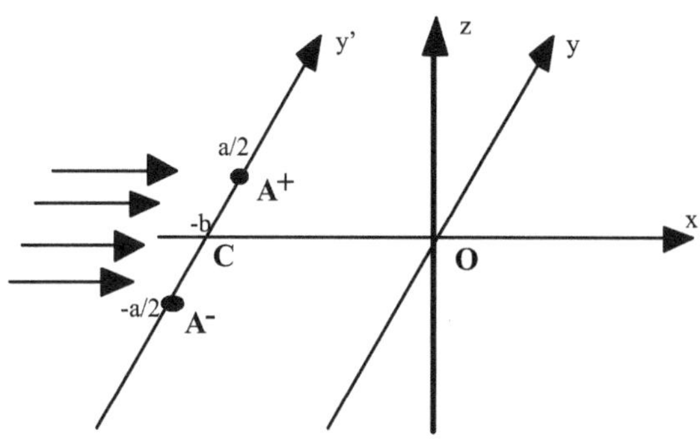

Figure 6.2 Aharonov-Bohm scheme.

calculations, we shall disregard the photon mass, which is only important in the symmetry laws, which are taken into account in all the formulas; therefore, to omit the photon mass only means to omit negligible corrections.

The electric charge of the diffracted electrons implies that they "see" the electromagnetism through the Lorentz potentials $(V, \mathbf{A})$, and thus through the equations (M): Eq. (6.21). These equations derive from the pseudo-invariant I_2. Now, there is an obvious invariant in the Aharonov-Bohm effect: the rotation angle $\varphi = \arctan y/x$ around the axis Oz. So we shall write

$$I_2 = \varepsilon k_0 \arctan (y/x), \tag{6.30}$$

where k_0 is the quantum wave number of the photon and ε a convenient dimensional constant, the value of which is not important for our calculation. Neverthless, something seems wrong here, because (y/x) is P-invariant so that, with the definition [Eq. (6.30)], I_2 seems to be a P-invariant and not a pseudoinvariant, as it needs to be in Eq. (6.22).

But this is not so because (y/x) is P-invariant only in the space $\mathbb{R}^2$: (x, y), not in the space $\mathbb{R}^3$ (x, y, z). In our case, the inversion is the P-transformation $(x, y, z) \rightarrow (-x, -y, -z)$, which implies the inversion of Oz and thus of the angle φ. So that (y/x) is really a pseudoinvariant in $\mathbb{R}^3$.

Thus, we have, by virtue of Eq. (6.22):

$$grad I_2 = k_0 \mathbf{B} \tag{6.31}$$

$$\mathbf{B}_x = -\varepsilon \frac{y}{x^2 + y^2}; \quad \mathbf{B}_y = \varepsilon \frac{x}{x^2 + y^2}; \quad \mathbf{B}_z = 0. \tag{6.32}$$

6.1.9 The Theory of the Effect

The commonly admitted theories are unnecessarily complicated (Olariu and Popescu, 1985). For the physical bases of the effect, the best is to start from the brillant book of Tonomura (1998). To find the formula of fringes, it is sufficient to take the geometrical optics approximation with the phase $\varphi = S/\hbar$ of de Broglie's wave and the principal Hamilton function S obeying the Hamilton-Jacobi equation with the potential [Eq. (6.32)]:

$$2m\frac{\partial S}{\partial t} = \left(\frac{\partial S}{\partial x} + \varepsilon \frac{y}{x^2 + y^2}\right)^2 + \left(\frac{\partial S}{\partial y} - \varepsilon \frac{x}{x^2 + y^2}\right)^2. \tag{6.33}$$

The electronic wave propagates from $x = -\infty$ to $x = +\infty$ and the Young slits A^+ and A^- (Figure 6.2) are on a parallel to Oy, at a distance $\pm a/2$ from the point C located at $x = -b$.

The pseudopotential **B** appearing in (6.30) and (6.31) is the gradient of I_2, so that **B** and I_2 satisfy up to μ_0 the equations (NM), Eq. (6.27). They are independent of t because $W = 0$.

Eq. (6.33) is immediately integrated, defining the phase as follows:

$$\Sigma = S - \varepsilon \arctan y/x\,, \tag{6.34}$$

which gives

$$2m\frac{\partial \Sigma}{\partial t} = \left(\frac{\partial \Sigma}{\partial x}\right)^2 + \left(\frac{\partial \Sigma}{\partial y}\right)^2. \tag{6.35}$$

Chosing a complete integral of Eq. (6.35) and thus of Eq. (6.32), owing to Eq. (6.33), we have

$$\Sigma = Et - \sqrt{2mE}\,(x \cos\theta_o + y \sin\theta_o) \tag{6.36}$$

$$S = Et - \sqrt{2mE}\,(x \cos\theta_o + y \sin\theta_o) + \varepsilon \arctan\frac{y}{x}, \tag{6.37}$$

or, in polar coordinates $x = r\cos\theta$, $y = r\sin\theta$:

$$S = Et - \sqrt{2mE}\, r \cos(\theta - \theta_o) + \varepsilon\,\theta. \tag{6.38}$$

The Jacobi theorem gives the trajectories (the wave rays):

$$\begin{aligned} \frac{\partial S}{\partial \theta_o} &= \sqrt{2mE}\,(x \sin\theta_o - y\cos\theta_o) = \mu; \\ \frac{\partial S}{\partial E} &= t - \sqrt{\frac{m}{2E}}\,(x\cos\theta_o + y\sin\theta_o) = t_o. \end{aligned} \tag{6.39}$$

Finally, with[17] $E = \frac{1}{2}mv^2$ we have the motion

$$x\cos\theta_o + y\sin\theta_o = v(t - t_o). \tag{6.40}$$

We see that the *rays* (electron trajectories), defined in Eq. (6.39) are orthogonal to the moving planes but they are not orthogonal to the equal phase surfaces [Eqs. (6.37)–(6.38)] except far from the magnetic string ($x \to \infty$), when the potential term of the order of ε becomes negligible.

Therefore, despite the presence of a potential, the electronic trajectories remain rectilinear and are not deviated, because the magnetic field equals zero by virtue of Eq. (6.22). The velocity $v = Const$ remains the one of the incident electrons because of the conservation of energy.

[17] We are obviously far from relativity.

But the diffraction of waves through the slits A^+ and A^- creates, for the electron trajectories, an interval of possible angles θ_o equal to the angles of the interference fringes, modified by the magnetic potential:

There is no deviation of the electrons, only a deviation of the angles of phase synchronization between the waves issued from A^+ and A^-. This is the Aharonov-Bohm effect, which is in accordance with the definition of the spin 0 photon [Eq. (6.22)].

It would be useless to reproduce the end of the theory of Aharonov-Bohm effect (see, for instance, Lochak, 1932b). Let us only recall the total phase-shift:

$$\Delta\varphi = \frac{\Delta S}{h} = \frac{a\theta_o}{\lambda} + \frac{2\varepsilon\xi}{h}. \tag{6.41}$$

The first term gives the standard Young fringes (the notations are those of Figure 6.2), while the second term is the Aharonov-Bohm effect: $\xi = \arctan a/2b$, which is equal to half the angle under which the Young slits are seen from the solenoid, which entails a dependence of the effect on the position of the string. One can assert that the effect decreases when the distance b increases.

We see that the theory of the Aharonov-Bohm effect is a simple consequence of the definition of the invariant in the system [Eq. (6.27)], as the invariant rotation angle around the axis of he solenoid.

6.1.10 Conclusions on the Theory of Light

We suggest a new theory of light based on four photons, as follows:

1. At first, the Einstein photon known in optics from 1905, and later identified by de Broglie (1922) as a vectorial spin 1 particle, which we call here the *electric photon,* because it interacts with the electric charges, principally with electrons.
2. A pseudovectorial spin 1 magnetic photon, analogous to the electric Einstein photon: it appears in the theory of leptonic magnetic monopoles (see: Chapters 2 and 3). The magnetic photon plays in the physics of monopoles a role exactly similar to the role played by the electric photon in the theory of electrons.
3. Two spin 0 photons (one electric and the other magnetic), related to 2 classes of respectively electric and magnetic fieldless phenomena; an example is the Aharonov-Bohm effect.
4. It must be added that in the four-photon theory of light, there are two Maxwell displacements: an electric displacement and a magnetic

displacement. Let us recall what is the Maxwell displacement[18]: at the beginning, he tried to unify the electromagnetism on the basis of several fundamental laws: the laws of Coulomb: $\nabla . \mathbf{E} = \mathbf{4}\pi\rho$, Ampere: $\nabla \times \mathbf{H} = \frac{4\pi}{c}\mathbf{J}$ and Faraday: $\nabla \times \mathbf{E} + \frac{1}{c}\frac{\partial \mathbf{H}}{\partial t} = 0$. But he found an incoherence between them because the third law depends on time and the other ones do not. The fact was well known and was objected to by Michael Faraday; but the critics of Maxwell went contrary to the unanimity of physicists: he considered the law of Faraday as the right one, and he decided to introduce a time dependence into the other two laws. He replaced the Coulomb law by a continuity law $\nabla . \ \mathbf{J} + \frac{\partial \rho}{\partial \mathbf{t}} = 0$, owing to which the Ampere law became $\nabla \times \mathbf{H} - \frac{1}{c}\frac{\partial \mathbf{E}}{\partial t} = \frac{4\pi}{c}\mathbf{J}$. So, he found the celebrated Maxwell equations in which appeared wave like solutions from which Maxwell found the electromagnetic theory of light and which later gave rise to the radio waves.

If we compare the (M) equations [Eq. (6.21)] of the electric photon with our (M) equations [Eq. (6.26)] of the magnetic photon, the analogy is evident: the terms of a Maxwell displacement are present in the magnetic photon, and it may be supposed that they lead to analogous physical consequences involving magnetic monopoles instead of electrons: the Aharonov-Bohm effect is a first example.

Now there is another fact that has been true for 70 years, without being pointed out until now, as far as I know. It is the fact that the de Broglie theory, based on the principle of fusion, implies automatically the displacement previously introduced by Maxwell through an external argument. The fact is hidden because the Maxwell equations make now a unit often abridged in different algebraic forms, while the displacements are more or less forgotten or rejected in the subtelties of history of science. Despite the fact that the postulate of fusion has an algebraic character, it has the advantage of unicity and of being a direct bridge between the problem of electromagnetism and the Dirac equation of the electron, the stronger equation of quantum mechanics.

It must be added that the de Broglie theory of the photon, being considered as a composite particle, gave rise to an extension to a general theory of spin particles, including gravitation (as discussed later in this chapter). We have already considered several generalizations of de Broglie's theory of

[18] See the excellent Chapter 6 of Jackson (1975). Our formalism is different from Jackson's because here, we are in the domain of quantum laws which are written in a vacuum.

light, as the magnetic photon linked to the magnetic monopole, and the Aharonov-Bohm effect, which gives rise to a new domain of electrodynamical phenomena.

The suggested theory of light is a generalization of Broglie's theory of light, with electric and magnetic photons. A new hypothesis of the present theory is that the spin 0 is condidered as a state of the photon with the same rights as the spin 1: there are not only two kinds of spin-1 photons, but also of two spin-0 photons. In other words, the photon world is divided into the same two categories as other composite quantum objects. There are orthophotons of spin 1 and paraphotons of spin 0, just as there are orthohydrogen and parahydrogen. But concerning the photon, it is a new idea, contrary to the case of orthohydrogen and parahydrogen, known for almost a century. This is why many questions still remain asked, such as the following:

- What happens with the spin-0 photons in the thermodynamical equilibrium?
- We have seen that paraphotons, being fieldless, are unable to create a force; so, are they able to produce something like a photoelectric effect? It seems not.
- More generally, are there true quantum wave-particle objects, or "pure-phases," pure potentials without particles? (*Pace* Louis de Broglie!)
- There are arguments in favor of some of these hypotheseses. For instance, the existence of a magnetic spin 1 photon is confirmed by the experiments on the leptonic monopole. Until now, the Aharonov-Bohm effect was a remarkable but isolated orphan effect. Here, it is integrated in a general theory. This is fine, but a question remains: is this effect exceptional, or is it a sample of a "class" of new phenomena? The equations define such a class mathematically, but it must be experimentally proved that such phenomena really exist as several physical effects. We have predicted at least one such effect: the Aharonov-Bohm effect with magnetic monopoles, but it is not yet observed and not yet calculated with all the details.

6.2 HAMILTONIAN, LAGRANGIAN, CURRENT, ENERGY, SPIN

6.2.1 The Lagrangian

Now, let us go back to the 16-line column wave function and the canonical form Eq. (6.12), keeping only (C) because (D) is deduced from it:

$$\frac{1}{c}\frac{a_4+b_4}{2}\frac{\partial\phi}{\partial t}=\frac{b_4a_k+a_4b_k}{2}\frac{\partial\phi}{\partial x_k}+i\frac{\mu_0 c}{\hbar}a_4b_4\phi. \tag{6.42}$$

Note the presence of $(a_4+b_4)/2$ in the factor of $\partial/\partial t$, so it seems unescapable that coherent definitions for tensor densities would be obtained. The Hamiltonian operator is

$$H=i\hbar\left[\frac{b_4a_k+a_4b_k}{2}\frac{\partial}{\partial x_k}+i\frac{\mu_0 c}{\hbar}a_4b_4\right] \tag{6.43}$$

and the Lagrangian density is (with $\phi^+=\phi\ (h.c.)$)

$$L=-i\hbar c\left[\phi^+\left(\frac{1}{c}\frac{a_4+b_4}{2}\frac{\partial\phi}{\partial t}-\frac{b_4a_k+a_4b_k}{2}\frac{\partial\phi}{\partial x_k}-i\frac{\mu_0 c}{\hbar}a_4b_4\phi\right)+c.c.\right]. \tag{6.44}$$

6.2.2 The Current Density Vector

The general formula

$$J_\mu=\frac{i}{\hbar}\left[\frac{\partial L}{\partial\phi_{,\mu}}\phi-\frac{\partial L}{\partial{\phi^+}_{,\mu}}\phi^+\right] \tag{6.45}$$

gives, with Eq. (6.44),

$$J_k=-c\phi^+\frac{b_4a_k+a_4b_k}{2}\phi;\ J_4=ic\rho;\ \ \rho=\phi^+\frac{a_4+b_{4k}}{2}\phi. \tag{6.46}$$

Therefore, $\int\rho dv$ *is not definite-positive.* But on the other hand, we shall find a *definite-energy* $\int\rho W dv\geq 0$, contrary to what happens in the Dirac electron. This result will be generalized in the general theory of particles with *spin* $=\frac{n}{2}$.

In terms of electromagnetic quantities, Eq. (6.45) is given by the Geheniau formulas, with two kinds of terms corresponding to spin 1 and spin 0 in the case of an electric photon (de Broglie, 1943). Here, until the end of the next section, we give only the translation of the formulas in the electric case (they were not translated until now in the magnetic case):

$$\begin{aligned}\mathbf{J}&=\frac{i}{\hbar c}[\mathbf{A}^*\times\mathbf{H}+\mathbf{H}^*\times\mathbf{A}+V^*\mathbf{E}-\mathbf{E}^*V]+\frac{c}{4}\left(I_2^*\mathbf{B}+\mathbf{B}^*I_2\right)\\ \rho&=\frac{i}{\hbar c}[(\mathbf{A}^*.\,\mathbf{E})-(\mathbf{E}^*.\,\mathbf{A})]+\frac{1}{4}\left(I_2^*W+W^*I_2\right).\end{aligned} \tag{6.47}$$

For the energy tensor, we have the general formula

$$T_{\mu\nu} = -\frac{\partial L}{\partial \phi_{,\mu}} \phi_{,\mu} - \frac{\partial L}{\partial \phi^+_{,\mu}} \phi^+{}_{,\mu} + L\delta_{\mu\nu} \,, \tag{6.48}$$

with the Lagrangian [Eq. (6.44)], which gives

$$\begin{aligned} T_{ik} &= -\frac{i\hbar c}{2}\left[\phi^+ \frac{b_4 a_k + a_4 b_k}{2}\frac{\partial \phi}{\partial x_k} + \frac{\partial \phi^+}{\partial x_k}\frac{b_4 a_k + a_4 b_k}{2}\phi\right] \\ T_{i4} &= \frac{\hbar}{2}\left[\phi^+ \frac{b_4 a_k + a_4 b_k}{2}\frac{\partial \phi}{\partial t} + h.c.\right]; \quad T_{4i} = -\frac{\hbar c}{2}\left[\phi^+ \frac{a_4 + b_4}{2}\frac{\partial \phi}{\partial x_i} + h.c.\right] \\ T_{44} &= -w = i\hbar\frac{i\hbar}{2}\phi^+\frac{a_4 + b_4}{2}\frac{\partial \phi}{\partial t} = -\phi^+ H\phi. \end{aligned} \tag{6.49}$$

In the electromagnetic form, we have

$$\begin{aligned} T_{\mu\nu} &= \frac{1}{2}\left(F_{\mu\lambda}\frac{\partial \mathbf{A}_\lambda}{\partial x_\nu} - \mathbf{A}_\lambda \frac{\partial F_{\lambda\mu}}{\partial x_\nu}\right) - \frac{i\hbar}{8}\left(I_2^* \frac{\partial \mathbf{B}_\lambda}{\partial x_\nu} + \mathbf{B}^*_\mu \frac{\partial I_2}{\partial x_\nu}\right) + \text{c.c.} \\ &\text{where}: \quad F_{\mu\lambda} = \frac{\partial \mathbf{A}_\mu}{\partial x_\nu} - \frac{\partial \mathbf{A}_\nu}{\partial x_\mu} \end{aligned} \tag{6.50}$$

In particular, the energy density ρW takes the form

$$\begin{aligned} T_{44} &= \frac{1}{2c}\left[\left(\mathbf{A}^*.\frac{\partial \mathbf{E}}{\partial t}\right)\left(\mathbf{E}^*.\frac{\partial \mathbf{A}}{\partial t}\right) + \left(\mathbf{A}.\frac{\partial \mathbf{E}^*}{\partial t}\right)\left(\mathbf{E}.\frac{\partial \mathbf{A}^*}{\partial t}\right)\right] \\ &+ \frac{i\hbar c}{2}\left[\left(I_2^*\frac{\partial W}{\partial t}\right)\left(W^*\frac{\partial I_2}{\partial t}\right) - \left(I_2\frac{\partial W^*}{\partial t}\right)\left(W\frac{\partial I_2^*}{\partial t}\right)\right]. \end{aligned} \tag{6.51}$$

The tensor $T_{\mu\nu}$ is often symmetrized, putting $T_{(\mu\nu)} = \frac{1}{2}(T_{\mu\nu} + T_{\nu\mu})$, but there are strong arguments in favor of the nonsymmetric tensor (Costa de Beauregard, 1943; de Broglie, 1943).

In addition, we can find other tensors, the integrals of which are equal to the integral of the precedings (they differ by a divergence). One of these tensors is[19]

[19] The factor μ_0 is surprising but according to Eq. (6.20), it disappears from the fields and potentials.

$$M_{ik} = M_{ki} = \mu_0 c^2 \phi^+ \frac{a_i b_k + a_k b_i}{2} \phi; \quad M_{i4} = M_{4i} = -\mu_0 c^2 \phi^+ \frac{a_i + b_i}{2} \phi;$$
$$M_{44} = \mu_0 c^2 \phi^+ \phi; \quad (i, k = 1, 2, 3). \tag{6.52}$$

This is a Maxwell-type tensor because we find, in electromagnetic terms, for the electric photon:

$$\begin{aligned} M_{i4} &= (\mathbf{E}.\,\mathbf{H}^*)_i + (\mathbf{E}^*.\,\mathbf{H})_i - k_0^2(V^*\mathbf{A}_i + V\mathbf{A}_i^*); \\ M_{44} &= |\mathbf{E}|^2 + |\mathbf{H}|^2 - k_0^2(|\mathbf{A}|^2 + |V|^2). \end{aligned} \tag{6.53}$$

We recognize the Maxwellian form, up to the mass terms, and we find

$$\int M_{\mu\nu}\, d\tau = \int T_{\mu\nu}\, d\tau. \tag{6.54}$$

6.2.3 The Photon Spin

Let us express the angular momentum with the nonsymmetric tensor $T_{\mu\nu}$:

$$m_{ik} = -\frac{i}{c} \int [x_i T_{4k} - x_k T_{4i}]\, d\tau \quad (i, k = 1, 2, 3), \tag{6.55}$$

where m_{ik} is *not* a constant of motion. But, as in Dirac's theory, we find a constant of motion $\mathrm{m}'_{\mathrm{ik}}$ if we add a convenient term of spin:

$$m'_{ik} = m_{ik} + S_{ik} \tag{6.56}$$

$$S_{ik} = i\hbar \int \phi^+ \frac{b_4 a_i a_k + a_4 a_i b_k}{2} \phi \quad (i, k = 1, 2, 3). \tag{6.57}$$

The dual $s_j = \varepsilon_{jik} S_{ik}$ of this tensor in $\mathbb{R}^3$ is a pseudovector. Analogous with the Dirac spin, we find a *space-time pseudovector,* by adding a time component:

$$s_4 = c\hbar \int \phi^+ \frac{b_4 a_1 a_2 a_3 + a_4 b_1 b_2 b_3}{2} \phi. \tag{6.58}$$

Now if we introduce into Eq. (6.55) the tensor $T_{(\mu\nu)} = \frac{1}{2}(T_{\mu\nu} + T_{\nu\mu})$ instead of the tensor $T_{\mu\nu}$, we find the new momentum, which is equivalent to Eq. (6.57):

$$m'_{ik} = -\frac{i}{c} \int \left[x_i T_{(4k)} - x_k T_{(4i)}\right] d\tau \quad (i, k = 1, 2, 3). \tag{6.59}$$

Of course, this is a conservative tensor. The difference between this and the theory of the electron is that the eigenvalues of the matrices in the integrals [Eq. (6.57)] are $-1, 0$, and 1, instead of $\pm\frac{1}{2}$. We have a particle of maximum spin 1. The space-time pseudovector $s_\mu = \{\mathbf{s}, s_4\}$ has the following form in terms of electromagnetic quantities in the case of the electric photon:

$$\mathbf{s} = \frac{1}{c}[\mathbf{E}^* \times \mathbf{A} - \mathbf{A}^* \times \mathbf{E} + V^*\mathbf{H} + \mathbf{H}^* V]; \ s_4 = \frac{1}{c}[\mathbf{A}^*.\,\mathbf{H} + \mathbf{H}^*.\,\mathbf{A}]. \tag{6.60}$$

Only terms corresponding to spin 1 appear in that formula. The terms corresponding to spin 0 vanish because $I_1 = 0$; this fact is not astonishing because $\mu_0 \neq 0$ [see Eq. (6.22)]. If, we had started from Eq. (6.11) instead of Eq. (6.12), we should be nearer to Dirac's theory. Now consider the orbital momentum operator:

$$\mathbf{M}_{op} = \mathbf{r} \times \mathbf{p}. \tag{6.61}$$

This operator is not an integral of the motion, but we can find a commutating operator by adding to M_{op} the new spin operators:

$$\mathbf{S} = \left\{ -i\hbar\left(\frac{a_2a_3 + b_2b_3}{2}\right),\ -i\hbar\left(\frac{a_3a_1 + b_3b_1}{2}\right),\ -i\hbar\left(\frac{a_1a_2 + b_1b_2}{2}\right)\right\}, \tag{6.62}$$

which must be completed by

$$S_4 = -\frac{i\hbar}{2}(a_1a_2a_3 + b_1b_2b_3), \tag{6.63}$$

which gives with $\mathbf{S}$ a relativistic quadrivector. The space components of $\mathbf{S}$ satisfy the spin commutation relations, and finally these definitions will be used in the generalized theory of fusion.

6.2.7 Relativistic Noninvariance of the Decomposition Spin 1—Spin 0

The spin operators $s_j = \varepsilon_{jik}S_{ik}$ satisfy the commutation rules of an angular momentum and they have the eigenvalues $\{-1,\ 0,\ 1\}$. The total spin $\mathbf{s}^2$ has the eigenvalues $l(l+1) = (2, 0)$, corresponding to $l = 1, 0$.

In the case of a plane wave in Eqs. (6.21) and (6.22) and Eqs. (6.29) and (6.24), one can show that the group of equations (M) is associated with $l = 1$, with projections $s = -1, 0, +1$ on the direction of propagation of the wave: $s = -1 \Leftrightarrow$ *right circular wave*, $s = +1 \Leftrightarrow$ *left circular wave*. For

$s = 0$, we have in both cases a small longitudinal *electric* wave (due to the mass) for the electric photon, and a small longitudinal *magnetic* wave for the magnetic photon. The group (NM) is associated with $l = 0$.

So we can speak of (M) as a "spin 1 particle" and of (NM) as a "spin 0 particle." However, de Broglie made an important distinction (de Broglie, 1943, Chapter 8): *although the equations (M) and (NM) are relativistically invariant, the separation between them is not covariant because it is based on the eigenvalues of the total spin-operator* $\mathbf{s}^2 = s_1^2 + s_2^2 + s_3^2$, which is not a relativistic invariant. The correspondence between the field values and the eigenvalues of $\mathbf{s}^2$ is as follows:

1. For the electric photon:

$$\begin{array}{cccc} \mathbf{A} & V & \mathbf{E} & \mathbf{H} \\ 2 & 0 & 2 & 2 \end{array} \quad \begin{array}{cccccc} I_1 & \mathbf{B} & W & I_2 & \mathbf{E}' & \mathbf{H}' \\ 0 & 2 & 0 & 0 & 2 & 2; \end{array} \tag{6.64}$$

2. For the magnetic photon:

$$\begin{array}{cccc} \mathbf{B}' & W' & \mathbf{H}' & \mathbf{E}' \\ 2 & 0 & 2 & 2 \end{array} \quad \begin{array}{cccccc} I_2 & \mathbf{A}' & V' & I_1 & \mathbf{H} & \mathbf{E} \\ 0 & 2 & 0 & 0 & 2 & 2, \end{array} \tag{6.65}$$

In both cases, the first group corresponds to the (M) equations and the second group to (NM). We can note, when passing from Eq. (6.64) to Eq. (6.65), the following exchanges:

- Between potentials $\mathbf{A}$, V and pseudopotentials $\mathbf{B}', W'$
- Between fields $\mathbf{E}$, $\mathbf{H}$ and anti-fields $\mathbf{E}'$, $\mathbf{H}'$ [we know that $\mathbf{E}', \mathbf{H}' = 0$ in Eq. (6.64) and $\mathbf{E}, \mathbf{H} = 0$ in Eq. (6.65)]
- Between I_1 and I_2, in the group (NM) ($I_1 = 0$ in Eq. (6.64) and $I_2 = 0$ in Eq. (6.65)

The most important fact is that there are in both groups (M) and (NM), field quantities with $\mathbf{s}^2 = 2$ and $\mathbf{s}^2 = 0$, and thus spin 1 and spin 0 components: there is no true separation between the values of spin. De Broglie has shown (for both photons) that the separation only occurs in the proper system, as follows:

1. Because for the electric photon, the potential $(\mathbf{A}, V)$ is *spacelike*, and the pseudopotential $(\mathbf{B}, W)$ is *timelike*, so that V and $\mathbf{B}$ disappear from (6.51), and only $\mathbf{s}^2 = 2$ remains in (M), conversely, only $\mathbf{s}^2 = 0$ remains in (NM) because we know that $\mathbf{E}' = \mathbf{H}' = 0$.
2. For the magnetic photon, the same thing happens because this case follows from the preceding by multiplying an electric solution by γ_5, exchanging polar and axial quantities:

$$(\mathbf{E}, \mathbf{H}) \leftrightarrow (\mathbf{H}', \mathbf{E}'); \quad (V, \mathbf{A}) \leftrightarrow (W, \mathbf{B}); \quad (I_1, I_2) \leftrightarrow (I_2, I_1). \tag{6.66}$$

Therefore, the potential $(\mathbf{A}, V)$ becomes *timelike* and the pseudopotential $(\mathbf{B}, W)$ becomes *spacelike*. And we have once more in the proper frame $\mathbf{s}^2 = 2$ in (M) and $\mathbf{s}^2 = 0$ in (NM), taking into account that we have $\mathbf{E} = \mathbf{H} = \mathbf{0}$ instead of $\mathbf{E}' = \mathbf{H}' = \mathbf{0}$.

In conclusion, the (M) and (NM) groups of equations cannot be rigorously separated, except in the proper frame, and they must be considered as forming one block, for two reasons:

1. The difficulty of separating spin 1 and spin 0 means that the composite photon cannot be considered as a spin 1 particle, but as a particle with a *maximum spin 1*, just as a two-electron atom or a two-atom molecule. It is noteworthy that the proper state in which the 1-components and 0-components are separated is obviously the same for both components.
2. On the contrary, the presence of two photons (electric and magnetic) is inscribed in the very structure of the theory; their separation is covariant and more radical than the separation of spin-states. The simultaneous presence in (M) and (NM) equations of potentials and pseudopotentials and of fields and anti-fields (even if half of them equal zero), and the "migration" of these quantities from one group of equations to the other according to the type of photon constitute another link.

Of course, at the present stage of the problem, a question remains: what is this spin 0 component, physically? It could seem that all these questions are raised by the hypothesis $\mu_0 \neq 0$. Of course, they could be avoided if we admit that $\mu_0 = 0$. But it would be certainly a bad idea to shield the theory from a physical difficulty by a formal condition, at the expense of a more synthetic structure, as was shown previously. A better answer will be given later in this chapter, by the simple fact that the spin 0 component is a photon state that plays a physical role, just as the spin 1 state, and they must be included in the same global theory of light.

6.2.8 The Problem of a Massive Photon

We have seen that many features of de Broglie's theory of the photon, including its logical coherence, are due to the hypothesis $\mu_0 \neq 0$. But even if μ_0 is small, it implies many differences with ordinary electromagnetism. These differences were examined in a number of papers (e.g., de Broglie, 1936, 1940–1942, 1943; Costa de Beauregard, 1997b, 1983; Borne, Lochak, & Stumpf, 2001; Lochak, 1995b, 2000, 2002, 2007b; Lochak and Costa de Beauregard, 2000).

6.2.9 Gauge Invariance

Obviously, the common phase invariance disappears if $\mu_0 \neq 0$, which then calls for some comments:

- First, why do we find in de Broglie's theory of light the *Lorentz gauge* as a field equation? Simply because it is the only relativistically invariant, linear differential law of the first order: it was the only possibility.
- There remain some practical problems. The relations between potentials and fields show that they are of the same order of magnitude. The mass terms are thus of k_0 order: that is, very small. Therefore, in general, the gauge symmetry remains, up to a negligible error, and we can still choose with good approximation the convenient gauge for most practical problems, provided that physics does not impose a particular choice.
- In the present theory, the potentials are deducible from the fields, thus from observable phenomena: they are no longer mathematical fictions, but physical quantities. It must be noted that such a conception was already devised by Maxwell himself (Maxwell, 1873).

This is important for zero-field phenomena only because of a potential, as is the case for the Aharonov-Bohm effect. The fact that this last effect is not gauge invariant is not an objection because we know other physical quantities that are only partially defined by some effects but exactly defined by others: for instance, energy is defined by spectral laws up to an additive constant, but exactly fixed by relativistic effects.

De Broglie gave another example of a physically defined potential: the electron gun (de Broglie, 1943), in which the potential V between the electrodes is exactly defined for several reasons:

1. The *measurable* velocity of the emerging electron is given by the increase of energy, which is equal to eV.
2. The phase of the wave associated with the electron is relativistically invariant only if the frequency and the phase velocity obey the classical de Broglie formula, which imposes the gauge of V (as already noted).
3. The fundamental reason is that the intertia of energy does not allow an arbitrary choice of the origin of electrostatic potentials, which actually are not gauge invariant. They are physical quantities, related to mesurable effects. More recently, Costa de Beauregard and Lochak published many other impressive experimental examples, in favor of the physical sense of electromagnetic potentials.

After several attempts, de Broglie and other authors supposed that the Dirac particles that were constituted by fusion photons and gravitons were *neutrinos*. For a long time, the neutrino was considered a massless particle, with arguments based on gauge invariance, separation of chiral components, etc. But new theoretical arguments based on hypothetical oscillations between different kinds of neutrinos, the subsequent need of coupling constants, and some experimental evidence pointed to a possible neutrino mass. If this is confirmed by facts, de Broglie's fusion theory will have as a consequence the prediction of a photon and a graviton mass, which will become in turn a credible idea. It must be confessed that the leptonic monopole theory (which is due to the author of these lines, who is a member of the same theoretical school) is not in agreement with the last opinion. Nevertheless, it must be remembered (see Chapter 4 of this part of the book) that there is also a theory of massive magnetic monopoles with the same symmetries, but it is a nonlinear theory, different from the present one.

6.2.10 Vacuum Dispersion

If $\mu_0 \neq 0$, we can write

$$h\nu = \frac{\mu_0}{\sqrt{1-\frac{v^2}{c^2}}} \rightarrow v = c\sqrt{1-\frac{\mu_0^2 c^2}{h^2\nu^2}} \quad (v = \text{group velocity}). \tag{6.67}$$

Thus, the vacuum must be dispersive, which was not yet observed, but it may be stressed that the supposed value $\mu_0 < 10^{-45}$g implies a Compton wavelength: $\lambda_c > 10^8$cm $= 10^3$km, so that the substitution of the Coulomb potential $\frac{1}{r}$ by the corresponding Yukawa potential $\frac{e^{-k_0 r}}{r}$ has a very small practical incidence, as with other numerical quantities. But the consequences of the symmetry laws are important.

Another question is that one could in principle observe a photon with a velocity smaller than c in the vacuum. In de Broglie's time, his estimations proved that it was impossible if $\mu_0 < 10^{-45}$g (de Broglie, 1936, 1940–1942). Nevertheless, with the progress of experimental physics, such a possibility must be reexamined.

6.2.11 Relativity

Practically, the velocity predicted for the photon is so near from c, that the difference has not any consequence (at least at the present level of knowledge). But the problem is: how shall we built the theory of relativity? De Broglie's answer was one of his favorite jokes: "Light is not obliged to go with the velocity of light." In other words: we need, in relativity, a maximum invariant velocity, but we do not need this velocity to be the velocity of light. It only happens that, in a vacuum, the velocity of light is close to it.

6.2.12 Blackbody Radiation

In a given unit-volume, there are $dn_\nu = \frac{4\pi\nu^2}{c^3} d\nu$ stationary waves of light in an elementary interval of frequencies, and we must have twice this number because of the transversality of light waves, which gives a factor of 8 in Planck's law of blackbody radiation. But if $\mu_0 \neq 0$, it seems that we must multiply by 3 (instead of 2) because there is a longitudinal electric-component that gives 12 in Planck's law.

But this is wrong. The answer is actually that if we apply the formula for energy, it is shown that the longitudinal part of the field (so as the one, corresponding to potentials) is of the order of k_0—that is, it is negligible (de Broglie, 1936, 1940–1942), so that it takes no part in the observed equilibrium and the factor 8 is the right one. This argument, given by de Broglie, was later independently confirmed by Bass and Schrödinger (1955).

6.2.13 A Remark on Structural Stability

A physical theory has (at least) three truth-criteria: experiment, logical consistency, and structural stability. The first two points are evident, while the third is less so. It means that a theory must have a sufficiant adaptability to withstand slight experimental deviations without its mathematical frame being destroyed.

Actually, most physical theories are too rigid and have structural *unstabilities:* for instance, Hamiltonian dynamics is structuraly unstable because its formalism does not allow the slightest dissipation. This means that the condition of structural stability, despite the strength of the argument and the high authority of the signatures, cannot be respected by all theories. But, at least, one must eliminate arithmetical conditions or too precise symmetries, which could not be verified experimentally.

An example is the mass of the photon. It is proved experimentally that the mass is *small,* but it cannot be proved that this mass is *exactly zero* because it would be an *arithmetical condition.* In other words, electromagnetic gauge invariance—as a law of symmetry—may be proved approximately, not exactly.

It would be extremely worrying if electromagnetism needed exactly zero mass and gauge invariance[20]. And this is not the case, but by virtue of Broglie's theory of photons, the smallness μ_0 implies negligible deviations in the experimental facts.

6.3 THEORY OF PARTICLES WITH MAXIMUM SPIN n

6.3.1 Generalization of the Theory

The general theory is the subject of the second part of de Broglie (1943). Here, we are giving only a short summary—even shorter than for the case of spin 1. The link with the monopole will appear later.

6.3.2 Generalized Method of Fusion

Extending Eq. (6.7), the fusion of n Dirac equations gives a generalization of Eq. (6.8):

$$\frac{1}{c}\frac{\partial \phi_{ikl\ldots}}{\partial t} = a_k^{(p)}\frac{\partial \phi_{ikl\ldots}}{\partial x_k} + i\frac{\mu_0 c}{\hbar}a_4^{(p)}\phi_{ikl\ldots} \quad (p = 1, 2, \ldots, n). \tag{6.68}$$

Thus, we have n equations instead of 2, and a 4^n component wave function (a spinor of nth rank) instead of 16 components for the photon. And there are $4n$ matrices $(a_r^{(p)})$ with 4^{2n} elements:

$$\left(a_r^{(p)}\right)_{ik\ldots opq\ldots,i'k',\ldots o'p'q'\ldots} = \delta'_{ii}\delta'_{kk}\ldots\delta'_{oo}(\alpha_r)'_{pp}\delta'_{qq}\ldots \tag{6.69}$$

They obey the Eq. (6.10) relations:

$$a_r^{(p)}a_s^{(p)} + a_s^{(p)}a_r^{(p)} = 2\delta_{rs};\ a_r^{(p)}a_s^{(q)} - a_s^{(q)}a_r^{(p)} = 0 \ (\text{if } p \neq q). \tag{6.70}$$

The same problem as in Eq. (6.8), occurs here: there are n times too many equations (whereas for the photon, we had twice as many). We have indeed $n4^n$ equations for 4^n components of the wave function. The answer is almost the same.

6.3.3 "Quasi-Maxwellian" Form

We shall proceed as in §**6.3.1)**. But we first put the following expression:

[20] A theory of A. Eddington was based on 16 degrees of freedom and needed the *exact formula* $\frac{1}{\alpha} = \frac{16(16+1)}{2} + 1 = 137$ (α = fine structure constant). Unfortunately, measurement gives $\frac{1}{\alpha} = 137.036\ldots$

$$F^{(p)} = a_k^{(p)} \frac{\partial}{\partial x_k} + i \frac{\mu_0 c}{\hbar} a_4^{(p)}. \tag{6.71}$$

We have the relation

$$F^{(p)} F^{(q)} = F^{(q)} F^{(p)}, \quad \forall p, q; \quad \left(F^{(p)}\right)^2 = \Delta - k_0^2, \tag{6.72}$$

which implies that the wave obeys the Klein-Gordon equation. Now, Eq. (6.68) takes the form

$$\frac{1}{c} \frac{\partial \phi}{\partial t} = F^{(p)} \phi; \quad p = 1, 2, \ldots n. \tag{6.73}$$

By adding these equations, we find a new evolution equation generalizing the (A) expression in Eq. (6.11):

$$(\text{A}) \ \frac{1}{c} \frac{\partial \phi}{\partial t} = F\phi; \quad F = \frac{1}{n} \sum_{p=1}^{n} F^{(p)}. \tag{6.74}$$

Now, subtracting the expressions Eq. (6.73) from each other in a convenient way, we can eliminate the time derivatives and find $(n - 1)$ "condition equations." This may be done in many ways. For instance, we can choose the following system, similar to the (B) expression in Eq. (6.11):

$$(\text{B}) \ B^{(p)} \phi = \frac{F^{(1)} - F^{(p)}}{2} \phi = 0 \ \ (p = 2, 3, \ldots, n). \tag{6.75}$$

It is easy to prove that the new systems (A) and (B) are equivalent to Eq. (6.68) or Eq. (6.72). Owing to Eq. (6.69), one can see that F and B commute, but their product doesn't equal zero, contrary to what happened with the operators on the right-hand side of Eq. (6.11) in the special case $n = 2$:

$$B^{(p)} F = F B^{(p)} \neq 0. \tag{6.76}$$

This means that, contrary to Eq. (6.11), we cannot use Eqs. (6.74) and (6.75) to prove that the conditions (B) are deducible from the evolution equation (A). However, as a consequence of Eq. (6.63), the left-hand sides $B^{(p)}\ \Phi$ of Eq. (6.75) are solutions of Eq. (6.73), so that, if the conditions (B) are satisfied at an initial time $t = 0$, they are satisfied at all time.

On the other hand, we can prove the compatibility of the $(n - 1)$ equations (B), so that the compatibility of the system (6.67)—or, equivalently of Eqs. (6.73) and (6.74)—is proved.

6.3.4 The Density of Quadri-current

Generalizing the case of maximum spin 1, de Broglie introduced another set of matrices (de Broglie, 1943):

$$B_4^{(p)} = a_4^{(1)} a_4^{(2)} \ldots a_4^{(p-1)} a_4^{(p+1)} \ldots a_4^{(n)}. \tag{6.77}$$

Each is the product of all the values of $a_4^{(i)}$ except the one corresponding to the index p. The quadri-current density is as follows, and it is easy to verify that it is conservative:

$$J_k = -c\phi * \frac{1}{n} \sum_{p=1}^{n} a_k^p B_4^p \phi; \quad \rho = \phi * \frac{1}{n} \sum_{p=1}^{n} B_4^p \phi; \quad \frac{\partial \rho}{\partial t} + \partial_k J_k = 0. \tag{6.78}$$

Generalizing a remark made in § 5.2, it is interesting to examine the ρ *density*. Following de Broglie, we shall do it in the case of the plane wave. Let us note, by the way, that it is not difficult to calculate a plane wave for a particle of maximum spin $n/2$: the phase is evident, and the amplitudes are given by the n products of 4 amplitudes of n Dirac plane waves, which gives 2^n constants restricted by the fusion conditions. The calculation is rather long (de Broglie, 1940–1942), but the result is simple. We find

$$\overline{Q} = \rho\, Q, \tag{6.79}$$

with

$$\rho = \left(\frac{\mu_0 c^2}{W}\right)^{n-1} |\phi|^2 \; (\mu_0 \text{ and } n = \text{mass and number of spin } 1/2 \text{ particles}) \tag{6.80}$$

From this, we see the following:

- If n is odd, the sign of ρ is definite-positive, as in the case $n = 1$ of a Dirac electron.
- If n is even, ρ has the same sign as energy, and it is indefinite: it was the case for a photon (spin 1), and it is the case for a graviton (spin 2).

It is interesting to note, with de Broglie, the curious presence, in Eq. (6.80), of the $(n - 1)$th power of the Lorentz contraction, which means that the density ρ, integrated over a volume ($\int \rho d\mathrm{v}$), will be contracted exactly n times (the number of elementary spin 1/2 particles). The exception is the Dirac particle, for which $n - 1 = 0$, so that the factor disappears and the integral is contracted only by the integration-volume itself. De Broglie conjectured that this factor is perhaps an echo of a hidden spatial structure of the composite particle, which we can describe only as a point in the present state of linear quantum mechanics.

6.3.5 The Energy Density

We begin with an elementary calculation of the *energy density* using the preceding density ρ for a plane wave. The definition of the density ρ means that all the mean values are obtained by the integration of a physical quantity multiplied by ρ.

The energy density is thus obtained (in the case of a plane wave) owing to Eq. (6.80):

$$\rho W = \left(\frac{\mu_0 c^2}{W}\right)^{n-1} W|\phi|^2. \tag{6.81}$$

Here, the power of W is not $(n-1)$ but $(n-2)$, so that we find a result opposite to the result for ρ:

- If n is odd, ρW has the same indefinite sign as energy: it was the case for $n=1$, for the Dirac electron.
- If n is even, the sign of ρW is definite-positive, as it was for the photon and as it will be for the graviton. This is confirmed by more sophisticated calculations using the energy tensor density.

We shall introduce two classes of tensors. The first, named "corpuscular" by de Broglie, is given by the receipts of quantum mechanics. The second class, called by de Broglie "of type M" (with M standing for "Maxwell"), is wider and is inspired by electromagnetism.

6.3.6 The "Corpuscular" Tensor

We use the B matrices definited in Eq. (6.77) with the following notation:

$$\rho U_i^{(p)} = a_i^{(p)} \quad (i = 1, 2, 3,), \quad U_i^{(p)} = 1; \quad (p = 1, 2, ..., n). \tag{6.82}$$

The tensor is then (de Broglie, 1943), generalizing the spin 1 case:

$$T_{\mu\nu} = T_{\nu\mu} = \frac{\hbar c}{4in} \sum_{p=1}^{n} \begin{bmatrix} \phi^* U_\mu^{(p)} B_4^{(p)} \dfrac{\partial \phi}{\partial x_\nu} - \dfrac{\partial \phi *}{\partial x_\nu} U_\mu^{(p)} B_4^{(p)} \phi \\ +\phi^* U_\nu^{(p)} B_4^{(p)} \dfrac{\partial \phi}{\partial x_\mu} - \dfrac{\partial \phi *}{\partial x_\mu} U_\nu^{(p)} B_4^{(p)} \phi \end{bmatrix}. \tag{6.83}$$

We verify its conservation by virtue of the equations

$$\partial_\nu T_{\mu\nu} = \partial_\mu T_{\mu\nu} = 0. \tag{6.84}$$

It is interesting to verify that the tensor takes the form that is to be expected for a plane wave, and we find indeed the following matrix for its components ($\mathbf{p}$ = momentum, $\mathbf{v}$ = group velocity):

$$\begin{Bmatrix} \rho p_1 v_1 & \rho p_1 v_2 & \rho p_1 v_3 & \rho p_1 c \\ \rho p_2 v_1 & \rho p_2 v_2 & \rho p_2 v_3 & \rho p_2 c \\ \rho p_3 v_1 & \rho p_3 v_2 & \rho p_3 v_3 & \rho p_3 c \\ \rho p_1 c & \rho p_2 c & \rho p_3 c & \rho W \end{Bmatrix}. \tag{6.85}$$

In particular T_{44} is the quantity given in Eq. (6.81).

6.3.7 The "type M" Tensors

At first, we shall generalize the formula [Eq. (6.77)] by the definition of a set of operators of rank m:

$$B_4^{(pq\ldots)} = a_4^1 a_4^2 \ldots a_4^{p-1} a_4^{p+1} \ldots a_4^{q-1} a_4^{q+1} \ldots. \tag{6.86}$$

This is the product of all the values of a_4^r ($r = 1, 2,\ldots,n$) except those for which r is equal to one of the m indices $p, q \ldots$ of B. Using of these operators and Eq. (6.81), we define a set of tensors of rank m (de Broglie, 1934c):

$$M_m = \mu_0 c^2 \phi^* \frac{\sum\limits_{pq\ldots} U_i^{(p)} U_j^{(q)} \ldots B_4^{(pq\ldots)}}{a_n^m} \phi; \quad \left(a_n^m = \frac{n!}{(n-M)!} \right). \tag{6.87}$$

These tensors are obviously symmetric, but we keep only those values of the rank $m = 2r$ that are even. Thus, we have defined (for a particle of maximum spin n) $n/2$ tensors if n is even and $(n-1)/2$ tensors if n is odd. Finally, we contract each tensor of rank $2r$, over $2r - 2$ indices, which gives a number equal to the half of the greatest even number contained in n of tensors of rank 2, according to the following formula:

$$M_{ij}^{(r)} = \sum_{ijkl\ldots}^{4} M_{ijkl\ldots}^{kl\ldots}. \tag{6.88}$$

We must remember that, applying the receipt to real space-cordinates, we must change the sign when indices 1, 2, and 3 go up or down.

These tensors were defined by de Broglie as tensors "of type M." By virtue of the general equations [Eq. (6.87)], we have, just as for the tensor T:

$$\partial_\nu M_{\mu\nu}^{(r)} = \partial_\mu M_{\mu\nu}^{(r)}, \tag{6.89}$$

and we have $n/2$ tensors $M^{(r)}$ of rank 2 if n is even and $(n - 2)/2$ tensors if n is odd.

A priori, each conservative tensor may be considered as an impulse-energy tensor, and it may be shown that, *for a plane wave*, $\forall\ r$ every tensor $M^{(r)}_{\mu\nu}$ gives exactly the table of components [Eq. (6.83)]. This is not true for other solutions, but it remains true of the integrals:

$$\int T_{\mu\nu} d\tau = \int M^{r}_{\mu\nu} d\tau; \quad \forall r. \tag{6.90}$$

6.3.8 Spin

Starting from Eq. (6.72)—the generalization of Eq. (6.11)—we have the same orbital operator, and the spin operators are now

$$S_i = \hbar \sum_{p=1}^{n} s_i^{(p)}; \quad (i = 1, 2, 3); \quad S_4 = \hbar \sum_{p=1}^{n} s_4^{(p)}. \tag{6.91}$$

It would be difficult to reproduce here the general nomenclature of spin states and (for an even number of spin 1/2 particles) the decomposition of wave functions in terms of tensor components. This nomenclature is based on the Clebsch-Gordan theorem for the product of irreducible representations, but it is completed in (de Broglie, 1934c), which defines the set of independent constants of a plane wave and the symmetry of tensors defined by an even number of particles.

These problems are treated in a different form by Fierz, whose work is based not on the fusion theory but on some conditions added to the field obeying the Klein-Gordon equation, to describe a spin $n/2$ particle. This point of view was developed by Fierz and Pauli (1939a, b) and on the basis of Dirac (1936) on the generalization of the equation of the electron, for higher spin-values.

6.4 THEORY OF PARTICLES WITH MAXIMUM SPIN 2

6.4.1 The Particles of Maximum Spin 2. Graviton

Fierz and Pauli (1939a, b) were the first to discover the connection between the equation of a particle of spin 2 and the linear approximation of the Einstein equation of a gravitation field. This approximation was given by Einstein himself (Einstein, 1916, 1918). It may be found, for instance, in Laue (1922) or Möller (1972). Einstein (1916) was the first study in which he formulated the idea of gravitational waves. He even alluded to a possible modification of gravitation theory by quantum effects, analogous to the modification of Maxwell's electromagnetism.

It must be stressed that the quantum theory of gravitation, developed by de Broglie and Tonnelat (de Broglie, 1934a,b,c; Tonnelat, 1942) on the basis of the fusion method, is not based on a particle of spin 2, but on the particle of *maximum* spin 2. This is an important point for two reasons:

1. The fusion theory raises the question: is the graviton a *composite particle*, just as the photon and all particles of spin higher than ½?
2. In the fusion theory, gravitons don't appear alone. They are linked to photons. This theory is actually a unitary theory of gravitation and electromagnetism (at least at the linear approximation), and the fields are not gathered by an extended geometry, but by the fusion of spins.

6.4.2 Why are Gravitation and Electromagnetism Linked?

When you ask why gravitation and electromagnetism are linked, formally you could say that fields are linked by Clebsch-Gordan's theorem because

$$D_{\frac{1}{2}} \times D_{\frac{1}{2}} \times D_{\frac{1}{2}} \times D_{\frac{1}{2}} = D_2 + 3D_1 + 2D_0. \tag{6.92}$$

Therefore, in the fusion of four spin 1/2 particles, we must find one particle of spin 2, three particles of spin 1, and two particles of spin 0. In particular, we have gravitons and photons. To this point we must add the spin 0 photons, the physical meaning of which is related to the Aharonov-Bohm effect, as was developed in the first part of §4.

De Broglie gave an interesting argument: he defined a particle of maximum spin 2 by the fusion of two particles of spin 1, described by the quadripotentials $A_\mu^{(1)} = \{A^{(1)}, V\}$ and $A_\mu^{(2)} = \{A^{(2)}, V\}$, and the invariants $I_2^{(1)}, I_2^{(2)}$ $(I_1^{(1)}, I_1^{(2)} = 0$. This was because $\mu_0 \neq 0$. We are only considering the electric case. The fusion gives

$$A_\mu^{(1)} \times A_\mu^{(2)};\ A_\mu^{(1)} \times I_2^{(2)};\ I_2^{(1)} \times A_\mu^{(2)};\ I_2^{(1)} \times I_2^{(2)}. \tag{6.93}$$

The first product is a tensor of rank 2 that defines a symmetric and an antisymmetric tensor:

$$A_{(\mu\nu)} = \frac{A_{(\mu\nu)} + A_{(\nu\mu)}}{2};\quad A_{[\mu\nu]} = \frac{A_{(\mu\nu)} - A_{(\nu\mu)}}{2}. \tag{6.94}$$

The products $A_\mu^{(1)} \times I_2^{(2)}$ and $I_2^{(1)} \times A_\mu^{(2)}$ are vectorlike quantities $P_\mu^{(1)}$, $P_\mu^{(2)}$, and it may be hoped that they will be photon potentials. The antisymmetric tensor $A_{[\mu\nu]}$ suggests the electromagnetic field.

The symmetric tensor $A_{(\mu\nu)}$ cannot be interpreted at this level of exposition, but actually, we can guess that it will be related to gravitation.

De Broglie shows, owing to a study of plane waves, that $P_\mu^{(1)}, P_\mu^{(2)}$ and the antisymmetric tensor $A_{[\mu\nu]}$ are related to spin 1; $A_{(\mu\nu)}$ is linked to spin 2 only if it is reduced to a zero-spur tensor because *spur* $A_{(\mu\nu)} = A_{(\mu\mu)}$ is an invariant; and it will be actually related to spin 0, just as is the invariant $I_2^{(1)} \times I_2^{(2)}$.

Now it must be remembered that, as was shown in the case of the photon, *the splitting between different spin states is not relativistically covariant* because it is based on the total spin operator which is not a relativistic invariant. Therefore, in the fusion theory, gravitation cannot appear without electromagnetism. Furthermore, it will be shown that, if $\mu_0 \neq 0$, splitting between spin 2 and spin 0 is impossible, and the interpretation of this fact is highly significant.

6.4.3 The Tensorial Equations of a Particle of Maximum Spin 2

We give only the tensorial form generalizing § 4.1. The total wave equations [Eq. (6.11) for $n = 4$] would have $4^4 = 256$ components with 168 independent quantities (de Broglie, 1934c):

$$\text{(A)} \quad \begin{aligned} \partial_\mu \phi_{(\nu\rho)} - \partial_\nu \phi_{(\mu\rho)} &= k_0 \phi_{[\mu\nu]\rho} \\ \partial_\rho \phi_{[\rho\mu]\nu} &= k_0 \phi_{(\mu\nu)} \\ \partial_\mu \phi_{[\rho\sigma]\nu} - \partial_\nu \phi_{[\rho\sigma]\mu} &= k_0 \phi_{[\mu\nu][\rho\sigma]} \\ \partial_\varepsilon \phi_{([\varepsilon\rho][\mu\nu])} &= k_0 \phi_{[\mu\nu]\rho} \end{aligned} . \tag{6.95}$$

Here, $\phi_{(\mu\nu)}$ is a symmetric tensor of rank 2, $\phi_{[\mu\nu]\rho}$ is a tensor of rank 3 antisymmetric with respect to the two first indices, and $\phi_{[\mu\nu][\rho\sigma]}$ is a tensor of rank 4 antisymmetric with respect to $\mu\nu$ and $\rho\sigma$, but symmetric with respect to these pairs. A consequence of Eq. (6.95) is

$$\begin{aligned} \partial_\nu \phi_{(\mu\nu)} &= \partial_\rho \partial_\nu \phi_{[\rho\mu]\nu} = 0 \\ \phi_{[\rho\rho]} &= \frac{1}{2} \phi_{[\mu\rho][\mu\rho]}; \quad \partial_\nu \phi_{(\rho\rho)} = k_0 \phi_{[\nu\rho]\rho}. \end{aligned} \tag{6.96}$$

The group (B) is divided in three subgroups where new tensors of rank 2, 3, and 4 appear:

$$(\mathrm{B}_1)\quad \begin{aligned} &\partial_\mu \phi^{(1)}_{(\nu\rho)} - \partial_\nu \phi^{(1)}_{(\mu\rho)} = k_0 \phi^{(1)}_{[\mu\nu]\rho} \\ &\frac{1}{2}\left(\partial_\rho \phi^{(1)}_{[\rho\mu]\nu} - \partial_\rho \phi^{(1)}_{[\rho\nu]\mu}\right) = k_0 \phi^{(1)}_{[\mu\nu]} \\ &\partial_\mu \phi^{(1)}_{[\rho\sigma]\nu} - \partial_\nu \phi^{(1)}_{[\rho\sigma]\mu} = k_0 \phi^{(1)}_{[\mu\nu][\rho\sigma]} \\ &\partial_\varepsilon \phi^{(1)}_{([\varepsilon\rho][\mu\nu])} = k_0 \phi^{(1)}_{[\mu\nu]\rho} \end{aligned}. \tag{6.97}$$

Note the antisymmetries (square brackets). From Eq. (6.97), we deduce the identities as follows:

$$\phi^{(1)}_{[\nu\mu]\nu} = \phi^{(1)}_{([\mu\nu][\rho\nu])} = 0. \tag{6.98}$$

The equations (B_2) and (B_3) are identical, and we have

$$(\mathrm{B}_2,\ \mathrm{B}_3)\quad \begin{aligned} &\partial_\mu \chi^{(1)}_\nu - \partial_\nu \chi^{(1)}_\mu = k_0 \chi^{(1)}_{[\mu\nu]} \\ &\partial_\rho \chi^{(1)}_{[\rho\nu]} = k_0 \chi^{(1)}_\nu \\ &\partial_\mu \chi^{(1)}_\nu = k_0 \chi^{(1)}_{\rho\nu} \\ &\partial_\rho \chi^{(1)}_{[\mu\nu]} = k_0 \chi^{(1)}_{[\mu\nu]\rho} \end{aligned}. \tag{6.99}$$

In the third equation, $\chi^{(1)}_{\rho\nu}$ is neither symmetric nor antisymmetric. Eq. (6.99) entails

$$\begin{aligned} &\chi^{(1)}_{\rho\rho} = 0;\ \chi^{(1)}_{\mu\nu} - \chi^{(1)}_{\nu\mu} = \chi^{(1)}_{[\mu\nu]} \\ &\chi^{(1)}_{[\nu\rho]\rho} = -\chi^{(1)}_\nu;\ \chi^{(1)}_{[\mu\nu]\rho} + \chi^{(1)}_{[\nu\rho]\mu} + \chi^{(1)}_{[\rho\mu]\nu} = 0. \end{aligned} \tag{6.100}$$

Finally, we find a last group of equations:

$$(\mathrm{C})\quad \begin{aligned} &\partial_\mu \phi^{(0)}_\nu = \partial_\nu \phi^{(0)}_\mu = k_0 \phi^{(0)}_{(\mu\nu)} \\ &\partial_\mu \phi^{(0)}_\mu = k_0 \partial_\mu \phi^{(0)} \\ &\partial_\mu \phi^{(0)} = k_0 \mu^{(0)}_\phi \end{aligned}. \tag{6.101}$$

The equations (B_1), (B_2), and (B_3) are three realizations of total spin 1. It is evident for (B_2) and (B_3) because putting

$$F_\mu = k_0 \chi^{(1)}_\mu;\ F_{[\mu\nu]} = k_0 \chi^{(1)}_{\mu\nu} \tag{6.102}$$

and defining potentials and fields as we did in Eq. (6.20), we find the Maxwell equations with mass (but we shall see that it needs some additional notation).

The correspondence is less evident for Eq. (6.97). Instead of Eq. (6.102), we must write

$$F_\mu = \frac{k_0}{6}\varepsilon_{\mu\lambda\nu\rho}\phi^{(1)}_{[\lambda\nu]\rho};\quad F_{[\mu\nu]} = k_0\phi^{(1)}_{\mu\nu}\,, \tag{6.103}$$

where $\varepsilon_{\mu\lambda\nu\rho}$ is the Levi-Civita symbol. Applying Eq. (6.20), we find the Maxwell equations.

Now, (C) is a realization of spin 0 as may be seen by comparing Eq. (6.101) with Eq. (6.22). But here we find a difficulty that justifies the preceding remarks: de Broglie (who did not know the magnetic case), considered only the electric photon [Eq. (6.21)] and he identified Eq. (6.101) with the non-Maxwellian equations [Eq. (6.22)]. But this implies the identity $\varphi^{(0)} = I_2$, where $\varphi^{(0)}$ is a *scalar* while I_2 is a pseudoscalar.

In de Broglie's time, people was less careful about parity than now, and he wrote that Eqs. (6.101) and (6.22) "are entirely equivalent (at least when vectors and pseudovectors are assimilated)." Today, we pay more attention to parity and we cannot neglect such a discrepancy: an equality like $\phi^{(0)} = I_2$ is unacceptable. There are two possible solutions:

1. We could admit that if $\phi^{(0)} = I_2 = 0$, $\phi^{(0)} = I_2$. Thus, the spin 0-component (C) vanishes. But there is a second spin 0-component, hidden in the equations (A) in the form of an invariant $\phi^{(0)}$, a vector $\phi^{(0)}_\mu$, and a symmetric tensor $\phi^{(0)}_{(\mu\nu)}$, that we can define as

$$\phi^{(0)} = \phi^{(0)}_{(\rho\rho)}\,;\quad \phi^{(0)}_\mu = \phi^{(0)}_{[\mu\rho]\rho};\quad \phi^{(0)}_{(\mu\nu)} = \phi_{([\mu\rho][\nu\rho])} - \phi_{(\mu\nu)}. \tag{6.104}$$

One can show using Eq. (6.94) that these tensors obey the group C of Eq. (6.101), but once more, if $\phi^{(0)}$ is a true scalar, we can write $\phi^{(0)} = I_2$ only if $\phi^{(0)} = I_2 = 0$. This implies that Eq. (6. 101) is submitted to the condition *sp* $\phi^{(0)}_{(\rho\rho)} = 0$ that was *a priori* supposed by Fierz and Pauli, who based their theory on a spin 2 (and not a maximum spin 2) particle. De Broglie criticized this postulate as artificial.

This suggestion, based on parity, could be considered as the justification of their hypothesis. However, it may be pointed out, as de Broglie did, that the splitting of spin components is not covariant. It is, at least, the case for the condition $\phi^{(0)} = I_2 = 0$, in spite of the fact that the equality *sp* $\phi^{(0)}_{(\rho\rho)} = 0$ is covariant. Thus, the problem remains unsolved. But there is a second proposition.

2. We can ask the question: Is $\phi^{(0)} = I_2$ a good equality? Perhaps instead it is $\phi^{(0)} = I_1$, which is covariant because I_1 is a true invariant. In such a

case, Eq. (6.101) must not be identified with Eq. (6.22), but rather with Eq. (6.27). Is this possible? It seems so.

Let us go back to Eq. (6.92). The products $A_\mu^{(1)} \times I_2^{(2)}$ and $I_2^{(1)} \times A_\mu^{(2)}$, denoted as $P_\mu^{(1)}$, $P_\mu^{(2)}$, were considered by de Broglie as vectors, but he said, more prudently, that they were "vectorlike". Actually, they are pseudovectors because they are the products of a polar-vector by a pseudoscalar. Therefore, $P_\mu^{(1)}$ and $P_\mu^{(2)}$ are not polar potentials but pseudopotentials of magnetic type as are those that appear in Eq. (6.26). On the contrary, the product $I_2^{(1)} \times I_2^{(2)}$ of two pseudoscalars is a *true* scalar of the same type as I_1, which appears in Eq. (6.27), and it can be identified.

The answer to the difficulty is that the third photon associated with the graviton is not electric but magnetic.

Let us suppose that, instead of introducing only electric photons, we introduce a magnetic photon in the symbolic formulas Eq. (6.92)] with pseudopotentials $\mathbf{B}_\mu^{(1)}, \mathbf{B}_\mu^{(2)}$, and pseudoscalars $I_2^{(1)}, I_2^{(2)}$. The fusion gives

$$B_\mu^{(1)}, B_\mu^{(2)};\ B_\mu^{(1)} \times I_\mu^{(2)};\ I_\mu^{(1)} \times B_\mu^{(2)};\ I_\mu^{(1)} \times I_\mu^{(2)}, \tag{6.105}$$

and we see the following:

- The spin 2 product: $B_\mu^{(1)} \times B_\mu^{(2)}$ has the same symmetry as $A_\mu^{(1)} \times A_\mu^{(2)}$ because the axial character of $B_\mu^{(1)} \times B_\mu^{(2)}$ is annihilated by the product.
- For the same reason, the spin 0 product $I_1^{(1)} \times I_1^{(2)}$ is a scalar, as was $I_2^{(1)} \times I_2^{(2)}$.
- The spin 1 products $B_\mu^{(1)} \times I_\mu^{(2)};\ I_\mu^{(1)} \times B_\mu^{(2)}$ are pseudovectors, as $A_\mu^{(1)} \times I_2^{(2)};\ I_2^{(1)} \times A_\mu^{(2)}$: they are products of a pseudovector by a scalar, while the latter were products of a polar vector by a pseudoscalar.

Thus, we find a magnetic photon whether we start from electric or from magnetic photons, and we can assert that one of the photons associated with the graviton is not electric but magnetic.

6.5 QUANTUM (LINEAR) THEORY GRAVITATION

6.5.1 The Particle of Maximum Spin 2. Graviton

Now, we shall follow de Broglie (1943) and Tonnelat (1942) and consider the general equations (A) when *spur* $\phi_{(\rho\rho)}^{(0)} \neq 0$. But we shall not be able to separate the spin 2 component from its spin 0 part.

We start from Eqs. (6.81) and (6.82) and the Klein-Gordon equation, verified by all the field quantities:

$$\Box\phi = -k_0^2\phi \ \left(\Box = -\partial_\rho\partial_\rho\right). \tag{6.106}$$

The metric tensor $g_{(\mu\nu)}$ will be taken at the linear approximation:

$$g_{(\mu\nu)} = \delta_{\mu\nu} + h_{(\mu\nu)} \quad \left(h_{(\mu\nu)} \ll 1\right). \tag{6.107}$$

At this limit, the propagation of gravitation waves is given by

$$\Box g_{(\mu\nu)} = -2R_{(\mu\nu)} \quad \left(R_{(\mu\nu)} = g^{\rho\sigma} R_{([\mu\rho][\nu\sigma])}\right), \tag{6.108}$$

where $R_{([\mu\rho][\nu\sigma])}$ is the tensor of Riemann-Christoffel; in the Euclidian regions of space-time, we have the d'Alembert equation $\Box g_{(\mu\nu)} = 0$ without a second member. This is true if we use "isothermic" coordinates x_μ, for which $D_2 x_\mu = 0$; D_2 is the second-order Beltrami differential parameter.

Now it seems that metrics could be defined by

$$g_{(\mu\nu)} = \phi_{(\mu\nu)}. \tag{6.109}$$

But Tonnelat remarked that, according to Eq. (6.98), this implies $\partial_\mu g_{(\mu\nu)} = 0$, which is wrong because "isothermic" coordinates obey the relation[21]

$$\partial_\mu g_{(\mu\nu)} = \frac{1}{2} \partial_\nu g_{(\rho\rho)} \quad \left(g_{(\rho\rho)} = g_{(\mu\nu)} \delta^{(\mu\nu)}\right), \tag{6.110}$$

and the second member is not equal to zero. Eq. (6.96) thus contradicts Eq. (6.95), which is why Tonnelat suggested the following metrics (which is possible because $k_0 \neq 0$):

$$g_{(\mu\nu)} = \phi_{([\mu\rho][\nu\sigma])} = \phi_{(\mu\nu)} + \frac{1}{k_0^2} \partial_\mu \partial_\nu \phi_{(\rho\rho)}. \tag{6.111}$$

From this, it follows immediately that

$$\partial_\mu g_{(\mu\nu)} = \partial_\mu \phi_{([\mu\rho][\nu\sigma])} = \partial_\nu \phi_{(\rho\rho)} \tag{6.112}$$

So we get from Eqs. (6.82), (6.111), and (6.112):

$$g_{(\rho\rho)} = 2\phi_{(\rho\rho)} \rightarrow \partial_\mu g_{(\mu\nu)} = \frac{1}{2} \partial_\nu g_{(\rho\rho)} \tag{6.113}$$

in accordance with Eq. (6.96).

Now, from Eq. (6.97), we deduce that $g_{(\mu\nu)}$ obeys the Klein-Gordon equation, as other field-quantities do:

$$\Box g_{(\mu\nu)} = -k_0^2 g_{(\mu\nu)}. \tag{6.114}$$

21 It must be noted that we do not have $g_\rho^\rho = g_{\rho\sigma} g^{\rho\sigma}$ because this quantity, *in the present case,* is equal to 4.

We have to identify Eq. (6.114) with Eq. (6.94), such that

$$R_{(\mu\nu)} = \frac{1}{2} k_0^2 g_{(\mu\nu)}. \tag{6.115}$$

Now, the tensor of Riemann-Christoffel may be deduced as the linear approximation from Eqs. (6.111), (6.81), and (6.82):

$$\phi_{([\mu\rho][\nu\sigma])} \cong \frac{2}{k_0^2} R_{([\mu\rho][\nu\sigma])}. \tag{6.116}$$

This formula is possible only if $\mu_0 \neq 0$, which imposes a curvature of the universe. Indeed, $k_0^2/2$ is nothing but the cosmological constant (which, unfortunately, Einstein disliked), defined by

$$R_{(\mu\nu)} = \lambda g_{(\mu\nu)}, \tag{6.117}$$

where λ is related to a natural curvature of space-time. In Euclidian space, $\lambda = 0$, in a de Sitter space of radius R, we have $\lambda = 3/R^2$. Therefore:

$$\lambda = \frac{k_0^2}{2} = \frac{\mu_0^2 c^2}{2\hbar^2}, \tag{6.118}$$

and the graviton mass is related to a natural curvature of radius R:

$$\mu_0 = \frac{\hbar\sqrt{6}}{Rc}. \tag{6.119}$$

If $R = 10^{26}$cm, the graviton (and photon) mass is

$$\mu_0 = 10^{-66}\text{g}. \tag{6.120}$$

The spin 0 may be eliminated from the equations of spin 2 only in one of two cases:

- By the *a priori* supposition that $\Phi^{(0)} = 0$ (Fierz equations)
- *At the limit case* $\mu_0 = 0$, when the radius of the universe is infinite—that is, the Euclidian case

In conclusion, the quantum theory of gravitation based on de Broglie's fusion theory raises the important question of a composite nature of photon and graviton, and above all, the theory furnishes the beginning of the quantum unitary field-theory of electromagnetism and gravitation. It only gives the beginning, however, because it is linear.

Two more remarks are relevant here:

- It could be asked whether the obstinate efforts of Einstein and other great physicists and mathematicians to find a unitary field theory had any basis, given that we are aware of hundreds of particles and it would seem that there is no reason to pay particular attention to only two of them: the photon and graviton. De Broglie's theory gives a reason for this, though: these particles are the only ones that are linked by spin properties in the fusion procedure. This argument is exterior to the geometrical path followed by Einstein.

- Concerning symmetry, the fact that a photon associated with the graviton could be magnetic instead of electric, as has been suggested, signifies the intrusion of duality, chirality, and magnetic monopoles instead of electric charges. *It is certainly of interest that a photon is perhaps not the one that was expected, and it must be stressed that there is another photon with zero spin.*

6.5.2 Comparison with Other Theories

First, we must emphasize the priority of Louis de Broglie in the quantum theory of the photon being considered as a *composite particle*. In his first paper on the subject (de Broglie, 1934b), the idea of a *fusion* of Dirac particles was the starting point of his theory of particles of higher spin. A second point is that, unlike others, de Broglie's initial aim was not a generalization of Dirac's equation, but a theory of light. This is why he did not introduce any electromagnetic interaction.

For these reasons, he was the only one to suppose a massive photon, unlike others who considered a massless photon as obvious. He never tried to extend his theory to massless particles and even scarcely alluded to this possibillity.

6.5.3 The "Proca Equation"

Eq. (6.21) and the very idea of a massive photon are often ascribed to Alexandru Proca. Actually, it is the result of a misunderstanding, if not a "misreading", as follows:

1. The "Proca equations" (Proca, 1936) appeared in 1936, two years after the Broglie equations (de Broglie, 1934b). Moreover, the paper of Proca was entitled: "On the ondulatory theory of positive and negative electrons". It was not a theory of the photon, but a theory of the electron: an attempt to avoid negative energies, as was frequent in that time.[22]

[22] Heisenberg and de Broglie were among the few who immediately adopted Dirac's equation, whatever the difficulties with negative energies were.

2. Rejecting spinorial wave functions, Proca suggests—for the electron—a *vectorial equation* deriving from the Lagrangian:

$$L = \frac{\hbar^2 c^2}{2} B^*_{rs} B_{rs} + m_0^2 c^4 \psi^*_r \psi_r$$
$$B_{rs} = (\partial_r - iA_r)\psi_s - (\partial_s - iA_s)\psi_r \quad (r, s = 1, 2, 3, 4). \tag{6.121}$$

The complex vectorial function ψ_r of the *electron* takes the place of de Broglie's *photon* potential (**A**, V); and A_r is a real potential of an *external* electromagnetic field acting on the electron ψ_r, and the electron—not the photon—was the object of the theory. From Eq. (6.121), Proca derived the following equations:

$$(\partial_r - iA_r)B_{rs} = k^2\psi_s, \quad (\partial_r + iA_r)B^*_{rs} = k^2\psi^*_s \quad \left(k = \frac{m_0}{\hbar c}\right), \tag{6.122}$$

and he remarked that "they have the form of Maxwell's equations [...], completed by an external potential (A_r)". But, in no way did he consider Eq. (6.122) as the equations of a photon.

Then, he gave a spin operator, but without calculating its eigenvalues and thus ignored the fact that his electron had a spin 1. This is astonishing because de Broglie worked one floor above Proca and had deduced this value 1 two years earlier in his equation of describing a massive photon (de Broglie, 1934b).

6.5.4 The Bargmann-Wigner Equation

The Bargmann-Wigner equation for higher values of spin was published in 1948 (Bargmann & Wigner, 1948) and was similar to de Broglie's equations (de Broglie, 1943). Not identical, indeed, because it lacked the idea of fusion and was restricted by an *a priori* condition of symmetry, so it had only half of the de Broglie solutions.

When the general theory is applied to the case of spin 1, Bargmann and Wigner found the equations identical with the equation taken from de Broglie's book p. 106 (de Broglie, 1943), with a difference: by virtue of their condition of symmetry, Bargmann and Wigner did not develop the wave on the 16 Clifford matrices, as we did in Eq. (6.19), but only on 10 of them: $\gamma_\mu, \gamma_{[\mu\nu]}$. The 6 others were forgotten, so the spin 1 Maxwell equations [Eq. (6.21)] were obtained, but not the non-Maxwellian ones [Eq. (6.22)], corresponding to spin 0, which clearly have an important physical meaning.

They would be unable to include the Aharonov-Bohm effect, as we did, and *a fortiori* to find the magnetic photon, of which we proved not only that it has a logical place in the theory, but that it was already hidden in de Broglie's theory and later experimentally observed. And it is the photon that automatically appears in the interaction between the leptonic monopole and the electromagnetic field.

CHAPTER 7

P, T, and C Symmetries, the Solutions with Negative Energy, and the Representation of Antiparticles in Spinor Equations

7.1 INTRODUCTION

In this chapter, we revisit the problem of P (Parity), T (Time), and C (Charge) symmetries in quantum electrodynamics, starting with the laws of Pierre Curie and classical electromagnetism rather than a priori postulating the formal covariance of quantum mechanics. It is only after having discussed these symmetries that we will assess how quantum mechanics agrees with them. In fact, it will turn out that the so-called Racah transformations are confirmed for the P and C symmetries, but not for the T symmetry. It thus evolves that there are *two* possible T transformations in the framework of classical electromagnetism, one of which is subsequently selected on the basis of a general physical argument and formal covariance. An examination of (linear) spinor equations will then reveal a large difference between the symmetries of an electric charge and those associated with a magnetic one.

It is an interesting historical point that Dirac, when first working on these questions, believed that the negative energy solutions of the Klein-Gordon equations stemmed from the fact that it is a *second*-order equation with respect to the time variable. In spite of all that he was eventually able to get from them, he was disappointed to find such solutions in his own *first*-order equation, and indeed, until the end of his life, he went on looking for an equation that would be free of such solutions. One may wonder whether these come from the *linearity* of the equation, but that is definitely not the case: negative energies, just as antiparticles are not associated with second derivatives or linearity, but rather with *symmetries* (i.e., relativistic covariance and *P, T, and C* symmetries). Recall that in special relativity, rotating bodies lead to negative *energies* [Möller (1972)], whereas in general relativity, Einstein showed, as early as 1925, that one cannot describe an electron without the appearance of a particle of opposite charge. He connected that property with *time reversal* and proved that the product *PT* reverses charges (Lochak, 1994; Einstein, 1925).

So in this chapter, our first goal will be to clarify some points pertaining to *P*, *T*, and *C* symmetries, using the work of Pierre Curie on the symmetries of the *electromagnetic field,* whose role cannot be overstated. We refrain from speaking, as some do [e.g., Berestetsky, Lifschitz, & Pitaevsky (1968)], of "charge conjugacy" about free-field equations because this leads to identifying the variance of the potentials with that of a world gradient (via gauge invariance). However, we would like to show that this variance can be *inferred* from the laws of electromagnetism. Most of this chapter is geared toward linear quantum equations; however, in Chapter 4, we already studied some *nonlinear* problems connected with chirality and the monopole, and here, we will return briefly to the nonlinear setting, with questions on the compatibility between nonlinearity and quantum mechanics. A more general and detailed study of the symmetries in nonlinear field equations can be found in Lochak (1997).

7.2 THE SPATIAL SYMMETRIES OF THE ELECTROMAGNETIC QUANTITIES

Few treatises of electromagnetism mention the *P*, *C*, and *T* symmetries. Jackson (1975) is an exception, but only formally so, as he *postulates* the invariance of the electric charge under space or time reversal[23]. We disagree about the time symmetry and, as far as space is concerned, although Jackson's choice does agree with the results of Pierre Curie, the latter does not impose them *a priori*, but rather deduces them from experimentation and his famous general symmetry law (Curie, 1894a,b):

> "Lorsque certaines causes produisent certains effets, les éléments de symétrie des causes doivent se retrouver dans les effets produits." *[When certain causes*

[23] Here is his only argument: "It is natural, convenient, and permissible to assume that charge is also a scalar under spatial inversion and even under time reversal" (Jackson, 1975, p. 249).

produce certain effects, the same patterns of symmetry should be found in the effects as in their causes.]

Conversely,

"Lorsque certains effets révèlent une certaine dissymétrie, cette dissymétrie doit se retrouver dans les causes qui lui ont donné naissance."[24] *[When certain effects betray a certain asymmetry, the latter must be present in the causes from which it originates.]*

Here are two applications of these principles, following Curie himself:

1. The spatial symmetry of the electric field: Consider an *electric* field created between two circular plates made of different metals and sharing a common axis. It displays the symmetries of the cause (i.e., of the set of the two plates): Rotation around their axes and planar symmetry with respect to any plane containing those axes, corresponding to the symmetry of a truncated cone. But there could be *more* symmetry—namely that of a cylinder[25] or a sphere. To decide this point, Curie imagines an electrically conducting sphere immersed in a uniform electric field. Then "a force will be exerted on the sphere along the direction of the field" and the asymmetry of the effect should be retraced to its cause. But the force (i.e., the effect) is not symmetric with respect to an axis orthogonal to its direction; hence, the field-sphere system admits no such axis either. Now the sphere does admit an infinity of axes of symmetry, which are included if viewed in the field because it is a conductor; so the cause of the asymmetry should lie in the field itself. The conclusion is that the electric field *exactly* admits the symmetry of a *cone* and thus can be represented as a *polar vector* of the space $\mathbb{R}^3$. The same holds true for an *electric current* or an *electric polarization*.
2. The spatial symmetry of the magnetic field: Consider now the magnetic field that is created at the center of a circular coil carrying an electric current. The axis of the coil is an *axis of isotropy* and its plane a *plane of symmetry*. So the magnetic field does possess a plane of symmetry normal to its direction. On the other hand, there is no normal axis associated with a binary symmetry. Indeed, think of a rod moving normally along its length; it does have a binary axis of symmetry spanned by the rod itself and the velocity

[24] However the effects may well be more symmetric than their causes, as some causes for asymmetry may not suffice to produce the expected effects.

[25] Curie does not elaborate, but he writes "'truncated cone'" rather than "cone," clearly because a cylinder can be seen as a particular case of the former.

vector. If one now creates a magnetic field normal to that plane, an electromotive force appears in the rod and the binary axis disappears. So it must be absent from the cause and the magnetic field admits no axis normal to its direction. The conclusion is that *the magnetic field has the symmetry of a rotating cylinder* and can be represented by an *axial vector* (in $\mathbb{R}^3$). The same can be said of a *magnetic current* or a *magnetic polarization*.

From Curie's reasonings about fields, one can deduce the symmetries of the *charges*. He does not do this himself,[26] but his reasoning can easily be adapted. Let us take up the parallel circular plates introduced here; they are swapped, as well as their charges, through a symmetry with respect to a parallel and equidistant plane. Since we know that this operation reverses the electric field, the signs of the charges do *not* change. For a magnetic charge, one can draw the opposite conclusion since in that case, the field is *not* reversed. In summary, parity reverses the sign of the *magnetic charge* (g), but not that of the *electric charge* (e):

$$P : \mathbf{E} \to -\mathbf{E};\ \ \mathbf{H} \to \mathbf{H};\ \ e \to e;\ \ g \to -g. \tag{7.1}$$

We will see in the upcoming discussion that although e is a scalar, $\mathbf{E}$ a polar vector, and $\mathbf{H}$ an axial vector, it would be wrong to conclude that the charge g is "pseudoscalar," as is sometimes asserted. In fact, *all* physical constants are *scalars*, no matter what physical quantities are characterized; no one would say that $\hbar$ is a component of an antisymmetric tensor because it is a unit of kinetic momentum. In the sequel, everything pertaining to the electric charge will remain as is, but we will have to modify our view of the magnetic charge in light of the quantum expression of chirality.

7.3 THE TIME SYMMETRY OF THE ELECTROMAGNETIC FIELD

Curie did not address the question of time reversal, which in those days was not an issue. We will first invoke Lorentz's force as exerted on an electric or magnetic charge [Jackson (1975)]:

$$\mathbf{F}_{\text{elec}} = \mathrm{e}\left(\mathbf{E} + \frac{1}{\mathrm{c}}\mathbf{v} \times \mathbf{H}\right);\ \ \mathbf{F}_{\text{magn}} = \mathrm{g}\left(\mathbf{H} - \frac{1}{\mathrm{c}}\mathbf{v} \times \mathbf{E}\right) \tag{7.2}$$

[26] However, he does deal with magnetic charges in his second memoir (*Curie, 1894a,b*).

Quantum mechanics, of course, should not contradict these formulas, which indeed are recovered in the semiclassical limit—especially since, although they are not enough to fix the symmetries, they should be coherent with them. Since **F** is T invariant (because $\mathbf{F} = m\boldsymbol{a}$) and v changes sign along with t, Eq. (7.2) forces the following transformation laws:[27]

$$T : e\mathbf{E} \rightarrow \mathrm{e}\mathbf{E};\, \mathrm{e}\mathbf{H} \rightarrow -\mathrm{e}\mathbf{H};\, \mathrm{g}\mathbf{H} \rightarrow \mathrm{g}\mathbf{H};\, \mathrm{g}\mathbf{E} \rightarrow -\mathrm{g}\mathbf{E}. \quad (7.3)$$

From there, one gets *two* possible sets of laws:

$$\begin{array}{l} (I)\, \mathbf{E} \rightarrow \mathbf{E};\, \mathbf{H} \rightarrow -\mathbf{H};\, \mathrm{e} \rightarrow \mathrm{e};\, \mathrm{g} \rightarrow -\mathrm{g} \\ (II)\, \mathbf{E} \rightarrow -\mathbf{E};\, \mathbf{H} \rightarrow \mathbf{H};\, \mathrm{e} \rightarrow -\mathrm{e};\, \mathrm{g} \rightarrow \mathrm{g}. \end{array} \quad (7.4)$$

Hence, there is an ambiguity which is enough to lift for *one* of these quantities since the others would follow, but that seems difficult (if not impossible) with the help of electromagnetic phenomena alone. All the ones that I have tried, whether classical or quantum, may accommodate both transformation laws. This difficulty is akin to the one that Curie himself encountered concerning spatial symmetry. In his words, *"Les phénomènes généraux de l'électricité et du magnétisme nous indiquent seulement une liaison entre les symétries du champ électrique et du champ magnétique [...]"* [The general electric and magnetic phenomena tell us only about a connection between the symmetries of the electric field and those of the magnetic field.]

In fact, these phenomena only showed him that one of the fields is polar and the other axial, but without specifying which is which. And he continues: *"Pour lever cette indétermination il faut faire intervenir d'autres phénomènes, les phénomènes electrochimiques ou d'electricite de contact, les phénomènes pyro ou piézoélectriques, ou encore le phénomène de Hall, ou celui de la polarisation rotatoire magnétique."* [In order to remove this indeterminacy, one needs to bring into play other types of phenomena, such as electrochemical phenomena or those involving contact electricity, or pyro or piezoelectric ones, or the Hall phenomenon or still rotating magnetic polarization.]

In our case, too, the general electromagnetic phenomena specify only the behavior of the three remaining quantities in Eq. (7.4), provided

[27] Note that following Eq. (7.2), one can also write $P : e\mathbf{E} \rightarrow -e\mathbf{E};\, e\mathbf{H} \rightarrow e\mathbf{H};\, g\mathbf{H} \rightarrow -g\mathbf{H};\, g\mathbf{E} \rightarrow g\mathbf{E}$, and by Curie's laws for fields, one recovers Eq. (7.1) for the charges.

that one of them is known[28]. We will now follow the route suggested by Curie.

Consider an electrochemical phenomenon (e.g., cations flowing toward an anode with a density of current $\mathbf{j} = \rho \mathbf{v}$, where ρ denotes the density of cations and $\mathbf{v}$ their velocity). Let us perform a time reversal; we do not know a priori whether the sign of the charges will be reversed or not, but in either case, the signs of the ions and that of the electrode will remain opposite, so the direction of the current will remain unaffected. But since the velocity v is reversed, the charge must change sign too: so it is the second possibility in Eq. (7.4)—namely, *(II)*—that is the correct one. This result is confirmed by the reasoning of Einstein (Einstein, 1925) and based on relativistic covariance. Indeed, relativity combines the two fields **E** and **H** into an antisymmetric tensor $F_{\mu\nu}$:

$$\mathbf{E} = \{iF_{k4}\}; \mathbf{H} = \{\mathbf{F}_{kl}\} \left(x_\mu = x_k, ict\right). \tag{7.5}$$

From Eq. (7.5), Einstein concludes that the electric field, regarded as the time component of the tensor, changes sign under P and T, as does the charge density since it occurs as the divergence of the field.

When Pierre Curie investigated the symmetries of the electric field, he assumed that the charged circular plates that he considered admit an infinite number of planes of symmetry through their common axis. He thereby implicitly assumed the P invariance of the electric charge, which he had not yet demonstrated. However, he did so in the sequel, using pyroelectricity and piezoelectricity. On the other hand, he never took up the hint that he gave about electrochemical phenomena, but one may imagine that he had in mind a line of reasoning similar to the one given here. Consider, in fact, two electrodes, anode and cathode, toward which two currents, cationic and anodic, are flowing. Let us perform a symmetry that permutes the electrodes. Whether the signs of the electrodes are permuted or not, those of the ions will also permute the electrodes, and they will thus aim for the same electrode. But since these have been permuted, the ionic currents must be reversed, as do the velocities, by parity. The conclusion is that the electric charge is unchanged (i.e., it is P invariant).

[28] This is probably the reason why Jackson admits that this choice is purely conventional.

7.4 *P*, *T*, AND *C* VARIANCE OF THE ELECTROMAGNETIC FIELD

This analysis must be supplemented by the effect of charge conjugacy, which offers no problem in classical physics: If one reverses the sign of a charge on which a force is applied, the *external* fields are left invariant and the sign of the force [as in Eq. (7.2)] is reversed:

$$C:\mathbf{E}\rightarrow\mathbf{E};\mathbf{H}\rightarrow\mathbf{H};e\rightarrow -e;g\rightarrow -g \tag{7.6}$$

However, we should keep in mind that the situation will be different for fields that are *emitted* by a given charge. Following Eqs. (7.1), (7.4), and (7.6), we can write the transformation laws as

$$\begin{cases} P:\mathbf{E}-\mathbf{E};\mathbf{H}\rightarrow\mathbf{H};e\rightarrow e;g\rightarrow -g \\ T:\mathbf{E}\rightarrow -\mathbf{E};\mathbf{H}\rightarrow\mathbf{H};e\rightarrow -e;g\rightarrow g. \\ C:\mathbf{E}\rightarrow\mathbf{E};\mathbf{H}\rightarrow\mathbf{H};e\rightarrow -e;g\rightarrow -g \end{cases} \tag{7.7}$$

7.5 TRANSFORMING THE POTENTIALS

The *P*, *T*, and *C* transformation laws for the potentials are obtained from the actual definitions of the fields. We will simultaneously introduce the Lorentz potentials *V* and **A** and the pseudopotentials *W* and **B** associated with the magnetic monopole (Lochak, 1995a,b); note, however, that **B** has nothing to do with induction), as follows:

$$\mathbf{E}=-\nabla V-\frac{1}{c}\frac{\partial\mathbf{A}}{\partial t};\mathbf{H}=curl\,\mathbf{A}\ \text{and} \tag{7.8}$$

$$\mathrm{E}=curl\,\mathbf{B};\mathbf{H}=\nabla W+\frac{1}{c}\frac{\partial\mathbf{B}}{\partial t}. \tag{7.9}$$

Following Eq. (7.7), one finds the following transformation laws for the potentials, which are to be satisfied by quantum equations:

$$\begin{cases} P:\mathbf{A}\rightarrow -\mathbf{A};\mathbf{V}\rightarrow\mathbf{V};B\rightarrow B;\mathbf{W}\rightarrow -\mathbf{W};e\rightarrow e;g\rightarrow -g \\ T:\mathbf{A}\rightarrow\mathbf{A};\mathbf{V}\rightarrow -\mathbf{V};\mathrm{g}\rightarrow -B;\mathbf{W}\rightarrow\mathbf{W};e\rightarrow -e;g\rightarrow g. \\ C:\mathbf{A}\rightarrow\mathbf{A};\mathbf{V}\rightarrow\mathbf{V};B\rightarrow B;\mathbf{W}\rightarrow\mathbf{W};e\rightarrow -e;g\rightarrow -g \end{cases} \tag{7.10}$$

Lorentz transformations combine the potentials (*V*, **A**) and (*W*, **B**) into two 4-vectors:

$$A_\mu=(\mathbf{A};iV);iB_\mu=\big(\mathbf{B};iW\big). \tag{7.11}$$

According to Eq. (7.7), A_μ is a polar vector, whereas B_μ is an axial vector in space-time. In Euclidean space $\mathbb{R}^3$, $\mathbf{A}$ is a polar vector and V is a scalar vector, whereas $\mathbf{B}$ is an axial vector and W is a pseudoscalar vector[29]. These are the transformation laws that are generally agreed upon. One can check that they fit well with other results, but note that they do not discriminate between the laws of type (I) and type (II) described previously. In particular, note the following:

1. Eq. (7.10) yields the right transformation laws for Lagrangian momenta.
If $\mathbf{P} = \mathbf{p} + \frac{e}{c}\mathbf{A}, E = mc^2 + eV$, or : $\mathbf{P} = \mathbf{p} + \frac{g}{c}\mathbf{B}, E = mc^2 + gW$, one gets

$$(P \text{ or } T) : \mathbf{P} \rightarrow -\mathbf{P}; \mathbf{E} \rightarrow -\mathbf{E}. \tag{7.12}$$

2. Here is a second property from pure electromagnetism. Namely, Eqs. (7.7) and (7.10), for the fields and potentials, respectively, ensure the covariance of Maxwell equations and that of de Broglie's equations for the photon (de Broglie, 1940–1942, 1943), in which potentials and fields appear on the same footing:

$$\begin{aligned}
&-\frac{1}{c}\frac{\partial \mathbf{H}}{\partial t} = curl\ \mathbf{E}; \quad \frac{1}{c}\frac{\partial \mathbf{E}}{\partial t} = curl\ \mathbf{H} + k_0^2\mathbf{A} \\
&div\,\mathbf{H} = 0; \quad div\,\mathbf{E} = -k_0^2 V \\
&\mathbf{H} = curl\,\mathbf{A}; \quad \mathbf{E} = -grad V - \frac{1}{c}\frac{\partial \mathbf{A}}{\partial t}; \quad \frac{1}{c}\frac{\partial V}{\partial t} + div\mathbf{A} = 0.
\end{aligned} \tag{7.13}$$

This also ensures the covariance of the equations for the "magnetic photon," which involve pseudopotentials (Lochak, 1995a,b):

$$\begin{aligned}
&-\frac{1}{c}\frac{\partial \mathbf{H}}{\partial t} = curl\,\mathbf{E} + k_0^2\mathbf{B}; \quad \frac{1}{c}\frac{\partial \mathbf{E}}{\partial t} = curl\,\mathbf{H} \\
&div\,\mathbf{H} = k_0^2 W; div\,\mathbf{E} = 0 \\
&\mathbf{H} = grad W + \frac{1}{c}\frac{\partial \mathbf{B}}{\partial t}; \quad \mathbf{E} = curl\,\mathbf{B}; \quad \frac{1}{c}\frac{\partial W}{\partial t} + div\mathbf{B} = 0.
\end{aligned} \tag{7.14}$$

[29] Recall that a polar vector in space-time has a spatial polar component and a time component that is P invariant because that transformation acts only on the space part. On the other hand, T acts only on the time component, as is visible from Eq. (7.10). The properties of an axial vector are the opposite of that of a polar vector: see again Eq. (7.10). It may also be useful to recall that an axial vector in $\mathbb{R}^3$ is dual to a second-order antisymmetric tensor: $B_i = \frac{1}{2}\in_{ijk} C_{[jk]}$. In much the same way, an axial vector in space-time is dual to an antisymmetric tensor of third order: $B_\mu = \frac{1}{6}\in_{\mu\alpha\beta\gamma} C_{[\alpha\beta\gamma]}$.

7.6 P, T, AND C INVARIANCE IN THE DIRAC EQUATION

The Dirac equation in the presence of an electromagnetic field reads

$$\gamma_\mu\left(\partial_\mu + i\frac{e}{\hbar c}A_\mu\right)\psi + \frac{m_0 c}{\hbar}\psi = 0 \quad (x_\mu = x_k, ict) \tag{7.15}$$

where A_μ is the potential four-vector defined as in Eq. (5.4) in Chapter 5, and one sets

$$\begin{aligned}\gamma_k &= i\begin{pmatrix} 0 & s_k \\ -s_k & 0 \end{pmatrix}; k = 1,2,3; \gamma_4 = \begin{pmatrix} I & 0 \\ 0 & -I \end{pmatrix};\\ \gamma_5 &= \gamma_1\gamma_2\gamma_3\gamma_4 = \begin{pmatrix} 0 & I \\ I & 0 \end{pmatrix}.\end{aligned} \tag{7.16}$$

Here, the s_k's are the Pauli matrices:

$$s_1 = \begin{pmatrix} 0 & 1 \\ 1 & 0 \end{pmatrix}; s_2 = \begin{pmatrix} 0 & -i \\ i & 0 \end{pmatrix}; s_3 = \begin{pmatrix} 1 & 0 \\ 0 & -1 \end{pmatrix}; I = \begin{pmatrix} 0 & 1 \\ 1 & 0 \end{pmatrix}. \tag{7.17}$$

In the sequel, one should keep in mind that the vector space components and the pseudovector time components, as well as s_1, s_3, γ_2, γ_4, and γ_5, are real, whereas the vector time components and the pseudovector space components, as well as s_2, γ_1, and γ_3, are purely imaginary.

We will use Weyl's spinorial representation, which diagonalizes γ_5 and displays the chiral two- components ξ and η:

$$\psi \to U\psi = \begin{pmatrix} \xi \\ \eta \end{pmatrix}; U = U^{-1} = \frac{1}{\sqrt{2}}(\gamma_4 + \gamma_5). \tag{7.18}$$

In this representation, the Dirac equation becomes (as mentioned already)

$$\begin{aligned}&\left[\frac{1}{c}\frac{\partial}{\partial t} - \mathbf{s}.\nabla - i\frac{e}{\hbar c}(V + \mathbf{s}.\mathbf{A})\right]\xi + i\frac{m_0 c}{\hbar}\eta = 0\\ &\left[\frac{1}{c}\frac{\partial}{\partial t} + \mathbf{s}.\nabla - i\frac{e}{\hbar c}(V - \mathbf{s}.\mathbf{A})\right]\eta + i\frac{m_0 c}{\hbar}\xi = 0.\end{aligned} \tag{7.19}$$

Let us now write out the *P*, *T*, and *C* invariances of Eq. (7.15), using Eqs. (7.7) and (7.10); the *P* and *T* invariances are expressed by the Racah formulas [Racah (1937)]; note that the tilde indicates transposition:

$$P : e \to e; x_k \to -x_k, x_4 \to x_4; A_k \to -A_k; A_4 \to A_4; \psi \to \gamma_4\psi$$
$$C : e \to -e; \psi \to \gamma_2\psi^* = \gamma_2\gamma_4\tilde{\overline{\psi}}(\overline{\psi} = \psi^+\gamma_4). \quad (7.20)$$

However, the Racah formula for time reversal has to be rejected because we now need to take into account the transformation formulas for the potentials and charges, which leads to writing

$$T^{(II)}_{Racah} : x_k \to x_k,\ x_4 \to -x_4; A_k \to A_k; A_4 \to -A_4$$
$$\psi \to -i\gamma_1\gamma_2\gamma_3\psi;\ e \to -e. \quad (7.21)$$

This is *not* the original Racah transformation, as the latter does not modify the sign of the charge; we have added a superscript (II) to make this clear. In order to apply this transformation, one should *first* change in Eq. (7.15) the signs of time, of the charge, and of A_4, getting

$$\left\{\gamma_1\partial_1 + \gamma_2\partial_2 + \gamma_3\partial_3 - \gamma_4\partial_4 - i\frac{e}{\hbar c}(\gamma_1A_1 + \gamma_2A_2 + \gamma_3A_3 - \gamma_4A_4) + \frac{m_0c}{\hbar}\right\}\psi = 0. \quad (7.22)$$

Applying then the operator $-i\gamma_1\gamma_2\gamma_3$, we find the Dirac equation again, but with an opposite charge, so that Eq. (7.21) does not express any invariance under time reversal. This is why we will look for an antiunitary solution by taking the complex conjugate of Eq. (7.22), namely

$$\left\{-\gamma_1\partial_1 + \gamma_2\partial_2 - \gamma_3\partial_3 + \gamma_4\partial_4 - i\frac{e}{\hbar c}(\gamma_1A_1 - \gamma_2A_2 + \gamma_3A_3 - \gamma_4A_4) + \frac{m_0c}{\hbar}\right\}\psi^* = 0. \quad (7.23)$$

If we now apply the matrix $-i\gamma_3\gamma_1$, which reverses γ_1 and γ_3 while leaving γ_2 and γ_4 invariant, we retrieve Eq. (7.15) with the correct sign (i.e., a plus sign) in front of the charge. The dual matrix, $-i\gamma_4\gamma_2 = -i\gamma_5\gamma_3\gamma_1$, would result in the same signs in front of the matrices $\gamma_1, \gamma_2, \gamma_3, \gamma_4$, but the sign of the mass would be reversed. Thus, the matrix $-i\gamma_3\gamma_1$ is the only possibility, whence the transformation laws for time reversal are as follows:

$$T : e \to -e; x_k \to x_k, x_4 \to x_4;$$
$$A_k \to A_k, A_4 \to -A_4; \psi \to -i\gamma_3\gamma_1\psi^*. \quad (7.24)$$

This leads to the following P, T, and C transformations associated with the Dirac equation:

$$\begin{cases} P : e \to e; x_k \to -x_k, x_4 \to x_4; \\ A_k \to -A_k; A_4 \to A_4; \psi \to \gamma_4 \psi \\ T : e \to -e; x_k \to x_k, x_4 \to -x_4; \\ A_k \to A_k; A_4 \to -A_4; \psi \to -i\gamma_3\gamma_1\psi^* \\ C : e \to -e; \psi \to \gamma_2\psi^* = \gamma_2\gamma_4\tilde{\overline{\psi}}\left(\overline{\psi} = \psi + \gamma_4\right). \end{cases} \tag{7.25}$$

These transformation laws conform to the Curie laws for electromagnetism, supplemented by those for charge and time reversal that we have added. They also comply with the objections which Costa de Beauregard raised concerning the Racah transformation, invoking relativity (Costa de Beauregard, 1983). The laws given here seem clearer, and especially better-grounded, than those of Jauch and Rohrlich (1955), with whom one would also disagree somewhat. The transformation of the wave function under the *T* transformation [see Eqs. (7.24) and (7.25)] coincides with those in Lochak, where it appears without the charge component because the reasoning there does not take interactions into account. In fact, the *T* transformation proposed here is sometimes called *weak* time reversal [Sokolov and Ternov (1974)], defined as the product of charge reversal by time reversal à la Racah. For us, however, Eq. (6.10) will feature pure time reversal *T*, and we will consider the original Racah transformation as representing the product *TC* as follows:

$$T_{\text{Racah}} = TC, \tag{7.26}$$

where T_{Racah} now leaves the charge invariant:

$$\begin{aligned} &T_{\text{Racah}} : e \to e; x_k \to x_k, x_4 \to -x_4; \\ &A_k \to A_k, A_4 \to -A_4; \psi \to -i\gamma_1\gamma_2\gamma_3\psi. \end{aligned} \tag{7.27}$$

This transformation expresses a law of invariance as a product of two such. How should we interpret it? We now have *two* time-reversal operations [namely, *T*, as in Eq. (7.24), and T_{Racah}], and two operations that reverse the charge (namely, *T* and *C*), but with different meanings because *C* associates with a negative energy solution of Eq. (7.15) a positive energy solution of that same equation with the opposite charge of that of the electron. Consider, indeed, a solution with negative energy, as displayed by the minus sign in the exponential ($\omega > 0$):

$$\psi = e^{-i\omega t}\phi(\mathbf{r}). \tag{7.28}$$

The charge reversal transformation C associates with that solution one with positive energy:

$$\psi' = \gamma_2\psi^* = e^{i\omega t}\gamma_2\phi^*(\mathbf{r}), \tag{7.29}$$

where the plus sign comes from complex conjugacy. Now, start instead from a solution with positive energy (again $\omega > 0$):

$$\psi = e^{i\omega t}\phi(\mathbf{r}) \tag{7.30}$$

and apply the T transformation of Eq. (7.24) or (7.25). Doing this will reverse both the sign of time and of the charge, and we get a solution with opposite charge but positive energy. The inversion of the sign in the exponential due to complex conjugacy is compensated for by time reversal:

$$\psi'' = -i\gamma_3\gamma_1\psi^*(-t,\mathbf{r}) = -ie^{(-i)\omega(-t)}\gamma_3\gamma_1\phi^*(\mathbf{r}) = -ie^{i\omega t}\gamma_3\gamma_1\phi^*(\mathbf{r}). \tag{7.31}$$

Time reversal associates with an electron with positive energy a positron with an equally positive energy: the positron can be thought of as an electron going back in time, as Richard Feynman liked to put it. This would not have been the case had we adopted the first transformation law (I) in Eq. (7.4). Let us now apply the Racah transformation to the positive energy solution in Eq. (7.30):

$$\psi''' = T_{\text{Racah}}\psi = TC\psi = -i\gamma_1\gamma_2\gamma_3\psi(-t,\mathbf{r}) = -ie^{-i\omega t}\gamma_1\gamma_2\gamma_3\phi(\mathbf{r}). \tag{7.31}$$

We find again a solution going backward in time, but this is obtained via the product of time reversal [in the sense of Eq. (7.25)] and charge conjugacy, producing a minus sign in the exponential.

Finally, we can transcribe [Eq. (7.25)] in the Weyl representation [Eq. (7.18)]:

$$\begin{aligned}
&P: e\to e; \mathrm{x}\to -\mathrm{x}, t\to t;\\
&\mathbf{A}\to -\mathbf{A}, V\to V; \xi\to\eta\\
&T: e\to -e; x\to x, t\to -t; \mathbf{A}\to\mathbf{A}, V\to -V;\\
&\xi\to s_2\xi^*; \eta\to s_2\eta^*\\
&C: e\to -e; \xi\to -is_2n^*; \eta\to is_2\xi^*.
\end{aligned} \tag{7.32}$$

It can be seen here, and it will appear more clearly later in this chapter in the case of magnetism that ξ and η are the *chiral components* of the Dirac wave function; they are permuted under parity (*P*) and charge conjugacy (*C*) but not under time reversal (*T*).

7.7 P, T, AND C INVARIANCE IN THE MONOPOLE EQUATION

Let us write down the linear equation for a magnetic monopole[30](Lochak, 1983, 1984, 1985, 1987a,b, 1995a,b):

$$\gamma_\mu(\partial_\mu - g\gamma_5 B_\mu)\psi = 0. \tag{7.33}$$

The equation being massless ensures its invariance with respect to the chiral gauge transformation:

$$\psi \to \exp(i\gamma_5\theta/2)\psi;\ B_\mu \to B_\mu + \partial_\mu\theta. \tag{7.34}$$

We have seen in Eq. (7.11) that B_μ combines the pseudopotentials W and B and these appear in the Weyl representation of Eq. (7.33), to wit:

$$\begin{aligned}
&\left[\frac{1}{c}\frac{\partial}{\partial t} - \mathbf{s}.\nabla - i\frac{g}{hc}(W + \mathbf{s}.B)\right]\xi = 0\\
&\left[\frac{1}{c}\frac{\partial}{\partial t} + \mathbf{s}.\nabla - i\frac{g}{hc}(W - \mathbf{s}.B)\right]\eta = 0.
\end{aligned} \tag{7.35}$$

This equation best displays the meaning of the Weyl representation. Indeed, comparing Eqs. (7.15) and (7.33) on the one hand with Eqs. (7.19) and (7.35) on the other makes quite clear the essential difference between an electric and a magnetic charge. In the Dirac equation [Eq. (7.15)], the charge operator is $E = eI$ (I denotes the identity matrix) with just one eigenvalue, whereas in the monopole equation [Eq. (7.33)], the operator reads $B = g\gamma_5$ with eigenvalues g and $-g$, which appear explicitly in Eq. (7.35). In fact, the main property of the Weyl representation is that it diagonalizes the charge operator B and thus separates the chiral components ξ and η:

$$UBU^{-1} = gU\gamma_5U^{-1} = g\gamma_4 = g\begin{pmatrix} I & 0 \\ 0 & -I \end{pmatrix}. \tag{7.36}$$

As was already mentioned in §2, the pseudoscalar character of the operator B is not to be ascribed to the charge constant g but to the operator γ_5.

[30] There is no factor i in front of the charge because $B\mu$ is a pseudovector; see Eq. (7.11).

That can only be understood in quantum mechanics and is simply ignored by its classical counterpart. In fact, chirality is related to the *polarization of the wave;* the wave itself disappears in classical mechanics and its trace is reduced to a phase in the action integral.

The chiral components ξ and η corresponding to the two eigenvalues of B satisfies the two *independent* equations [Eq. (7.35)], in contrast with Eq. (7.19) for an electron, which is coupled via the mass term. The magnetic charge g occurs with opposite signs in the two equations [Eq. (7.35)] (these are the eigenvalues of the charge operator $B = g\gamma_5$), whereas the electric charge e enters with the same sign in both equations [Eq. (7.19)] (operator $E = eI$)—an essential difference between magnetism and electricity.

This entails that, at variance with what happens in classical physics, a change of sign of the magnetic charge may a priori have two different meanings in quantum mechanics:

1. It can denote a transition between two monopoles with opposite signs of their charge constants, in analogy with a transition between an electron and a positron.
2. At the same time, it can describe a transition between the two chiral components of the *same* monopole, with a given charge constant, but opposite eigenvalues of the charge operator B.

Therefore, at least in principle, we are confronted with four cases (see the first reference in Lochak, 1985):

$$\begin{aligned} &m^+: \text{ left monopole } (\xi-\text{ component}) \text{ charge } g > 0 \\ &\overline{m}^+: \text{ right antimonopole } (\eta-\text{ component}) \text{ charge } g > 0 \\ &m^-: \text{ left monopole } (\xi-\text{ component}) \text{ charge } g < 0 \\ &\overline{m}^-: \text{ right antimonopole } (\eta-\text{ component}) \text{ charge } g < 0. \end{aligned} \tag{7.37}$$

Let us express the P invariance, which is explicit in Eq. (7.33), changing the signs of the components of B_μ according to Eq. (7.10) but leaving the charge constant fixed g [contrary to the prescription of Eq. (7.10)—we will see why in a moment]:

$$\left\{-\gamma_1\partial_1 - \gamma_2\partial_2 - \gamma_3\partial_3 + \gamma_4\partial_4 - \frac{g}{\hbar c}(\gamma_1 B_1 + \gamma_2 B_2 + \gamma_3 B_3 - \gamma_4 B_4)\gamma_5\right\}\psi = 0. \tag{7.38}$$

Since δ_μ and B_μ transform in opposite ways and the matrices γ_4 and γ_5 anticommute, Eq. (7.38) is equivalent to Eq. (7.33) up to global multiplication by γ_4. We get the following explicit form of the P invariance for Eq. (7.33):

$$P : g \rightarrow g; x_k \rightarrow -x_k, x_4 \rightarrow x_4; \\ B_k \rightarrow B_k; W \rightarrow -W; \psi \rightarrow \gamma_4 \psi, \tag{7.39}$$

where the magnetic charge constant is invariant which looks as if it contradicts (7.10) but this is deceptive. In fact, starting from the Weyl representation, Eqs. (7.35) and (7.39) can be rewritten as

$$P : g \rightarrow g; x_k \rightarrow -x_k, x_4 \rightarrow x_4; \\ B_k \rightarrow B_k; B_4 \rightarrow -B_4; \xi \leftrightarrow \eta. \tag{7.40}$$

This makes it plain that whereas the charge constant is invariant, the parity operator swaps the chiral components ξ and η, hence the eigenvalues $\pm$g of the charge operator B. It thus changes the sign of the charge of the monopole, as predicted in Eq. (7.10), although this is effected not via changing the sign of the charge constant but rather a change of chirality, in conformity with Curie's memoir (see §7.2). Following the terminology in Eq. (7.37), we will get, according to the sign of g (for a priori, both values $+g$ and $-g$ could exist in nature):

$$P : m^+ \leftrightarrow \overline{m}^+ \text{ or } m^- \leftrightarrow \overline{m}^-. \tag{7.41}$$

Let us now take a look at the charge conjugation operator C. To this end, we take the complex conjugate of Eq. (7.33), keeping Eqs. (7.11) and (7.16) in mind:

$$\left\{-\gamma_1\partial_1 + \gamma_2\partial_2 - \gamma_3\partial_3 - \gamma_4\partial_4 - \frac{g}{\hbar c}(\gamma_1 B_1 - \gamma_2 B_2 + \gamma_3 B_3 + \gamma_4 B_4)\gamma_5\right\}\psi^* = 0. \tag{7.42}$$

Multiplying by γ_2, we fall back onto Eq. (7.33), which is thus invariant under C, but again *without* changing the sign of the constant g:

$$C : g \rightarrow g; \psi \rightarrow \gamma_2\psi^* = \gamma_2\gamma_4 \tilde{\overline{\psi}} \ \left(\overline{\psi} = \psi^+\gamma_4\right). \tag{7.44}$$

Here again, this may seem to contradict Eq. (7.10) but using the Weyl representation, Eq. (7.43) reads

$$C : g \rightarrow g; \xi \rightarrow -is_2\eta^*; \eta \rightarrow -is_2\xi^*. \tag{7.44}$$

In other words, charge conjugacy leaves Eq. (7.33) invariant, as well as the equivalent *system* of equations [Eq. (7.35)] without changing the sign of the constant g; but inside Eq. (7.35), it leads to permuting the chiral components; that is, the left and right monopoles with respective eigenvalues $\pm g$ of the charge operator. The charge changes sign much as in Eq. (7.41):

$$C: m^{+} \leftrightarrow \overline{m}^{+} \; or \; m^{-} \leftrightarrow \overline{m}^{-}. \tag{7.45}$$

Finally we come to the time reversal transformation T, by introducing Eq. (7.10) into Eq. (7.33):

$$\left\{\gamma_1\partial_1 + \gamma_2\partial_2 + \gamma_3\partial_3 - \gamma_4\partial_4 - \frac{g}{\hbar c}(-\gamma_1 B_1 - \gamma_2 B_2 - \gamma_3 B_3 + \gamma_4 B_4)\gamma_5\right\}\psi = 0. \tag{7.46}$$

Apply $-i\gamma_1\gamma_2\gamma_3 i$, which commutes with $\gamma_1, \gamma_2, \gamma_3$, and anticommutes with γ_4 et γ_5 to retrieve Eq. (7.33) again and a Racah transformation, which looks compatible with Eq. (7.10):

$$\begin{aligned} T_{\text{Racah}} &: g \rightarrow g;\, x_k \rightarrow x_k,\, x_4 \rightarrow -x_4; \\ B_k &\rightarrow -B_k,\, B_4 \rightarrow B_4;\, \psi \rightarrow -i\gamma_1\gamma_2\gamma_3\psi. \end{aligned} \tag{7.47}$$

We check what happens to the chiral components by transcribing Eq. (7.47) in the Weyl representation:

$$\begin{aligned} T_{\text{Racah}} &: g \rightarrow g;\, x_k \rightarrow x_k,\, x_4 \rightarrow -x_4; \\ & B_k \rightarrow -B_k,\, B_4 \rightarrow B_4 \\ & \xi \rightarrow -i\eta;\, \eta \rightarrow i\xi. \end{aligned} \tag{7.48}$$

One finds that the charge constant is left invariant but that again chirality is not, permuting the ξ and η components, hence the charge of the monopole, which violates Eq. (7.10), which we took as our starting point. So that Racah transformation is *not* admissible for representing time reversal for a magnetic charge. This conclusion contradicts my own previous papers on the subject, where I had overlooked (and so apparently did everyone else) the P, T, and C transformation laws for electromagnetic quantities, as embodied by Eq. (7.10). On the other hand, the unitary transform of §7.6 still holds true here. Indeed, consider the complex conjugate of Eq. (7.46), taking Eq. (7.10) into account:

$$\left\{-\gamma_1\partial_1 + \gamma_2\partial_2 + \gamma_3\partial_3 - \gamma_4\partial_4 + \frac{g}{\hbar c}(\gamma_1 B_1 - \gamma_2 B_2 + \gamma_3 B_3 - \gamma_4 B_4)\gamma_5\right\}\psi^* = 0. \tag{7.49}$$

Multiplying by $-i\gamma_3\gamma_1$ changes the sign of γ_1 and γ_3, leaving γ_2, γ_4, and γ_5 invariant and it takes us back to Eq. (7.33), leading to the following transformation law:

$$\begin{aligned} T &: g \rightarrow g;\, x_k \rightarrow x_k,\, x_4 \rightarrow -x_4; \\ B_k &\rightarrow -B_k,\, B_4 \rightarrow B_4;\, \psi \rightarrow -i\gamma_1\gamma_2\gamma_3\psi^*, \end{aligned} \tag{7 50}$$

but now in the Weyl representation, we find:

$$\begin{aligned} &T: g\rightarrow g; x_k\rightarrow x_k, t\rightarrow -t; \\ &B_k\rightarrow -B_k, W\rightarrow W; \\ &\xi\rightarrow -s_2\xi^*; \eta\rightarrow s_2\eta^*, \end{aligned} \tag{7.51}$$

which makes it clear that, at variance with the Racah transformation [Eq. (7.48)], the chiral components are *not* permuted and the magnetic charge stays invariant, in conformity with our starting point [Eq. (7.10)]. Summarizing, and in parallel with the transformation laws [Eq. (7.25)] for the electron, we arrive at the following table for the monopole:

$$\begin{aligned} &\mathrm{P}: g\rightarrow g; x_k\rightarrow -x_k, x_4\rightarrow x_4; \\ &B_k\rightarrow B_k; B_4\rightarrow -B_4; \psi\rightarrow \gamma_4\psi \\ &\mathrm{T}: g\rightarrow g; x_k\rightarrow x_k, x_4\rightarrow -x_4; \\ &B_k\rightarrow B_k; B_4\rightarrow B_4; \psi\rightarrow -i\gamma_3\gamma_1\psi^* \\ &C: g\rightarrow g; \psi\rightarrow \gamma_2\psi^* = \gamma_2\gamma_4\tilde{\overline{\psi}}(\overline{\psi} = \psi^+\gamma_4), \end{aligned} \tag{7.52}$$

which can be rewritten in the Weyl representation as

$$\begin{aligned} &P: g\rightarrow g; x_k\rightarrow -x_k, t\rightarrow t; \\ &B_k\rightarrow B_k; W\rightarrow -W; \xi\leftrightarrow \eta \\ &T: g\rightarrow g; x_k\rightarrow x_k, t\rightarrow -t; \\ &B_k\rightarrow -B_k; W\rightarrow W; \xi\rightarrow s_2\xi^*; \eta\rightarrow s_2\eta^* \\ &C: g\rightarrow g; \xi\rightarrow -is_2\eta^*; \eta\rightarrow is_2\xi^*. \end{aligned} \tag{7.53}$$

As was mentioned in §7.6 already, chirality is seen much more clearly with magnetism than it is with electricity. The system [Eq. (7.35)] is made of two independent equations for ξ and η, respectively, and formulas [Eq. (7.53)] show that they are exchanged under parity and under charge conjugacy. There are two monopoles, left and right, which form a particle-antiparticle pair. Eq. (7.35) also displays the neutrino as a particular case or, conversely, suggests that monopoles can be considered as "magnetically excited" neutrinos. This leads to a question that we have raised previously: Could it be that such monopoles are produced in weak interactions, in which case the fact that they interact strongly with matter, in contrast with neutrinos, could explain the deficit of solar neutrinos in terrestrial observations?

Let us stress one more time that in Eq. (7.53), the P and C transformations leave the magnetic charge constant *invariant,* which does not contradict Eq. (7.10) because *the charge of a monopole changes via its chirality*. Chirality remains invariant under time reversal, and thereby so does the magnetic charge, contrary to what happens with the electric charge. In a figurative way, and in sharp contrast with the case of the electron, one can say that *an antimonopole is not a monopole going back in time, but rather its mirror image*. It thus seems that one should consider a monopole with a charge constant $-g$ not as the antiparticle partner of one of charge $+g$, but rather as another particle altogether.

7.8 *P, T,* AND *C* TRANSFORMATION LAWS FOR TENSOR QUANTITIES

Consider the 16 tensor quantities associated with the Dirac equation:

$$\begin{aligned}&\omega_1 = \overline{\psi}\psi; J_\mu = i\overline{\psi}\gamma_\mu\psi; M_{\mu\nu} = -i\overline{\psi}\gamma_\mu\gamma_\nu\psi;\\&\Sigma_\mu = -i\overline{\psi}\gamma_\mu\gamma_5\psi; \omega_2 = -i\overline{\psi}\gamma_5\psi,\end{aligned} \tag{7.54}$$

or in the Weyl representation:

$$\begin{aligned}&\omega_1 = \xi^+\eta + \eta^+\xi; \omega_2 = i(\xi^+\eta - \eta^+\xi)\\&\omega_1^2 + \omega_2^2 = 4(\xi^+\eta)(\eta^+\xi)\\&J_\mu = \{J_4,\mathbf{J}\} = \{\mathrm{i}(\xi^+\xi + \eta^+\eta), -(\xi^+ s\xi - \eta^+ s\eta)\\&\Sigma_\mu = \{\Sigma_4, \mathbf{\Sigma}\} = \{i(\xi^+\xi - \eta^+\eta), -(\xi^+ s\xi + \eta^+ s\eta)\}\\&M_{\mu\nu} = \{M_{j4}, M_{jk}\} = \{\xi^+ s\eta - \eta^+ s\xi), (\xi^+ s\eta + \eta^+ s\xi)\}.\end{aligned} \tag{7.55}$$

We know that ω_1 and ω_2 are Lorentz invariants, J_μ and S_μ are vectors, and $M_{\mu\nu}$ is an antisymmetric tensor. Their behaviors under P, T, and C are displayed in Table 7.1, together with that of the *chiral currents* (Lochak, 1959, 1983), which are consequences of Eq. (7.55):

$$X_\mu = \{i\xi^+\xi, -\xi^+ s\xi\}; Y_\mu = \{i\eta^+\eta, \eta^+ s\eta\}. \tag{7.56}$$

From Eq. (7.55), it is plain that

$$J_\mu = X_\mu + Y_\mu; \Sigma_\mu = X_\mu - Y_\mu. \tag{7.57}$$

Now there is a table of the P, T, and C transformation laws as derived from Eqs. (7.25), (7.32), (7.52), and (7.53), and in accordance with (*II*) in Eq. (7.4):

Let us now introduce the physical dimensions of these quantities, taking into account the P, T, and C transformation rules for the charges. We denote by $\mathbf{P} = -\frac{e\hbar}{2m_0c}M_{4k}$ and $\mathbf{M} = -\frac{e\hbar}{2m_0c}M_{kl}$ the electric and magnetic polarizations of the electron, respectively. The transformation rules for the electric charge are taken from Eq. (7.10), but those for the magnetic charge have been modified according to quantum mechanics, as in Eqs. (7.52) and

Table 7.1

		P	T	C
ψ	$\rightarrow$	$\gamma_4\psi$	$-i\gamma_3\gamma_1\psi^*$	$\gamma_2\psi^*$
ξ	$\rightarrow$	η	$s_2\xi^*$	$-is_2\eta^*$
η	$\rightarrow$	ξ	$s_2\eta^*$	$is_2\eta^*$
ω_1	$\rightarrow$	ω_1	ω_1	$-\omega_1$
J_4	$\rightarrow$	J_4	J_4	J_4
J_k	$\rightarrow$	$-J_k$	$-J_k$	J_k
M_{4k}	$\rightarrow$	$-M_{4k}$	M_{4k}	M_{4k}
M_{jk}	$\rightarrow$	M_{jk}	$-M_{jk}$	M_{jk}
$\sum_4$	$\rightarrow$	$-\sum_4$	$\sum_4$	$-\sum_4$
$\sum_K$	$\rightarrow$	$\sum_K$	$-\sum_K$	$-\sum_K$
ω_2	$\rightarrow$	$-\omega_2$	$-\omega_2$	$-\omega_2$
$\xi^+\xi$	$\rightarrow$	$\eta^+\eta$	$\xi^+\xi$	$\eta^+\eta$
$\xi^+s_K\xi$	$\rightarrow$	$\eta^+s_K\eta$	$-\xi^+s_K\xi$	$-\eta^+s_K\eta$
$\eta^+\eta$	$\rightarrow$	$\xi^+\xi$	$\eta^+\eta$	$\xi^+\xi$
$\eta^+s_K\eta$	$\rightarrow$	$\xi^+s_K\xi$	$-\eta^+s_K\eta$	$-\xi^+s_K\xi$

(7.53). The sign changes do not come from the constant g, but from the changes of chirality.

We will now give the transformation laws for the fields and the potentials, but at variance with Eqs. (7.7) and (7.10), the fields are not external anymore; rather, they are caused by the currents. The P and T rules are identical, but this is not so for C. In order to check the rules for parity and time reversal, it is enough to check that the rules previously noted for the fields are in accordance with those for the currents. But for charge conjugacy, the currents (i.e., the cause) determine the rule for the fields (i.e., the effect) via the covariance that Curie's laws imply for Maxwell equations (as discussed earlier in this chapter); we thus add the subscript *em* (meaning "emitted") to the fields in the table:

$$A_\mu = \frac{4\pi}{c} eJ_\mu; \, B_\mu = \frac{4\pi}{c} g\Sigma_\mu, \tag{7.58}$$

or, more explicitly:

$$\begin{aligned} V &= \frac{4\pi}{c} eJ_4; \, \mathbf{A}_\mu = \frac{4\pi}{c}\mathbf{J}; \\ W &= \frac{4\pi}{c} g\Sigma_4; \, \mathbf{B} = \frac{4\pi}{c} g\boldsymbol{\Sigma}. \end{aligned} \tag{7.59}$$

So here is the revised version of the table, given as Table 7.2:

Table 7.2

		P	T	C
e	$\rightarrow$	e	$-e$	$-e$
g	$\rightarrow$	g	g	g
eJ_4	$\rightarrow$	eJ_4	$-eJ_4$	$-eJ_4$
$e\mathbf{J}$	$\rightarrow$	$-e\mathbf{J}$	$e\mathbf{J}$	$-e\mathbf{J}$
$\mathbf{P}$	$\rightarrow$	$-\mathbf{P}$	$-\mathbf{P}$	$-\mathbf{P}$
$\mathbf{M}$	$\rightarrow$	$\mathbf{M}$	$\mathbf{M}$	$-\mathbf{M}$
$g\sum_4$	$\rightarrow$	$-g\sum_4$	$g\sum_4$	$-g\sum_4$
$g\sum$	$\rightarrow$	$g\sum$	$-g\sum$	$-g\sum$
$\mathbf{E}$	$\rightarrow$	$-\mathbf{E}$	$-\mathbf{E}$	$-\mathbf{E}_{(\mathrm{em})}$
$\mathbf{H}$	$\rightarrow$	$\mathbf{H}$	$\mathbf{H}$	$-\mathbf{H}_{(\mathrm{em})}$
V	$\rightarrow$	V	$-V$	$-V_{(\mathrm{em})}$
$\mathbf{A}$	$\rightarrow$	$-\mathbf{A}$	$\mathbf{A}$	$-\mathbf{A}_{(\mathrm{em})}$
W	$\rightarrow$	$-W$	W	$W_{(\mathrm{em})}$
$\mathbf{B}$	$\rightarrow$	$\mathbf{B}$	$-\mathbf{B}$	$\mathbf{B}_{(\mathrm{em})}$

By using this table, one can check easily the covariance of the polarizations and fields, as well as that of the currents and potentials; it makes plain that the correct transformation rules hold true for the quantities with physically meaningful coefficients. Notice also that the T invariance of $e\mathbf{J}$ in Table 7.2, which results from Eq. (7.24) goes along with the discussion in §7.3 leading to the choice of the second possibility (*II*) in Eq. (7.4).

Finally, Table 7.1 provides a new argument in favor of the T transformation and against the original Racah transformation. Indeed, one finds from the table that the first invariant ω_1 is a true invariant in space-time, being both P and T invariant, whereas ω_2 appears as a *pseudo*invariant, changing sign under P and T. The T invariance of ω_1 is especially important, as it ensures the invariance of the Dirac Lagrangian equation:

$$L = \hbar c\left\{\overline{\psi}\gamma_\mu\left(\frac{1}{2}\left[\partial_\mu\right] + i\frac{e}{\hbar c}A_\mu\right)\psi + \frac{m_0 c}{\hbar}\overline{\psi}\psi\right\}$$
$$\left\{\left[\partial_\mu\right] = (\partial_\mu\rightarrow) - (\leftarrow\partial_\mu)\right\}, \qquad (7.60)$$

and thereby the T invariance of the energy density:

$$E = \frac{\partial L}{\partial\left(\frac{\partial\psi}{\partial t}\right)}\left(\frac{\partial\psi}{\partial t}\right) + \left(\frac{\partial\overline{\psi}}{\partial t}\right)\frac{\partial L}{\partial\left(\frac{\partial\overline{\psi}}{\partial t}\right)} - L. \qquad (7.61)$$

By contrast, using the Racah version of the T transformation, we would get

$$T_{\text{Racah}} : \omega_1 \to -\omega_1 \text{ hence} : E \to -E, \tag{7.62}$$

and this property suffices to rule out this transformation law; indeed, recall that an energy density appears as the T_{44}-component of the energy-momentum tensor and varies as the square of the time variable, which immediately entails its T invariance.

7.9 NONLINEARITY AND QUANTUM MECHANICS: ARE THEY COMPATIBLE?

We start from a question that seems to be simple enough: What are the main features of the Dirac equation that ensure that it complies with the general principles of quantum mechanics, and in particular, produces the correct semiclassical approximation? Recall the original Dirac equation [Eq. (7.15)] and its Weyl representation [Eq. (7.19)], which diagonalize the matrix γ_5, displaying the chiral components. In order to derive Planck's law from the Dirac equation, it is best to write the corresponding Lagrangian equation as follows:

$$\begin{aligned} L &= \hbar c\left\{\overline{\psi}\gamma_\mu\left(\frac{1}{2}[\partial_t] + i\frac{e}{\hbar c}A_\mu\right)\psi + \kappa_0\overline{\psi}\psi\right\}; \quad \{[\partial] = (\partial\rightarrow) - (\leftarrow\partial)\} \\ &= \frac{\hbar c}{i}\left\{\xi^+ + \left(\frac{1}{2c}[\partial_t] - \frac{ie}{\hbar c}V\right)\xi - \xi^+ s.\left(\frac{1}{2}[\nabla] + \frac{ie}{\hbar c}\mathbf{A}\right)\xi \right. \\ &\quad \left. + \eta^+\left(\frac{1}{2c}[\partial_t] - \frac{ie}{\hbar c}V\right)\eta + \eta^+ s\left(\frac{1}{2}[\nabla] + \frac{ie}{\hbar c}\mathbf{A}\right)\eta + i\kappa_0\left(\xi^+\eta + \eta^+\xi\right)\right\}, \end{aligned} \tag{7.63}$$

with $\kappa_0 = \frac{m_0 c}{\hbar}$. This determines in turn the energy-momentum tensor, and thus the following energy density:

$$\begin{aligned} E &= \frac{\partial L}{\partial(\partial_t\psi)}(\partial_t\psi) + \left(\partial_t\overline{\psi}\right)\frac{\partial L}{\partial\left(\partial_t\overline{\psi}\right)} - L \\ &= \frac{\partial L}{\partial(\partial_t\xi)}(\partial_t\xi) + \left(\partial_t\xi^+\right)\frac{\partial L}{\partial\left(\partial_t\xi^+\right)} + \frac{\partial L}{\partial(\partial_t\eta)}(\partial_t\eta) + \left(\partial_t\eta^+\right)\frac{\partial L}{\partial\left(\partial_t\eta^+\right)} - L. \end{aligned} \tag{7.64}$$

Since the equations of the motion imply that L vanishes (a point to which we will return later in this discussion), this density can be rewritten as

$$E = \frac{\hbar}{2i}\left(\overline{\psi}\partial_t\psi - \partial_t\overline{\psi}\psi\right) = \frac{\hbar}{2i}\left\{\left(\xi^+\partial_t\xi - \partial_t\xi^+\xi\right) + \left(\eta^+\partial_t\eta - \partial_t\eta^+\eta\right)\right\}. \tag{7.65}$$

Consider a stationary wave

$$\psi = e^{i\omega t}\phi(\mathbf{r}), \tag{7.66}$$

with energy density

$$E = \phi^+\phi \ \hbar\omega = \left(\xi^+\xi + \eta^+\eta\right)\hbar\omega. \tag{7.67}$$

Assuming that the wave function is normalized, we find after integrating over the whole space,

$$E = \hbar\omega, \tag{7.68}$$

confirming that Planck's laws follow from the Dirac equation. Let us now compute the plane waves in the Weyl representation [Eq. (7.19)] with no external field; to this end, set

$$\xi = ae^{i(\omega t - \mathrm{k.r})}; \quad \eta = be^{i(\omega t - \mathrm{k.r})}, \tag{7.69}$$

in which ω, **k**, *a,* and *b* are constants (*a* and *b* being spinors). We then get

$$\begin{aligned} \left(\frac{\omega}{c} + \mathbf{s.k}\right)a + k_0 b = 0 \\ \left(\frac{\omega}{c} - \mathbf{s.k}\right)a + k_0 b = 0. \end{aligned} \tag{7.70}$$

Setting the determinant equal to 0 yields the dispersion relation in the following form:

$$\left(\frac{\omega}{c}\right)^2 = k^2 + k_0^2; \quad \omega = 2\pi\nu; \quad \mathbf{k} = \left(\frac{2\pi}{\lambda}\right)\mathbf{n}, \tag{7.71}$$

with frequency ν and wavelength λ (with **n** as a unit vector). Multiplying the phase in Eq. (7.69) by Planck's constant and using Eq. (7.68), we find

$$\hbar\omega t - \hbar\mathbf{k.r} = Et - \left(\frac{h}{\lambda}\right)\mathbf{n.r}. \tag{7.72}$$

However, by definition, plane waves belong to the domain of geometrical optics, and the connection with quantum mechanics is effected by identifying this phase with the action integral of Hamiltonian classical mechanics. In other words, $\hbar\mathbf{k}$ represents a Lagrange momentum $\mathbf{p}$, which immediately yields de Broglie's wavelength formula:

$$\lambda = \frac{h}{p}. \tag{7.73}$$

This formula is thus included, along with Planck's law, in the Dirac equation. Finally, let us multiply the dispersion relation [Eq. (7.71)] by $\hbar^2$, taking Eq. (7.73) into account. We retrieve the expression of the energy E, whence come the Hamilton-Jacobi equation and the semiclassical approximation, expressing E and $\mathbf{p}$ as the space-time gradient of the action:

$$\left(\frac{E}{c}\right)^2 = \mathbf{p}^2 + m_0^2c^2. \tag{7.74}$$

At this point, these results depend heavily on the *linearity* of the Dirac equation. In particular, in order to get Planck's law, we used the vanishing of the Lagrangian expression in Eq. (7.64), which is a consequence of linearity; or, equivalently, that this Lagrangian is quadratic. Besides, one then needs to integrate Eq. (7.71) over space, which requires the wave function to be normalized—another consequence of linearity.

One could be led to think that linearity is not really necessary and that one could accommodate a first-degree homogeneous equation, which makes it possible to normalize the wave function and entails the vanishing of the Lagrangian along any solution. However, this is not sufficient to recover de Broglie's wavelength formula. Indeed, to make sense of Eq. (7.73) and recover the Hamilton-Jacobi equation, it is first necessary that Eq. (7.71) give the general dispersion relation and that it be identified (to within a term of order h^2) with the expression [Eq. (7.74)] of the energy. But Eq. (7.74) is imposed by relativistic considerations, and Eq. (7.71) agrees with it for the Dirac equation because iterating the latter leads to the Klein-Gordon equation, which was built out of Eq. (7.74). Note that this is not a fortuitous coincidence—rather, it was built in from the very beginning. It is thus clear that this will not occur for a nonlinear equation, if only because the dispersion relation will be different: it may happen that some solutions are admissible in the semiclassical approximation, but that will almost never be the case for the general solution.

Besides, being in accordance with quantum mechanics requires more than Planck's law and de Broglie's formula. One still needs the so-called superposition principle, which by definition will not hold true in the nonlinear case, except asymptotically in regions with vanishingly small wave function and a weak nonlinearity. This is what de Broglie was counting on in his theory of what he called "*onde à bosse*" (one-bump wave), whose goal was to describe the link between the wave and the particle by representing the latter as an intense region of the wave.

We are thus led to the conclusion that in general, when looking for *nonlinear* wave equations, one is drifting away from quantum mechanics, even in the semiclassical approximation. It thus seems naive to look for a nonlinear version of quantum mechanics that could replace and improve the one we are used to. Nonlinear equations tell another story altogether, and their connection with quantum or even classical mechanics can be at best asymptotic. The search for such equations can thus be meaningful only inasmuch as one is determined to make a foray into unchartered territories where quantum mechanics itself does not venture, such as the structure of particles, the connection between waves and particles, the description of quantum transitions, etc.

7.10 NONLINEAR SPINORIAL EQUATIONS AND THEIR SYMMETRIES

Now we are going to partially extend to nonlinear equations the results of this chapter thus far, confining our attention to relativistic Lagrangian equations in which only the mass term is nonlinear while a linear differential part is retained. We start with the general form of the Lagrangian equation in the electric case:

$$\begin{aligned} L_e &= L_D + \hbar c F(\omega_1, \omega_1) \\ &= \hbar c \left\{ \overline{\psi} \gamma_\mu \left(\frac{1}{2} [\partial_\mu] + i \frac{e}{\hbar c} A_\mu \right) \psi + F(\omega_1, \omega_2) \right\}, \end{aligned} \tag{7.75}$$

in which L_D is the differential part of Dirac's Lagrangian and $F(w_1, W_2)$ is an as-yet-arbitrary function, with dimension L^{-1} being the inverse of a length. We insist that this nonlinear term is the most general one possible and that it subsumes all terms of the following form:

$$F(J_\mu J_\mu); \ F\Big(\sum_\mu \sum_\mu\Big); \ F\big(M_{\mu\nu}\tilde{M}_{\mu\nu}\big); \ F\Big(J_\mu \sum_\nu M_{\mu\nu}\Big) \text{etc.} \tag{7.76}$$

We now define *chiral invariance* via the magnetic gauge transform, which we have already discussed at length, namely

$$\psi \to e^{i\gamma_5\theta/2}\psi \quad (\theta = \text{Const.}). \tag{7.77}$$

Now recall the Weyl transform diagonalizing the matrix γ_5 and splitting Eq. (7.77) into two transformations that exchanged the chiral components (ξ, η) of ψ:

$$\xi \to e^{i\theta/2}\xi; \quad \eta \to e^{-i\theta/2}\eta. \tag{7.78}$$

Performing this transformation leaves the tensor quantities X_μ, Y_μ, J_μ, and $\sum_\mu$ invariant since they do not contain mixed ξ-η terms. But the ω_i's $(i = 1, 2)$ do, and that gauge transformation indeed induces a rotation of the (ω_1, ω_2)-plane by an angle θ. Namely, we have (Lochak, 1983):

$$\begin{pmatrix} \omega'_1 \\ \omega'_2 \end{pmatrix} = \begin{pmatrix} \cos\theta & -\sin\theta \\ \sin\theta & \cos\theta \end{pmatrix} \begin{pmatrix} \omega_1 \\ \omega_2 \end{pmatrix}. \tag{7.79}$$

This plane is chiral because whereas ω_1 is a relativistic invariant, ω_2 is pseudo-invariant such that the sense of rotation changes with parity, and θ is also a pseudo-invariant, in contrast with the phase angle of the electron. This is a fundamental difference between electricity and magnetism, with which Maxwell and Pierre Curie were fully familiar in the framework of classical physics.

Define the polar coordinates as ρ and the angle A (not to be confused, of course, with the Lorentz potential); here again, ρ is invariant and A is pseudo-invariant:

$$\rho = \sqrt{\omega_1^2 + \omega_2^2} = 2\sqrt{(\xi + \eta)(\eta + \xi)}; A = \arctan\frac{\omega_2}{\omega_1} \tag{7.80}$$

The transformation [Eq. (7.78)] now amounts to the rotation $A \to A + \theta$, and we see that the chiral-invariant equations are derived from a Lagrangian expression that is invariant under rotations of the chiral plane: the nonlinear term depends on ρ only, not on A:

$$F(\omega_1, \omega_2) = F\left(\sqrt{\omega_1^2 + \omega_2^2}\right) = F\left(2\sqrt{(\xi + \eta)(\eta + \xi)}\right) = F(\rho). \tag{7.81}$$

Now recall the linear equations [Eqs. (7.33) and (7.35)] for the magnetic monopole. We thus find that the general nonlinear chiral-invariant Lagrangian expression in the Weyl representation reads as follows (Lochak, 1983, 1984, 1985, 1987a,b):

$$
\begin{aligned}
L_m &= \hbar c\left[\overline{\psi}\gamma_\mu\left(\frac{1}{2}[\partial_\mu] - \frac{g}{\hbar c}\gamma_5 B_\mu\right)\psi + F(\rho)\right] \\
&= \frac{\hbar c}{i}\left\{\xi^+\left(\frac{1}{2}\frac{1}{c}[\partial_t] - \frac{g}{\hbar c}W\right)\xi - \xi^+ s.\left(\frac{1}{2}[\nabla] + \frac{g}{\hbar c}\mathbf{B}\right)\xi\right. \\
&\quad \left. + \eta^+\left(\frac{1}{2}\frac{1}{c}[\partial_t] + \frac{g}{\hbar c}W\right)\eta + \eta^+ s.\left(\frac{1}{2}[\nabla] - \frac{g}{\hbar c}\mathbf{B}\right)\eta + iF(\rho)\right\},
\end{aligned}
\tag{7.82}
$$

with ρ as in Eq. (7.80) and an arbitrary function F. From this Lagrangian, one first derives in the Dirac representation the following equation:

$$
\gamma_\mu\left(\partial_\mu - \frac{g}{\hbar c}\gamma_5 B_\mu\right)\psi + k(\rho)\frac{\omega_1 - i\gamma_5\omega_2}{\sqrt{\omega_1^2 + \omega_2^2}}\psi = 0, \tag{7.83}
$$

where k is the derivative of F: $k = F'(\rho)$. In the Weyl representation, this reads as

$$
\begin{aligned}
&\frac{1}{c}\frac{\partial\xi}{\partial t} - \mathbf{s}.\nabla\xi - i\frac{g}{\hbar c}(W + \mathbf{s}.\mathbf{B})\xi + ik(\rho)\sqrt{\frac{\eta + \xi}{\xi + \eta}}\eta = 0 \\
&\frac{1}{c}\frac{\partial\eta}{\partial t} + \mathbf{s}.\nabla\eta + i\frac{g}{\hbar c}(W - \mathbf{s}.\mathbf{B})\eta + ik(\rho)\sqrt{\frac{\xi + \eta}{\eta + \xi}}\xi = 0.
\end{aligned}
\tag{7.84}
$$

So these are the most general possible equations for a magnetic monopole. Note that the mass term of each of the equations in Eq. (7.84) has the same phase as the corresponding differential term, indicating a phase independence between the chiral components ξ and η, which is absent from the Dirac equation. This extra degree of freedom comes from the chiral invariance, which the Dirac equation does not possess, and this is a fundamental difference. Eq. (7.84) leaves magnetism invariant, whereas the Dirac equation does not, at least in general. Only a subset of its solutions enjoys this property—namely, those that satisfy the so-called Majorana condition (Lochak, 1992). These solutions live on the light cone and leave the norms of the isotropic currents X_μ and Y_μ invariant.

On the other hand, if one introduces a term representing magnetic interaction in the Dirac equation, one does not get an equation for a magnetic monopole (except in special cases). And if one replaces the magnetic interaction by an electric one in an equation of the same type as Eq. (7.83), the result does not represent an electron [Daviau (2005), Daviau & Lochak (1991)].

We will now examine two particular cases of Eqs. (7.83) and (7.84), starting with the *cubic equation* that is derived from the Lagrangian [i.e., Eq. (7.82)] with a fourth-degree term that is, in the absence of an external field:

$$\begin{aligned} L &= \frac{\hbar c}{2}\left(\overline{\psi}\gamma_\mu[\partial_\mu]\psi \pm l^2\rho^2\right) \quad (l = \text{Const.}) \\ &= \frac{\hbar c}{2i}\left(\xi^+\frac{1}{c}[\partial_t]\xi - \xi^+\mathbf{s}.[\nabla]\xi + \eta^+\frac{1}{c}[\partial_t]\eta + \eta^+\mathbf{s}.[\nabla]\eta \pm il^2\rho^2\right). \end{aligned} \tag{7.85}$$

Bearing in mind the definitions of ω_1 and ω_2, this leads to several equivalent forms of the corresponding equation:

$$\gamma_\mu\partial_\mu\psi ml^2\left(\overline{\psi}\gamma_5\gamma_\mu\psi\right)\gamma_5\gamma_\mu\psi = 0 \tag{7.86a}$$

$$\gamma_\mu\partial_\mu\psi \pm l^2\left(\overline{\psi}\gamma_\mu\psi\right)\gamma_\mu\psi = 0 \tag{7.86b}$$

$$\gamma_\mu\partial_\mu\psi \pm l^2\left[\overline{\psi}\psi - \left(\overline{\psi}\gamma_5\psi\right)\gamma_5\right]\psi = 0 \tag{7.86c}$$

$$\begin{aligned} &\frac{1}{c}\frac{\partial\xi}{\partial t} - \mathbf{s}.\nabla\xi \pm 2il^2\left(\eta^+\xi\right)\eta = 0 \\ &\frac{1}{c}\frac{\partial\eta}{\partial t} - \mathbf{s}.\nabla\eta \pm 2il^2\left(\xi^+\eta\right)\xi = 0. \end{aligned} \tag{7.86d}$$

The first one appears in Heisenberg, (1953,1954), and the second in Finkelstein, Lelevier, and Ruderman (1951), both in particle physics. Heisenberg's equation was later obtained in Rodichev (1961) in a space with torsion. The first three equations are particular cases of the monopole equation (Lochak 1987a,b, 2007a,b). The last equation is nothing but the Weyl representation of any of the other three.

We now come to our second special case of Eqs. (7.83) and (7.84)—namely, the *homogeneous equation*. Here, we choose a linear (with respect to ρ) mass term known as $m_0(\rho)$, but the equation is still nonlinear in ψ:

$$F(\rho) = \kappa_0\rho; \quad \kappa(\rho) = \kappa_0 = \text{Const.} \tag{7.87}$$

So we can rewrite Eq. (7.84) with $k = k_0$, a constant:

$$\begin{aligned}
&\frac{1}{c}\frac{\partial \xi}{\partial t}\xi - \mathbf{s}\cdot\nabla\xi - i\frac{g}{\hbar c}(W + \mathbf{s}\cdot\mathbf{B})\xi + i\kappa_0\sqrt{\frac{\eta + \xi}{\xi + \eta}}\eta = 0 \\
&\frac{1}{c}\frac{\partial \eta}{\partial t} - \mathbf{s}\cdot\nabla\eta + i\frac{g}{\hbar c}(W - \mathbf{s}\cdot\mathbf{B})\eta + i\kappa_0\sqrt{\frac{\xi + \eta}{\eta + \xi}}\xi = 0.
\end{aligned} \tag{7.88}$$

This is homogeneous of degree 1, with a corresponding second-degree Lagrangian:

$$L_m = \frac{\hbar c}{i}\left\{\begin{array}{l}
\xi^+\left(\frac{1}{2}\frac{1}{c}[\partial t] - \frac{g}{\hbar c}W\right)\xi - \xi^+ s\cdot\left(\frac{1}{2}[\nabla] + \frac{g}{\hbar c}\mathbf{B}\right)\xi \\
+\eta^+\left(\frac{1}{2}\frac{1}{c}[\partial t] + \frac{g}{\hbar c}W\right)\eta + \eta^+ s\cdot\left(\frac{1}{2}[\nabla] + \frac{g}{\hbar c}\mathbf{B}\right)\eta \\
+2i\kappa_0\sqrt{(\xi + \eta)(\eta + \xi)}
\end{array}\right\}.$$

Adding in an electric interaction, the equation reads as

$$\begin{aligned}
&\frac{1}{c}\frac{\partial \xi}{\partial t} - \mathbf{s}\cdot\nabla\xi - i\frac{e}{\hbar c}(V + \mathbf{s}\cdot\mathbf{A})\xi + i\kappa_0\sqrt{\frac{\eta + \xi}{\xi + \eta}}\eta = 0 \\
&\frac{1}{c}\frac{\partial \eta}{\partial t} - \mathbf{s}\cdot\nabla\eta + i\frac{e}{\hbar c}(V - \mathbf{s}\cdot\mathbf{A})\eta + i\kappa_0\sqrt{\frac{\xi + \eta}{\eta + \xi}}\xi = 0.
\end{aligned} \tag{7.90}$$

This particular case was studied by Daviau and Lochak (1991). The Lagrangian equation vanishes by virtue of the equations of the motion, just as in the linear case, and one gets Planck's law again. Some—but not all—solutions also yield de Broglie's wavelength formula and a correct semiclassical approximation. However, the equation does not account for particle-antiparticle pairs.

In closing, we return to a recurrent theme of this chapter: namely, *symmetries,* which we now study in the nonlinear framework, starting with the general Lagrangian equations [Eqs. (7.75) and (7.82)], with linear parts L_D and L_M. We first obtain Table 7.3, which is derived from Tables 7.1 and 7.2.

One finds that the electromagnetic potentials are now C invariant because they are *external* to the system instead of being *emitted*, as in Table 7.2 where they changed signs. The variance of the linear parts L_D and L_M of the Dirac and monopole Lagrangians are given in Table 7.4.

Taking into account the variances of ω_1 and ω_2 as displayed in Table 7.3, we will find those of the nonlinear terms $F(\omega_1,\omega_2)$. Here are some examples;

Table 7.3

		P	T	C
ψ	$\rightarrow$	$\gamma_4\psi$	$-i\gamma_3\gamma_1\psi^*$	$\gamma_2\psi^*$
ξ	$\rightarrow$	η	$s_2\xi^*$	$-is_2\eta^*$
η	$\rightarrow$	ξ	$s_2\eta^*$	$is_2\xi^*$
X_4	$\rightarrow$	Y_4	X_4	Y_4
$\mathbf{X}$	$\rightarrow$	$-\mathbf{Y}$	$-\mathbf{X}$	$\mathbf{Y}$
Y_4	$\rightarrow$	X_4	Y_4	X_4
$\mathbf{Y}$	$\rightarrow$	$-\mathbf{X}$	$-\mathbf{Y}$	$\mathbf{X}$
J_4	$\rightarrow$	J_4	J_4	J_4
$\mathbf{J}$	$\rightarrow$	$-\mathbf{J}$	$-\mathbf{J}$	$\mathbf{J}$
Σ_4	$\rightarrow$	$-\Sigma_4$	Σ_4	$-\Sigma_4$
Σ	$\rightarrow$	Σ	$-\Sigma$	$-\Sigma$
ω_1	$\rightarrow$	ω_1	ω_1	$-\omega_1$
ω_2	$\rightarrow$	$-\omega_2$	$-\omega_2$	$-\omega_2$
e	$\rightarrow$	e	$-e$	$-e$
g	$\rightarrow$	g	g	g
V	$\rightarrow$	V	$-V$	V
$\mathbf{A}$	$\rightarrow$	$-\mathbf{A}$	$\mathbf{A}$	$\mathbf{A}$
W	$\rightarrow$	$-W$	W	W
$\mathbf{B}$	$\rightarrow$	$\mathbf{B}$	$-\mathbf{B}$	$\mathbf{B}$

Table 7.4

		P	T	C
L_D	$\rightarrow$	L_D	L_D	$-L_D$
L_M	$\rightarrow$	L_M	L_M	$-L_M$

the variances of the corresponding equations are obtained by combining the information from Tables 7.4 and 7.5.

From Table 7.5, one finds that the three invariances C, P, and T are rarely simultaneously respected; for this to happen, the nonlinear term must be P and T invariant and change sign under C, as do ω_1 and the differential part of the Lagrangian (see Table 7.3). In particular, Heisenberg's equation and that of the magnetic monopole are not C invariant because this invariance is actually *incompatible* with chiral invariance, for the somewhat paradoxical reason that this last invariance entails too much symmetry. Indeed, the mass term of the nonlinear Lagrangian expression is chiral invariant, as a function of the radius ρ, and is thus P, T, and C invariant. For the *equation* to be C invariant, however, one would need the mass term to change sign in the C transform, as previously mentioned. In that sense, this mass term can indeed be called too symmetric. But the P and T

Table 7.5

		P	T	C
ω_1	$\rightarrow$	ω_1	ω_1	$-\omega_1$
ω_2	$\rightarrow$	$-\omega_2$	$-\omega_2$	$-\omega_2$
$F = \omega_1$ (Dirac)	$\rightarrow$	$+F$	$+F$	$-F$
$F = \omega_1^2$	$\rightarrow$	$+F$	$+F$	$+F$
$F = \omega_1^3$	$\rightarrow$	$+F$	$+F$	$-F$
$F = G(\omega_1, \omega_2)\,\omega_1, (G > 0)$	$\rightarrow$	$+F$	$+F$	$-F$
$F = F(\omega_1^2 + \omega_2^2)$	$\rightarrow$	$+F$	$+F$	$+F$
$F = F(\omega_1^2 + \omega_2^2) = -(i\overline{\Psi}\gamma_5\gamma_\mu\Psi)(i\overline{\Psi}\gamma_5\gamma_\mu\Psi)$	$\rightarrow$	$+F$	$+F$	$+F$
$F = (\omega_1, \omega_2)$	$\rightarrow$	$-F$	$-F$	$+F$

invariances are still verified, with the definition of T as in Table 7.3. As a result, these chiral-invariant equations do *not* obey the CPT theorem.

The fact that an equation is not C invariant does not by itself preclude the existence of solutions with both signs for the energy, but these do not describe particle-antiparticle pairs as they do in the Dirac equation. Curiously enough, such pairs could exist because of T invariance, with one element going forward in time and the other backward. But the pairs defined via charge conjugation and time inversion, respectively, are not of the same nature. In the first case, the two elements have opposite chiralities and move in the same direction with respect to the course of time, whereas in the second case, both elements have the same chirality but move in opposite time directions (see Table 7.3).

Finally, we would like to conclude by stating that nonlinearity is a complicated matter (an assertion that few would dispute). More information on this subject can be found in Lochak (1997a,b).

CHAPTER 8

A Catalytic Nuclear Fusion Arising from Weak Interaction

8.1 MAIN IDEAS

In this chapter, we shall examine a possible way to avoid the superhigh temperatures that are generally introduced in nuclear fusion experiments in order to overcome the Coulomb barrier between electrically charged particles of the same sign. We start from the observation that although temperatures are very high in the middle of stars, they are lower than the temperatures used in terrestrial experiments in nuclear fusion. We suggest the possibility of a catalyst that could be present in stars and absent from our experiments. It will be shown that neutrinos in the stars could play the role of this catalyst since they are abundant and subject to weak interactions, which makes them able to play this role. But they cannot do so in terrestrial experiments, first because there are too few neutrinos, and second because since they have no charge, they are scattered in all directions. Nevertheless, instead of neutrinos, we have at our disposal leptonic magnetic monopoles, which were at first theoretically discovered and described and have now for many years been experimentally produced, observed, and applied. These leptonic monopoles have the same weak interaction properties as neutrinos and could accelerate such phenomena as a proton-proton fusion at temperatures that are far lower than the temperatures used in other attempts at nuclear fusion. In addition, they can easily be focused and accelerated thanks to their magnetic charge, rather than being scattered in space. Thus, they are potentially able to modify the problem of nuclear fusion.

8.2 INTRODUCTION

One of the most difficult problems in applied physics today is the industrial development of nuclear fusion energy. As is well known, the current attempts in this area are based on a race to achieve giant temperatures of plasma (several hundreds of million degrees) in order to increase the velocity of nucleons with the same charge sign (positive or negative), to the extent that they are brought close enough to each other to overcome Coulomb repulsion. But there are problems with this approach.

First, there is no material enclosure that can survive such temperatures. The tokamak, a device invented by Russian physicists Igor Tamm and Andrei Sakharov, seemed to provide an answer: in it, electrically charged particles are forced to rotate around a magnetic field to prevent them from approaching the walls. But the tokamak becomes unstable and works only over brief intervals of time; thus, the problem is not solved.

The message of this chapter is that high temperatures must be abandoned. But how can we do this? Let us start with a few opening remarks:

1. Astrophysicists have discovered that although temperatures inside stars are very high, they are lower than those used in the terrestrial research installations for nuclear fusion. So it is natural to ask whether there is a catalyst in the stars that boosts fusion, but which would be absent from terrestrial laboratories.
2. What could such a catalyst be? It is interesting to observe that weak interactions constitute, in some manner, an obstacle to the strong interactions that provide the energy of stars. For instance, they are in a position to slow down the carbon and hydrogen cycles. Thus, one can ask if, conversely, they could speed up the strong interactions, and whether we could use this property that results from this.
3. The weak interactions that occur in the known astronomical cycles lead to the emission of neutrinos. Conversely, many antineutrinos in the stars are produced by the β disintegration of free neutrons:

$$n \rightarrow p + e^- + \tilde{\nu}. \tag{1}$$

Consequently, these antineutrinos can be absorbed in inverse β disintegration:

$$p + \tilde{\nu} \rightarrow n + e^+. \tag{2}$$

In this last reaction, an antineutrino coming from outside gives rise to the reaction in Eq. (2): its absorption is equivalent to the forced emission of a neutrino. Such a reaction can boost an entire cycle, the antineutrino playing the role of a quasi-catalyst (which is only "quasi" because it disappears in the reaction).

4. Unfortunately, even if this hypothesis is justified, it cannot be applied in laboratories, which do not have sufficient neutrinos. Furthermore, since neutrinos are neutral, they are diffused in all directions and cannot be focused. This is one of the reasons why the famous 1996 experiment of Reines and Cowan, which proved the existence of the neutrino, was very difficult.
5. Nevertheless, a result found by Ivoilov (2006) lends support to this hypothesis; he showed that when irradiating a β radioactive source with leptonic magnetic monopoles, the lifetime of the source decreases; in other words, the β radioactivity is accelerated.

8.3 A POSSIBLE CATALYST FOR NUCLEAR FUSION

Now, we shall show that a possible catalyst for nuclear fusion may be the leptonic nuclear monopoles that were theoretically predicted and then observed in our group (see Lochak, 1983, 1985, 1997a,b; Lochak, 2007; and this entire book). For now, let us say briefly that these monopoles are very light (even massless in the present theory), contrary to the monopoles described by other authors, which are supposed to be very heavy. Furthermore, they have two main properties:

- They have a *magnetic charge* g which is equal to $137/2$ times the electron charge (in the same Gaussian units).
- They are magnetically excited neutrinos that are subject to the same weak interactions, and are thus able to influence the same nuclear phenomena. The most important observation is that unlike neutrinos, leptonic monopoles can be focused by electromagnetic forces and their energy can be increased in magnetic fields.

Therefore, we can substitute the following reaction to the reaction in Eq. (2):

$$p + \tilde{m} \rightarrow n + e^{+}, \tag{3}$$

where $\tilde{m}$ is an anti-monopole with the same weak interactions as the antineutrino in Eq. (2). Next we suggest a test experiment in order to verify the reaction (3).

8.3.1 Some Remarks

- *Concerning the reaction in Eq. (3)*: It may be objected that energy is not conserved because a proton is lighter than a neutron. But that was already the case for the β inverse formula [Eq. (2)]: such formulas are only written in conformity with quantum rules, and the conservation of energy is admitted for other reasons. Our case is simpler because the monopole may be accelerated in a magnetic field.
- *Concerning temperature:* The temperature required must be enough to create a plasma, which is much lower than the several hundred million degrees needed for other experiments in nuclear fusion.

8.4 A TEST-EXPERIMENT

Leptonic magnetic monopoles generally appear in disruptive electric phenomena in water, such as the following[31]:

1. An explosion in water of a stepped-up electric conductor (Urutskoiev, Moscow, described at a conference in Nantes, France)
2. An electric arc in water (Ivoilov, Kazan)
3. Strong electric sparks in water (Bergher, Fondation Louis de Broglie, Paris)

The monopoles are recorded in different manners, but most often on photographic film. Here are three characteristic tracks: (a) The first is greatly enlarged (characteristic caterpillar form). (b) The second is a track with its image in a germanium mirror; this image is identical to the original rotated by an angle of π, in accordance with the theoretical predictions. (c) The third track was observed on the magnetic north pole; a solar monopole created by a β decay in a strong solar magnetic field. We have hundreds of other different examples of tracks.

By definition, the recorded monopoles were originally present in the source. Thus, these same monopoles were able to accelerate the creation of deuterons, which is the first and principal stage of the hydrogen cycle:

$$p + p \rightarrow {}^2D + e^+ + \nu. \tag{4}$$

But at this point, there occurs a forced reaction due to an antimonopole coming from outside, which then disappears in the reaction, according to the following formula:

$$p + p + \tilde{m} \rightarrow {}^2D + e^+. \tag{5}$$

A possible experiment could be to create, in a container of water, the explosion of an electric conductor (as in Urutskoiev's experiments), or an electric arc (as in Ivoilov's experiments). Then, the proportion of heavy water must increase as the water cools in the container.

I am aware that I am making several hypotheses here, but there is no science without hypotheses. Nevertheless, apart from other possible objections, one important question concerns the number of produced monopoles, and thus the probability of the purported nuclear phenomena.

[31] There are other examples, such as β radioactivity in a magnetic field.

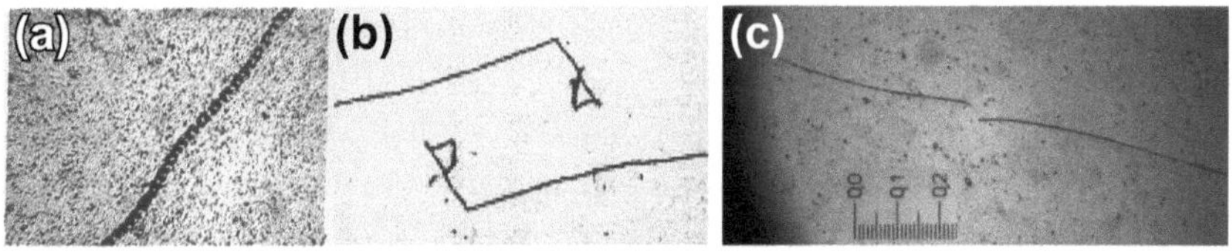

Figure 8.1

Despite that, at first glance, Eq. (5) seems to be easy to realize. However, there is a problem: the number of monopoles (and consequently, the probability of the reaction). If the number of monopoles produced by a source corresponded to the number of trajectories observed in the previous images, this number would be very small and the probability would be negligible.

But an important observation by Ivoilov, confirmed more recently by Daniel and Sonia Fargue (Daviau et al., 2013a,b), shows that apart from the rare large tracks that are generally observed, there are many other tracks which, contrary to the preceding statements, come directly from the direction of the source. These tracks are very thin and are easy to miss. Nevertheless, they have the charge of a monopole, and they probably constitute the hidden but most significant emission of monopoles.

The first to observe these thin tracks (indirectly) was Urutskoiev. When he obtained the first photographs of large monopole tracks on photographic film (curiously, in a plane perpendicular to the direction of the monopole source), he tried to find three-dimensional (3D) tracks in a bubble chamber. But he was disappointed because instead of 3D tracks, he obtained a large white cloud. I now believe that this cloud consisted of the extremely numerous thin tracks later identified by Ivoilov. Actually, the large tracks correspond to deviated monopoles that strike the limit between the plastic film support and the photosensitive coating, and are strongly deviated from their initial trajectory. The large tracks are thus due to a kind of rare accidents, which explains why they are so rare. On the contrary, "true" tracks (which is called "true" because they are free) seem to be the numerous thin tracks. Therefore, we can assert that the emitted monopoles are actually extremely numerous. Thus, it is possible that my test-experiment might be not so difficult to perform. Some other arguments support this idea, among which the macroscopic phenomena observed by Urutskoiev's group after the Chernobyl catastrophe[32], such as the following:

1. In the same hall, there were two parallel conductors separated by some meters: one was a water pipe coming from the nuclear reactor, and the other an electric cable in a concrete box. During the disaster, this concrete protection was broken by the strength of the electric cable strongly attracted by the water pipe. Puzzled by this phenomenon, a young physicist raised the audacious hypothesis that there could be magnetic monopoles in the water coming from the reactor. He meant it as a joke, but actually, it was later confirmed; this implies an enormous quantity of monopoles.
2. The reactor had a concrete cover weighing about 3,000 tons. During the accident, it was pushed aside and fell vertically against the wall of the reactor. If this was due to gas pressure inside the reactor, then a calculation proves that the walls would have exploded, whereas in fact the walls of the reactor were in perfect condition. The paint on the inside of these walls was not even burned, and it would have burned at a temperature of 300°C. Two conclusions emerged from these observations:

   There was no fire in the reactor except in a very limited volume (this was, of course, later verified directly) and thus there was no strong pressure.

   If the cover was lifted, this means that its weight was far less than 3,000 tons during the catastrophe. It was assumed that monopoles have modified gravitation. This was confirmed experimentally by Urutskoiev and theoretically by myself on the basis of my theory of monopoles and of de Broglie's theory of light, which gives a quantum theory of gravitation (de Broglie proved the influence of electromagnetism on gravitation—i.e., Einstein's unitary theory, which he presented in 1942 in his *General Theory of Spin Particles*). All this would be impossible unless an enormous number of monopoles was produced.

[32] It must be emphasized that we absolutely disagree with the official Soviet interpretation of the catastrophe as the result of "errors" by engineers. We have strong arguments in favor of the hypothesis of an explosion of an electrical machine (probably a transformer) that produced a stream of monopoles that then activated a chain of disintegrations never seen before. An important point is that we came up with this hypothesis independently: we met only years later. It must be added that two completely independent groups were sent to Chernobyl to explain the disaster: one by a scientific center, the Kurchatov Institute, under the direction of Urutskoiev, and the other, the official one, in order to find a guilty party. And this is the origin of the official "explanation."

CHAPTER 9

Conclusion

The Foreword of this book presented a brief history of the magnetic monopole. From this, it follows that the monopole is not a particle like others—just another constituent particle of matter. Without its name ever being mentioned, this particular particle was long awaited because it is necessary to the entire concept of electromagnetism. We saw in Chapter 6 that this monopole even plays a role in universal gravitation, in the form of a graviton. If we now take a bird's-eye view of the whole of physics over the last four centuries, we can say that it is dominated by three great ideas: Isaac Newton's universal gravitation, James Clerk Maxwell's electromagnetism, and quantum theory.

And if we want to identify the greatest unachieved dream of the twentieth century, it is most certainly Einstein's Grand Unified Theory, which would have unified the two theories of gravitation and electromagnetism in a single geometric vision of the universe. To this day, such a theory has never been established in the form imagined by its creator. However, although this is little known, Louis de Broglie and Marie-Antoinette Tonnelat[33] gave a different version, based not on geometry but on quantum theory, which emerges from de Broglie's theory of light (see Chapter 6 and de Broglie, 1940–1942, 1949) and his general theory of particles with spin (de Broglie, 1950).

Recall that the idea on which these theories are based is that there exists a fundamental particle of spin ½, defined by Dirac's equation[34], and that the particles of greater spin arise through the fusion of several of these. In particular:

- Two particles of spin ½ give Einstein's photon, of spin 1, which is thus no longer an elementary particle but a composite particle; and de Broglie showed that when the photon is defined this way, it obeys Maxwell's equations. Heisenberg gave as much importance to this discovery of de Broglie's as to the wave properties of the electron (Heisenberg, 1953).

[33] M. A. Tonnelat, an ex-student of de Broglie, was a great specialist in relativity was invited by Schrödinger to Dublin and by Einstein to Princeton.

[34] *Proceedings of the Royal Society, 187,* 610; and *118,* 34; 1928.

- Four particles of spin ½ give the graviton of spin 2, which obeys Einstein's equations of general relativity, unfortunately only in the linear approximation, given the linearity of quantum mechanics. But the most important thing is that the equations of gravitation come together naturally with three of Maxwell's equations of electromagnetism, as in Einstein's vision: it is indeed a version of the Grand Unified Theory, but it is a quantum rather than a geometric theory.
- Furthermore, I recently showed that only two of these photons are Einstein's electric photons (Lochak, 1995a, and Chapter 6 of this volume): the third photon is a magnetic photon that arises in the theory of the monopole. This establishes a link between the magnetic monopole and gravitation.

This book is based on two ideas that form the essence of its content:

- Dirac's electron must be accompanied by another particle, the *leptonic magnetic monopole.*
- Einstein's photon is not unique: it is the fundamental element of a sequence of elecphotons of spins 0, 1, 2, etc... and each electric photon is paired with a magnetic photon.

We already know that the idea of the magnetic monopole—although it was never explicitly named—was expressed by Maxwell and by Pierre Curie. My contribution will be to propose several wave equations, of which one is linear and very simple (1983). At the start of this research, I submitted to Louis de Broglie in his final years (he died in 1987) some preliminary ideas, on the subject of which I will share here a personal anecdote that I have kept to myself until now. One day, he listened to me in silence and then made this single comment: "Too bad Einstein is dead—this would have interested him."

To me, this memory at least partially makes up for the blindness of the scientific community, which is locked into the idea of a very heavy monopole, which has never been observed, and has put up a wall of resistance to my idea, even though it has been repeatedly confirmed by experiments[35]. I can only refer to the saying of Heraclitus: "If you don't seek for the unexpected, you will never know the truth."

Let us now make two remarks on this subject. The idea of the leptonic monopole is hidden in Dirac's equation, in the form of two equations that I

[35] The silence is not absolute, however, since I received from a great institute of nuclear physics the "mathematical proof" that my monopole doesn't exist. The "proof" only had two flaws: (1) It did not start with the correct equation; and (2) It totally neglected the experiments, but that, of course, is a mere detail...

deduced in 1956 thanks to a different formulation of the equation. One of these equations is obvious: it expresses the conservation of electric charge. The other one was incomprehensible at the time: even de Broglie could not decode it, and it took me twenty years to understand that it expresses the conservation of a magnetic quantity which yielded the equation of the monopole.

This first equation leads to a zero mass for the monopole, which must then move in the vacuum at the speed of light, which is not the case for the other two equations found in this book. But it is remarkable that it suffices to write Dirac's usual equation (with a nonzero mass) on the light cone (see Chapter 5), while requiring the current to be isotropic, to render the equation ambivalent, representing both an electron and a monopole.

The second remark concerns de Broglie's theory of light. The original form of the equation that he found is very complicated, but thanks to an algebraic transformation, he was able to deduce (de Broglie, 1950) Maxwell's equation from it with the definition of the fields using potentials and Lorentz's condition. Impressed by this, de Broglie did not look for further algebraic transformations.

It was only recently that I showed that in reality, there are two, and only two, such transformations: de Broglie's and another one (described in Chapter 6). This other transformation also yields Maxwell-like equations, which do not correspond to the electron but to the monopole whose equation I already knew. And they define new covariant derivatives that represent the action of electromagnetism not on an electric charge, but on a magnetic one.

In other words, de Broglie's equations for the photon do not only represent Einstein's photon, but also a second photon of spin 1: namely, a magnetic photon. This second photon introduced by me (and also, independent of me, by Dominique Spehler) is thus not an artificial invention since it was hidden in de Broglie's original equations. It just needed to be found.

It is thus very strange that these two new particles, the magnetic monopole and the magnetic photon, already existed in a hidden form in Dirac's equation for the electron and in de Broglie's equation for light. But I could not show this to de Broglie, who would have understood it in a second and leapt up, interrupting my explanation. Sadly, he had died by then.

Let me add to this what I showed in Chapter 6 concerning the photon of spin 0, which is to my mind a fundamental aspect of de Broglie's theory of light. The application of this to the Aharonov-Bohm effect that I describe shows that this photon of spin 0 represents a kind of closure of the entire theory of light. This circle of ideas constitutes a general theory of electromagnetism, which englobes Einstein's gravitation in an entirely new manner. But it is necessary to state one important criticism: the simple fact—already acknowledged in the theory of de Broglie-Tonnelat—that Einstein's equations of gravitation appear only in the form of linear approximations of general relativity indicates a fundamental incompleteness of the theory.

Unlike the well-known quantum theory, a future theory should start from nonlinear quantum concepts, and the first proof of success would obviously be to find the exact form of general relativity, which should serve as a guide in this research.

As a "conclusion to the conclusion," I would like to add a few further remarks.

The magnetic monopole reveals a duality between electricity and magnetism that Maxwell has already noted in his treaty, since he relied on a double Coulomb's law for electric and magnetic poles. Note that Charles Augustin de Coulomb really established these two laws by measures on electrically charged bodies and at the extremities of long magnetic wires.

We find this duality in what precedes, under two different forms:

- The two different gauges of Dirac's equation, one electric and the other magnetic, which yield the equation of the electron and the equation of the monopole, respectively.
- The second form in which the duality appears is, as we have seen, de Broglie's theory of light and the two photons, electric and magnetic. It must be noted that the magnetic photon already implicitly arose via the second gauge, which gave the equation of the monopole, so that this photon is confirmed by all the experiments on interactions of the monopole with an electromagnetic field, just as Einstein's electric photon is confirmed by the corresponding interactions.

Certainly, there are many less proofs of this duality on the magnetic side than on the electric side, but then they are separated by a century. And it must also be said that when we speak of duality, it is never absolutely symmetric. Our world, at least in the present state of knowledge, is much more electric than magnetic.

This was already visible in wave-particle duality. The material world is much more particle than wave. We know that Heinrich Hertz thought that cathode rays were waves (and his choice is understandable). But Jean Perrin showed the particle properties of these rays, and that characteristic is at the origin of the discovery of the electron. Indeed, de Broglie showed later that both Perrin and Hertz were right about wave-particle duality.

Nevertheless, the electron really is "more particle than wave," whatever Erwin Schrödinger may think. But for the photon, the contrary holds: light is more wave than particle. It is a question of mass—much smaller for the photon than for the electron—and of spin: ½ for the electron, which is an individualistic fermion, and 1 for the photon, which is a boson prone to collective states.

But what happens with a monopole? The equation suggested in Chapters 1, and 3 describes the essential properties of a leptonic monopole. This equation is inspired by the Dirac equation and the change from the electric to the magnetic particle is characterized by the presence of a γ_5 matrix, absent from the Dirac equation. So that the problem of the kinetic moment is very different from the electric case, especially for the Coulomb law.

We saw in Chapter 3 that when a monopole interacts with an electric Coulomb center, we find no more the Laplace functions with integer idexes, as in the case of a hydrogen atom, but generalized spherical functions, the quantum numbers of which are either integers or half-integers. In the case of an integer, we must add the spin value ½: thus, the kinetic moment becomes a half-integer and the monopole is in a fermion state. But if the quantum number of the kinetic moment is a half-integer (which is possible because of the top-symmetry of the monopole) we must, once more, add a spin value ½, so that the total moment will be an integer, and the monopole is now in a boson state: that is a wholly different case.

This is why the interaction between a monopole and an electric charge is a particular case of electromagnetic interaction because, owing to the presence of half-integer kinetic moments of monopoles, due to their top-symmetry, there is the possibility of transitions between fermion and boson states, which is absolutely forbidden. One example of this is between hydrogen quantum states with photon interactions because the latter are bosons. The possible existence of monopoles in boson states implies the possibility of phenomena of the type of magnetic supraconductivity, which could be very interesting.

As was shown in Chapter 3, another important consequence is that in such an electron-monopole interaction, the conservation of the magnetic current is lost because of the quantification of the product (e.g., of both charges, electric and magnetic), and possible jumps between the quantum values. It signifies that, if we admit the general conservation of electricity as seems to be true in all the experiments to date, we must admit that the magnetic charge is not conserved in this special case. It is quantized, and there must be quantum transitions between quantum states. In particular, if the magnetic charge was initially equal to zero, we shall observe the birth of a magnetic monopole. The description of such a phenomenon needs a new equation, which does not seem to be hard to build.

REFERENCES

Aharonov, Y., & Bohm, D. (1959). Significance of electromagnetic potentials in the quantum theory. *Physical Review, 115*, 485–491.

Akhiezer, A. I., & Berestetsky, V. B. (1965). *Quantum Electrodynamics*. New York: Interscience.

Bachelot, A. (1988a). *Global Existence of Large-Amplitude Solutions to Nonlinear Wave Equations in inkowski Space*. Preprint N°8802 UER de Math Bordeaux I.

Bachelot, A. (1988b). *Mémoire d'Habilitation*. Université de Bordeaux I.

Bardout, G., Lochak, G., & Fargue, D. (2007). Sur la présence de monopôles magnétiques légers au Pôle nord. *Annales de la Fondation Louis de Broglie, 32*, 551–553.

Bargmann, V., & Wigner, E. P. (1948). Group theoretical discussion of relativistic wave equations. *Proceedings of the National Academy of Sciences of the United States of America, 34*, 211–223.

Barrett, T. W. (1989). On the distinction between fields and their metrics. *Annales de la Fondation Louis de Broglie, 14*, 37–75.

Bass, L., & Schrödinger, E. (1955). Must the photon mass be zero? *Proceedings of the Royal Society of London A, 232*, 1–6.

Berestetsky, V. B., Lifschitz, E. M., & Pitaevsky, L. P. (1968). Théorie quantique relativiste, *Nauka, French translation Mir, Moscou 1972.*

Birkeland, K (1896). Sur les rayons cathodiques sous l'action de forces magnétiques intenses. *Archives Sciences Physiques et Naturelles, 1*, 497–512.

Bohm, D. (1960). *Quantum Theory*. London: Prentice Hall.

Borne, T., Lochak, G. H., & Stumpf, H. (2001). *Nonperturbative Quantum Field Theory and the Structure of Matter*. Dordrecht: Kluwer.

Broglie, L. de (1922). Rayonnement noir et quanta de lumière. *Journal de Physique Radium Série, 3*, 422–428.

Broglie, L. de (1932a). Sur une analogie entre l'électron de Dirac et l'onde électromagnétique. *Comptes rendus de l'Academie des sciences, 195*, 536–537.

Broglie, L. de (1932b). Sur le champ électromagnétique d l'onde lumineuse. *Comptes rendus de l'Academie des sciences, 195*, 862–864.

Broglie, L. de (1932c). Sur la densité de l'énergie dans la théorie de la lumière. *Comptes rendus de l'Academie des sciences, 197*, 1377–1380.

Broglie, L. de (1934a). Sur la nature du photon. *Comptes rendus de l'Academie des sciences, 198*, 135–138.

Broglie, L. de (1934b). L'équation d'onde du photon. *Comptes rendus de l'Academie des sciences, 199*, 445–448.

Broglie, L. de (1934c). *Une nouvelle conception de la lumière*. Paris: Hermann.

Broglie, L. de (1936). *Nouvelles recherches sur la lumière*. Paris: Hermann.

Broglie, L. de (1940–1942). Une nouvelle théorie de la lumière, la mécanique ondulatoire du photon *(Tome I:* La lumière dans le vide*)* (1940–1942). Paris: Hermann (Tome II: *L'interaction entre les photons et la matière*); Hermann, Paris (1942).

Broglie, L. de (1943). *La théorie générale des particules à spin*. Paris: Gauthier-Villars.

Broglie, L. de (1949). *Mécanique ondulatoire du photon et théorie quantique des champs*. Paris: Gauthier-Villars.

Broglie, L. de (1950). *Optique ondulatoire et corpusculaire*. Paris: Hermann.

Cabibbo, N., & Ferrari, G. (1962). Quantum electrodynamics with Dirac monopoles. *Nuovo Cimento, 23*, 1147–1154.

Cabrera, B. (1982). First results from a superconductive detector for moving magnetic monopoles. *Physics Review Letters, 48*, 1378–1381.

Cartan, E. (1938). *Leçons sur la théorie des spineurs*. Paris: Hermann.

Case, K. M. (1957). Reformulation of the Majorana theory of the neutrino. *Physics Review, 107*, 307–316.
Costa de Beauregard, O. (1943). *Contribution à l'étude de la théorie de l'électron de Dirac*. Paris: Thèse.
Costa de Beauregard, O. (1997b). Induced electromagnetic inertia and physicality of the four-vector potential. *Physics Essays, 10*, 646–650.
Costa de Beauregard, O. (1983). CPT revisited. In S. Diner, D. Fargue, G. Lochak, & F. Selleri (Eds.), *The Wave–Particle Dualism* (pp. 495–499). Dordrecht: Reidel.
Curie, P. (1894a). Sur la symétrie dans les phénomènes physiques, symétrie d'un champ électrique et d'un champ magnétique. *Journal de Physique, 3*, 393–415.
Curie, P. (1894b). Sur la possibilité d'existence de la conductibilité magnétique et du magnétisme libre. *Journal de Physique, 3*, 415–417.
Curie, P. (1994). *Preceding papers republished in the Annales de la Fondation Louis de Broglie. 19* pp. 137–157). and 159–160.
Daviau, C. (2005). Equation d'onde relativiste linéaire du second ordre pour un monopôle magnétique et ses solutions dans le cas coulombien. *Annales de la Fondation Louis de Broglie, 30*, 71–84.
Daviau, C., & Lochak, G. (1991). Sur un modèle d'équation spinorielle non linéaire. *Annales de la Fondation Louis de Broglie, 16*, 43–71.
Daviau, C., Fargue, D., Priem, D., & Racineux, G. (2013a). Tracks of magnetic monopoles. *Annales de la Fondation Louis de Broglie, 38*, 139–153.
Daviau, C., Priem, D., & Racineux, G. (2013b). Experimental report on magnetic monopoles. *Annales de la Fondation Louis de Broglie, 38*, 189–194.
Dirac, P. A. M. (1931). Quantised singularities in the electromagnetic field. *Proceedings of the Royal Society of London A, 133*, 60–72.
Dirac, P. A. M. (1936). Relativistic wave equations. *Proceedings of the Royal Society of London A, 155*, 447–459.
Dürr, H., Heisenberg, W., Mitter, H., Schlieder, S., & Yamakaki, K. (1959). Zur Theorie der Elementarteilchen. *Zeitschrift für Naturforschung, 14a*, 441–485.
Ehrenberg, W., & Siday, R. E. (1949). The Refractive index in electron optics and the principles of dynamics. *Proceedings of the Physical Society B, 62*, 8–21.
Einstein, A. (1916). Näherungsweise Integration der Feldgleichungen der Gravitation. *Sitzungsberichte der Preussische Akademie Wissenschaften, 1*, 688–696.
Einstein, A. (1918). Über Gravitationswellen. *Sitzungsberichte der Preussische Akademie Wissenschaften, 1*, 154–167.
Einstein, A. (1925). Elektron und allgemeine Relativitätstheorie. *Physica, 5*, 330–334.
Fierz, M., & Pauli, W. (1939a). On relativistic field equations for arbitrary spin particles in an electromagnetic field. *Helvetica Acta, 12*, 297–300.
Fierz, M., & Pauli, W. (1939b). On relativistic field equations for arbitrary spin particles in an electromagnetic field. *Proceedings of the Royal Society of London A, 173*, 211–232.
Finkelstein, R., Lelevier, P., & Ruderman, M. (1951). Nonlinear spinor fields. *Physics Review, 16*, 326–332.
Gelfand, I. M., Minlos, R. A., & Shapiro, Z. Y. (1963). *Representation of the rotation and Lorentz groups and applications*. New York: Pergamon Press.
Heisenberg, W. (1953). Remarques sur la théorie neutrinienne de la lumière. *Louis de Broglie: Physicien et penseur* (pp. 283–286). Paris: Albin Michel.
Heisenberg, W. (1954). Zur Quantentheorie nichtrenormalisierter Wellengleichungen. *Zeitschrift für Naturforschung, 9a*, 292–303.
Heisenberg, W. (1966). *Introduction to the Unified Field Theory of Elementary Particles*. London: Interscience.
Ince, E. L. (1956). *Ordinary Differential Equations*. York: Dover, New.

Ivoilov, N. G. (2006). Low-energy generation of "strange" radiation. *Annales de la Fondation Louis de Broglie, 31*(1), 115–124.
Jackson, J. D. (1975). *Classical Electrodynamics*. New York: Wiley.
Jakobi, G., & Lochak, G. (1956a). Introduction des paramètres relativistes de Cayley-Klein dans la représentation hydrodynamique de la mécanique ondulatoire. *Comptes rendus de l'Academie des sciences, 243*, 234–238.
Jakobi, G., & Lochak, G. (1956b). Décomposition en paramètres de Clebsch de l'impulsion de Dirac et interprétation physique de l'invariance de jauge des équations de la mécanique ondulatoire. *Comptes rendus de l'Academie des sciences, 243*, 357–360.
Jauch, J. M., & Rohrlich, F. (1955). *The Theory of Electrons and Photons*. New York: Addison Wesley.
Kazama, Y., Yang, C. N., & Golhaber, A. S. (1977). Scattering of a Dirac particle with charge *Ze* by a fixed magnetic monopole. *Physics Review D, 15*, 2287–2299.
Kramers, H. A. (1964). *Quantum Mechanics*. New York: Dover.
Laue, M. von (1922). *Die Relativitätstheorie, Vieweg, I, II, Braunschweig* (2nd ed.) French translation: *La théorie de la relativité*, Gauthier-Villars, Paris, I, 1922; 2, 1924.
Lochak, G. (1959). Quelques problèmes sur le groupe des rotations et la toupie quantique. *Cahiers de Physique, 13*, 41–81.
Lochak, G. (1983). Sur un monopôle de masse nulle décrit par l'équation de Dirac et sur une équation générale non linéaire qui contient des monopôles de spin ½ (1ère partie. *Annales de la Fondation Louis de Broglie, 8*, 345–372.
Lochak, G. (1984). Sur un monopôle de masse nulle décrit par l'équation de Dirac et sur une équation générale non linéaire qui contient des monopôles de spin ½ (2ème partie). *Annales de la Fondation Louis de Broglie, 9*, 5–30.
Lochak, G. (1985). Wave equation for a magnetic monpole. *International Journal of Theoretical Physics, 24*, 1019–1050.
Lochak, G. (1987a). The symmetry between electricity and magnetism and the wave equation of spin ½ magnetic monopole. In A. Blaquière, S. Diner, & G. Lochak (Eds.), *Information, Complexity, and Control in Quantum Physics*. Wien: Springer Verlag.
Lochak, G. (1987b). Electric and magnetic states of the Majorana field. *Annales de la Fondation Louis de Broglie, 12*, 135–164.
Lochak, G. (1992). A magnetic monopole in the Dirac field. *Annales de la Fondation Louis de Broglie, 17*, 203–216.
Lochak, G. (1994). *The Geometrisation of Physics*. Paris: Flammarion.
Lochak, G. (1995a). A second photon in de Broglie's theory of light. *Annales de la Fondation Louis de Broglie, 20*, 111–114.
Lochak, G. (1995b). Magnetic monopole. The symmetry between electricity and magnetism and the problem of the existence of a magnetic monopole. In T. L. W. Barrett, & D. M. Grimes (Eds.), *Advanced Electromagnetism* (pp. 105–147). Singapore: World Scientific.
Lochak, G. (1997a). *P, T, C* symmetries Part I. *Annales de la Fondation Louis de Broglie, 22*, 1–22.
Lochak, G. (1997b). *P, T, C* symmetries Part II. *Annales de la Fondation Louis de Broglie, 22*, 187–217.
Lochak, G., & Costa de Beauregard, O. (2000). The vector potential mesured in the Tonomura experiment. *Annales de la Fondation Louis de Broglie, 25*, 303–308.
Lochak, G. (2000). The shifting of an interference electronic field in a fieldless magnetic potential,. *Annales de la Fondation Louis de Broglie, 25*, 107–127.
Lochak, G. (2002). A new theory of the Aharonov-Bohm effect. *Annales de la Fondation Louis de Broglie, 27*, 529–547.
Lochak, G. (2003). The Dirac equation on the light cone: Majorana electrons and magnetic monopoles. *Annales de la Fondation Louis de Broglie, 31*, 193–206.

Lochak, G. (2004). Electric photons and magnetic photons in the de Broglie theory of photon. *Annales de la Fondation Louis de Broglie, 29*, 297–316.
Lochak, G. (2006). Magnetic monopoles in the Dirac field (magnetic states of the Majorana field). *Annales de la Fondation Louis de Broglie, 28*, 403–439.
Lochak, G. (2007a). The equation of a light leptonic magnetic monopole and its experimental aspects,. *Zeitschrift für Naturforschung, 62a*, 231–246.
Lochak, G. (2007b). Twisted space, chiral gauge and magnetism. *Annales de la Fondation Louis de Broglie, 32*, 125–136.
Lochak, G. (2014). *A theory of light with 4 different photons: Electric and magnetic, with spin 1 and 0, Prikladnaya Fizika (to appear, in Russian).*
McLennan, J. A. (1957). Parity nonconservation and the theory of the neutrino. *Physics Review, 106*, 821–822.
Majorana, E. (1937). Teoria simmetrica dell'elettrone, neutrino e del positrone. *Nuovo Cimento, 14*, 171–184.
Maxwell, J.-C. (1873). *A Treatise on Electricity and Magnetism* (3rd ed.). New York: Dover, 1954.
Mignani, R., & Recami, E. (1975). Complex electromagnetic four-potential and the Cabibbo–Ferrari relation for magnetic monopoles. *Nuovo Cimento, 30A*, 533–540.
Mikhaïlov, V. F. (1985). Observation of magnetic monopoles in the field of a line conductor with a current. *Journal of Physics A, 18*, L903.
Mikhaïlov, V. F. (1987). Observation of the magnetic charge effect in experiments with ferromagnetic aerosols. *Annales de la Fondation Louis de Broglie, 12*, 491–523.
Mikhaïlov, V. F. (1993). *Light, microparticles and magnetic charge phenomenon* (pp. 279–290). Paris: In Courants, Amers, Ecueils en Microphysique, Fondation Louis de Broglie.
Möller, C. (1972). *The Theory of Relativity* (2nd ed.). Oxford, UK: Clarendon Press.
Olariu, S., & Popescu, I. I. (1985). The quantum effects of electromagnetic fluxes. *Review of Modern Physics, 57*, 339–436.
Pauli, W. (1936). Contribution mathématique à la théorie des matrices de Dirac. *Annales de l'Institut Henri Poincaré, 6*, 109–136.
Peshkin, M., & Tonomura, A. (1989). *The Aharonov-Bohn Effect.* Berlin: Springer.
Poincaré, H. (1896). Remarques sur une expérience de M. Birkeland. *Comptes rendus de l'Academie des sciences, 123*, 530–533.
Price, P. B., Shirk, E. K., Osborne, W. Z., & Pinsky, L. S. (1975). Evidence for detection of a moving monopole. *Physics Review Letters, 35*, 487–490.
Proca, A. (1936). Wave theory of positive and negative electrons. *Journal de Physique et le Radium, 7*, 347–353.
Racah, G. (1937). Sulla simmetria tra particelle e antiparticelle. *Nuovo Cimento, 14*, 322–328.
Recami, E., & Mignani, R. (1976). Magnetic monopoles and tachyons in special relativity. *Physics Letters, 62B*, 41–43.
Rodichev, V.I. (1961). *Soviet Physics JETP*, 1029.
Salam, A. (1966). Magnetic monopole and two photon theories of C-violation. *Physics Letters, 22*, 683–684.
Sokolov, A. A., & Ternov, J. M. (1974). *The Relativistic Electron.* Moscow [in Russian]: Nauka.
Takabayasi, T. (1957). Relativistic hydrodynamics of the Dirac matter. *Progress in Theoretical Physics, Suppl., 4*, 2–80.
Tonnelat, M.-A. (1938). Relations entre la théorie du photon et la théorie de l'électron lourd. *Comptes rendus de l'Academie des sciences, 207*, 1180–1182.
Tonnelat, M.-A. (1942). Une nouvelle forme de théorie unitaire: Etude de la particule de spin 2. *Annales de Physique, 17*, 158–208.
Tonomura, A. (1998). *The Quantum World Unveiled by Electron Waves.* Singapore: Singapore.
Thomson, J. J. (1904). *Elements of the Mathematical Theory of Electricity and Magnetism.* Cambridge: Cambridge University Press.

Weyl, H. (1950). A remark on the coupling of gravitation and electron. *Physics Review, 77,* 699–701.
Whittaker, E. T., & Watson, G. N. (1958). *A Course of Modern Analysis.* Cambridge University Press.
Wu, T. T., & Yang, C. N. (1976). Concept of nonintegrable phase factors and global formulation of gauge fields. *Nuclear Physics, 107B,* 365–380.
Wu, T. T., & Yang, C. N. (1975). Dirac monopole without strings: Monopole harmonics. *Physics Review, 12D,* 3845–3857.

FURTHER READING

Ampère, A.-M. (1958). *Théorie mathématique des phénomènes électrodyamiques uniquement déduite de l'expérience.* Paris: Blanchard (original edition published 1827).
Born, M., & Wolf, E. (1964). *Principles of Optics.* Oxford, UK: Pergamon.
Broglie, L. de, & Winter, J. (1934). Sur le spin du photon. *Comptes rendus de l'Academie des sciences, 199,* 813–816.
Costa de Beauregard, O. (1995). Electromagnetic gauge as integration condition: Einstein's mas–energy equivalence law and action–reaction opposition. In T. W. Barrett, & D. M. Grimes (Eds.), *Advanced Electromagnetism* (pp. 77–104). Singapore: World Scientific.
Costa de Beauregard, O. (1997a). A new c^{-2} effect: A hidden angular momentum in magnets. *Physics Essays, 10,* 492–496.
Costa de Beauregard, O., & Lochak, G. (1999). The action of an electrostatic potential on the electron mass. Comment on Mikhailov's article. *Annales de la Fondation Louis de Broglie, 24,* 159–160.
Daviau C. (1993). *Thèse.* Nantes.
Daviau, C. (1994). Remarques sur une équation de Dirac non linéaire. *Annales de la Fondation Louis de Broglie, 19,* 349–360.
Dirac, P. A. M. (1948). The theory of magnetic poles. *Physics Review, 74,* 817–830.
Dirac, P. A. M. (1978a). The monopole concept. *International Journal of Theoretical Physics, 17,* 235–247.
Dirac, P. A. M. (1978b). *Directions in Physics.* New York: John Wiley.
Einstein, A. (1905). Zur Elektrodynamik bewegter, Körper. *Annalen der Physik, 17,* 891–921.
Goldstein, H. (1980). *Classical Mechanics* (2nd ed.). Reading, MA: Addison-Wesley.
Hooft, G.'t (1974). Magnetic monopoles in unified gauge theories. *Nuclear Physics B, 79,* 276–284.
Lochak, G. (1957). Signification mécanique de l'invariance de jauge. *Comptes rendus de l'Academie des sciences, 245,* 2023–2026.
Lochak, G., & Costa de Beauregard, O. (1999). The action of an electrostatic potential on the electron mass,. *Annales de la Fondation Louis de Broglie, 24,* 159–160.
Lochak, G. (2008). Electric and magnetic photons, a possible new theory of the Einstein unitary field. *Annales de la Fondation Louis de Broglie, 33,* 107–127.
Lochak, G. (2010). A new electromagnetism based on 4 photons: Electric, magnetic, with spin 1 and spin 0. Part I. *Annales de la Fondation Louis de Broglie, 35,* 1–21.
Lochak, G. (2011). A new electromagnetism with 4 different photons. Part II: Gravitation. *Annales de la Fondation Louis de Broglie, 36,* 1–17.
Pauli, W. (1957). On the conservation of the lepton charge. *Nuovo Cimento, 6,* 204–215.
Planck, M. (1906). *Vorlesung über die Theorie der Wärmestrahlung 6^e^ ed; 1959, eng. trans. The theory of heat radiation.* New York: Dover.
Spehler, D., & Marques, G. C. (2003). Magnetic monopoles and chiral asymmetry. *International Journal of Modern Physics A, 18,* 2457–2475.
Touschek, B. (1957). The mass of the neutrino and the non-conservation of parity. *Nuovo Cimento, 5,* 1281–1291.

PART TWO

Symmetry Breaking by Electric Discharges in Water and Formation of Lochak's Light Magnetic Monopoles in an Extended Standard Model

Harald Stumpf
Institute of Theoretical Physics, University of Tübingen

Contents

Advances in Imaging and Electron Physics, Volume 189
ISSN 1076-5670
http://dx.doi.org/10.1016/bs.aiep.2015.01.002

INTRODUCTION

Most particle physicists anticipate that the LHC will produce Higgs bosons. Many believe it will create supersymmetric particles. Some think it will yield evidence of extra dimensions. Perhaps. But there is another possibility that very few take seriously: substructure—the idea that quarks and leptons (and maybe also the mediatoirs) are composite particles, made of even more elementary constituents. This would change everything, just as the quark model changed everything 40 years ago, and Rutherford's atomic model changed everything a century ago. In any event, we almost certainly stand at the threshold of a fundamental revolution of elementary particle physics.

D. Griffiths (2008),* Introduction to Elementary Particles, *2nd ed. Wiley-VCH, Weinheim, p. 417.

In the past decade, electric discharges in fluids (mainly water) were reported that obviously induce low-energy nuclear reactions, particularly nuclear transmutations (Urutskoiev *et al.*, 2002; Filippov, Rukhadze, & Urutskoiev, 2004; Filippov & Urutskoiev, 2004; Urutskoiev, 2004). These results were confirmed by various groups and carefully analyzed but could not be explained in the usual way (Urutskoiev, 2004).

In view of this situation by Lochak and Urutskoiev (Urutskoiev, 2004; Lochak 2005; Lochak & Urutskoiev, 2006), it has been suggested that during discharges in water, light (leptonic) magnetic monopoles could be created which by their strong magnetic fields could trigger weak nuclear reactions. An idea that was formulated in analogy to the hypothetical catalysis of proton decay in high-energy physics by (hypothetical) heavy magnetic monopoles.

However, the heavy nonabelian (topological) monopoles do not seem to be very suitable for physical applications. They are too heavy to play any role in the current accessible energy ranges, their constructions are based only on approximate solutions of the associated field equations (Cheng & Li, 1986), and their topological stability (Georgi & Glashow, 1972) might not be suitable as possibly transient phenomena. A transfer of the catalysis idea to low-mass magnetic monopoles and to low-energy phenomena, therefore, seems to be neither possible nor desirable and should not be attempted for mathematical reasons.

A more economical and physically reasonable way of treating magnetic monopoles was pioneered by Lochak. Thirty years ago, Lochak showed that massless Dirac fermions can couple to magnetic vector potentials and can in this way reveal their magnetic monopole properties (Lochak, 1983, 1984, 1985). As neutrinos have small masses and do not show strong magnetic interactions with matter, Lochak's massless fermionic magnetic monopoles cannot be directly identified with them. Softening the zero mass condition, Lochak thus proposed that his fermionic monopoles were identical to excited neutrinos (Lochak, 2003, 2007).

Such an assumption has an immediate consequence: As point particles cannot be internally excited, the magnetically excited neutrinos must be composites. This is in accordance with another discovery by Lochak. After having formulated his equation for massless (light) magnetic monopoles, he showed that from de Broglie's photon theory, either electric photons or magnetic photons can be derived, both of which are composed of two fermions (Lochak, 1995). Therefore, in the discussion of monopole phenomena, the application of de Broglie's fusion idea (de Broglie, 1940, 1942) should or even must be included.

Furthermore, in electrodynamics, the fields of electric and magnetic bosons are linked by duality transformations that forestall the independence of these bosons and their associated charges (Lehner, 1990). In contrast, nonabelian theories generally do not admit self-duality (Deser and Teitelboim, 1976). So, in that case, electric charges and magnetic charges and their associated bosons should be truly independent physical quantities. This is of great importance because photons are represented by mixtures of $U(1)$ and $SU(2)$ fields in the Standard Model, and thus inclusion of the weak isospin is unavoidable. The extension of de Broglie's fusion idea to the full electroweak theory, therefore, is imperative.

As in de Broglie's theory, magnetic and electric photon states cannot exist simultaneously, and the whole theory refers only to single composite particle states (Borne, Lochak, and Stumpf, 2001, Chapter 1). The theoretical formalism of de Broglie's fusion idea must be generalized. In this way, a field theoretic version of de Broglie's and Lochak's discoveries is required, which should lead to an extended electroweak Standard Model as an effective theory for electric and magnetic electroweak bosons as well as for fermions. This was advocated by Lochak (2004).

To treat these problems, we use a model that is based on a relativistically invariant nonlinear spinor field theory with local interaction, canonical quantization, self-regularization, and probability interpretation (Stumpf, 2000). This model implies that in agreement with the ideas of de Broglie and Heisenberg, the present elementary particles are assumed to possess a fermionic substructure. The model is expounded in Borne, Lochak, and Stumpf (2001) and Stumpf (2000, 2004b). It allows the performance of nonperturbative calculations, and its mathematical treatment is in accordance with the basic ideas of the algebraic representation theory of quantum fields. In this approach, the Standard Model is considered an effective theory derived by weak mapping theorems (Stumpf, 2004d), which implies the chance to study processes beyond the present Standard Model.

For the successful application of this model to the monopole problem, an essential prerequisite is the theoretical incorporation of *symmetry breaking,* which is caused by electric discharges in water.

By purely theoretical reasoning, it was demonstrated in Stumpf (2004b) that in the spinor field model, electric and magnetic electroweak boson states can coexist if the CP symmetry of the vacuum is violated. These findings are in accordance with the phenomenological observation that the existence of magnetic monopoles implies CP-symmetry breaking (Cabibbo & Ferrari, 1962; Ellis, 1983; Flambaum & Murray, 2000), and it was shown in a subsequent paper (Stumpf, 2008) that the formal way of theoretical symmetry breaking discussed by Stumpf (2004b) *mathematically* reflects the *experimental situation.*

As monopoles are to be identified by their fields, the analysis of fields generated by symmetry breaking is a first step in the investigation of a possible extension of the effective Standard Model. *Again, as in the case of the derivation of the effective dynamics by weak mappimg for conserved symmetries, the calculations for symmetry breaking must be done nonperturbatively because otherwise, no meaningful results can be obtained.*

In the algebraic formulation of quantum field theory, the behavior of a system can be expressed by an infinite set of inequivalent representations that are generated by an infinite set of inequivalent vacuum states (Bratelli &

Robinson, 1979). This algebraic method has been successfully applied to the quantum field theory of solids. There, the various ground states of matter constitute the set of inequivalent vacuum states generated by different symmetry properties of the system (Volovik, 2003; Benner & Rieckers, 2005; Bogolubov, 1991). The algebraic treatment frequently leads to a system with broken symmetry exhibiting completely different behavior than that of the corresponding system with conserved symmetry, a result that cannot be obtained by application of the quantum mechanical Fock representation, cf. Emch (1972, section 1e).

Finally, one should answer three elementary questions that could possibly be asked by the reader of this article.

The first question is: *What is the concept of the extended Standard Model?*

The concept certainly was not the idea of the substructure itself. Anyone who studies the Standard Model will agree that it needs a more fundamental foundation. Indeed, several decades ago, there was a real boom in composite particle models. However, all these models ended up with the old problems of quantum field theory; see Buchmueller (1985), Lyons (1983), and D'Souza and Kalman (1992). Nevertheless, the need for new ideas is obvious; cf. Ellis (2007).

The spinor field model breaks with tradition, in that no new gauge groups or new physical subparticles are introduced. The spinor field model describes substructures without physical subparticles, and the extended Standard Model is an effective theory that includes substructures. The name *subfermion* or *parton* is only an illustration of mathematical quantities. Because the formalism is new, in Chapter 2 of this part, the theory of relativistic two- and three-particle equations is reviewed in detail, which is not only instructive for the formalism, but also crucial for the monopole concept.

Another question is: Why is this article based on discharges in water *only?*

First, Urutskoiev and his collegues have preferred to use water for numerous experiments. On the other hand, however, there have been observations of accidents in power plants that suggest that magnetic monopoles have been created in electric cables during discharges and transmitted over considerable distances; cf. Lochak (2005, 2008). Urutskoiev and Lochak suppose that insulating materials of the cables (e.g., oil) are responsible for these processes. Therefore, the discussion of monopole effects in oil or other insulating materials would be equally interesting. But this is not the topic of this article.

Second, water plays an extraordinary role in discharges, which is well documented (cf. section 4.1 of this part), and one can easily refer to its thermodynamic properties.

Finally, another question: Are magnetic charges conserved?

In the research literature, the conservation of magnetic charge seems to be self-evident (Shnir, 2005).

For heavy magnetic monopoles that after spontaneous symmetry breaking result from gauge theories, this premise is satisfied by the derivation of topological quantum numbers.

In contrast to the heavy magnetic monopoles that are bosons [cf. Cheng and Li (1986, section 15.3)], the light magnetic monopoles are leptons and emerge during low energy electric discharges under symmetry breaking conditions. For these low energy reactions the influence of strong forces can be excluded Urutskoiev (2004, p. 1154), cf. also Lochak, (2005, 2008). Hence the existence of these leptonic monopoles must be connected with electroweak forces and no conservation of magnetic charge can be proven.

CONVENTIONS AND EXPLANATORY NOTES

- **Units**: For practical purposes, in Chapter 4 of this part, partly Gaussian or natural units are used, while in Chapters 1,2, and 3, only natural units are applied. Conversion formulas between these unit systems can be found in Mandl and Shaw (1993, section 6.1). In order to avoid mixing up the summation index i with the imaginary unit i the later is denoted by ι if necessary.
- **Bold lettering**: For example, **j** and **d** stand for elements of functional algebra or for three-dimensional (3D) space vectors (such as **r** and **k**) and vectorial functions [such as **A**(**r**)].
- **Temporal gauge**: The electroweak sector of the conventional Standard Model is an $SU(2)_L \otimes U(1)$ gauge theory. For the comparison of the phenomenological gauge theory with the effective theory resulting from the spinor field model it is important to know that the temporal gauge is often considered as the "**physical gauge**" which sets the standard for the treatment of nonabelian gauge theories, cf. Cheng (1990); Tsai (1990); Haller (1990); Huang (1992); Jackiw (1980); Goldstone and Jackiw (1978); Leibbrandt (1987). Of interest is that by the use of this gauge an unambiguous formulation of the theory for canonical quantization in the Hamiltonian formalism is ensured, cf. Huang (1992, p. 79) and Stumpf and Pfister (1996, 1997), whereas the difficulties

produced by using this gauge in quantum field theoretic calculations are not relevant in this context.

- **Relativistic covariance**: By Stumpf and Borne (1994), Borne, Lochak, Stumpf, (2001, chapts. 2) a complete spinor field dynamics has been formulated in a covariant way by using multi-time ordered matrix elements. However, against such a formulation serious objections can be raised, cf. Borne, Lochak, Stumpf (2001, sect. 3.7). This can be illustrated by considering hard-core states, cf. Stumpf, Borne (1996, sect 4.2) which are part of the covariant formalism. While two particle hard-core equations can be explicitly solved already the three particle equations are too complicated for practical use cf. Stumpf (2003b) and this situation becomes more critical with increasing particle numbers. As a way out single time algebraic energy equations can be treated which are compatible with the covariant formalism, cf. Borne, Lochak, Stumpf (2001, chapts. 3,4).

CHAPTER 1

Elements of the Spinor Field Quantum Theory

1.1 ALGEBRAIC REPRESENTATION OF THE SPINOR FIELD

The algebraic representation is the basic formulation of the spinor field model. In order to avoid lengthy deductions, we give only some basic formulas of this formalism; for details, refer to Borne, Lochak, and Stumpf (2001); Stumpf and Borne (1994). The associated Lagrangian density reads as follows [see Borne, Lochak, and Stumpf (2001, Eq. (2.52)]:

$$\begin{aligned}\mathcal{L}(x) := & \sum_{i=1}^{3} \lambda_i^{-1} \overline{\psi}_{A\alpha i}(x)\left(\iota\gamma^{u}\partial_\mu - m_i\right)_{\alpha\beta}\delta_{AB}\psi_{B\beta i}(x) \\ & -\frac{1}{2}g\sum_{h=1}^{2}\delta_{AB}\delta_{CD}v^h_{\alpha\beta}v^h_{\gamma\beta}\sum_{i,j,k,l=1}^{3}\overline{\psi}_{A\alpha i}(x)\psi_{B\beta j}(x)\overline{\psi}_{C\gamma\kappa}(x)\psi_{D\delta l}(x),\end{aligned} \tag{1.1}$$

with $v^1 := 1$ and $v^2 := \iota\gamma^5$. The field operators are assumed to be Dirac spinors with index $\alpha = 1, 2, 3, 4$ and additional isospin with index $A = 1, 2,$

as well as auxiliary fields with index $i = 1, 2, 3$ for nonperturbative regularization; cf. Borne, Lochak, and Stumpf [(2001, Eq. (2.51)] and Stumpf (2000). The algebra of the field operators is defined by the anticommutators

$$\left[\psi^{+}_{A\alpha i}(\mathbf{r}, t)\psi_{B\beta j}(\mathbf{r}', t)\right]_{+} = \lambda_i \delta_{ij}\delta_{AB}\delta_{\alpha\beta}\delta(\mathbf{r}-\mathbf{r}'), \tag{1.2}$$

which result from canonical quantization. All other anticommutators vanish. To obtain a uniform transformation property with respect to Lorentz transformations, the adjoint spinors are replaced by formally charge-conjugated spinors that are defined by

$$\psi^{c}_{A\alpha i}(x) = C_{\alpha\beta}\overline{\psi}_{A\beta i}(x), \tag{1.3}$$

and the index Λ is introduced by

$$\psi_{A\Lambda\alpha\iota}(x) = \begin{pmatrix} \psi_{A\alpha i}(x); & \Lambda = 1 \\ \psi^{c}_{A\alpha i}(x); & \Lambda = 2 \end{pmatrix} \tag{1.4}$$

Then the set of indices is defined by

$$Z := (A,\ \Lambda,\ \alpha,\ i).$$

In order to derive definite results from the theory, a state space is needed in which the dynamical equations can be formulated. This is achieved by the use of the algebraic Schroedinger representation. For a detailed discussion, see Borne, Lochak, and Stumpf (2001) and Stumpf and Borne (1994).

To ensure transparency of the formalism, we use the symbolic notation

$$\psi_{I_1}\ldots\psi_{I_n} := \psi_{Z_1}(\mathbf{r}_1, t)\ldots\psi_{Z_n}(\mathbf{r}_n, t), \tag{1.5}$$

with $I_k := (Z_k, \mathbf{r}_k, t)$. Then in the algebraic Schroedinger representation, a state $|a\rangle$ is characterized by the following set of matrix elements:

$$\tau_n(a) := \langle 0|\mathcal{A}(\psi_{I_1}\ldots\psi_{I_n})|a\rangle, \quad n = 1\ldots\infty, \tag{1.6}$$

where $\mathcal{A}$ means antisymmetrization in $I_1 \ldots I_n$.

From this definition, the calculation of an eigenstate $|a\rangle$ is transferred to the calculation of the set of matrix elements [Eq. (1.6)] that characterizes this state. For a compact formulation of this method, generating functionals are introduced and the set [Eq. (1.6)] is replaced by the functional state

$$|\mathcal{A}(\mathbf{j}; a)\rangle := \sum_{n=1}^{\infty}\frac{i^n}{n!}\sum_{I_1\ldots I_n}\tau_n(I_1\ldots I_n|a)\ \mathbf{j}_{I_1}\ldots\mathbf{j}_{I_n}|0\rangle_f, \tag{1.7}$$

where $\mathbf{j}_I := \mathbf{j}_Z(\mathrm{r})$ are the generators of a CAR-algebra with corresponding duals $\mathbf{d}_I := \mathbf{d}_Z(\mathrm{r})$ that satisfy the anticommutation relations

$$[\mathbf{j}_I, \mathbf{d}_{I'}] = \delta_{ZZ'}\delta(\mathbf{r} - \mathbf{r}'), \tag{1.8}$$

while all other anticommutators vanish.

With $\mathbf{d}_I|0\rangle_f = 0$, the basis vectors for the generating functional states can be defined. The latter must not be confused with the creation and annihilation operators of particles in physical state spaces because generating functionals are formal tools for a compact algebraic representation of the states. We define: With each state $|a\rangle$ in the physical state space, we associate a functional state $|\mathcal{A}(\mathbf{j}; a)\rangle$ in the corresponding functional space. The map is biunique, and the symmetries of the original theory are conserved. For details, see Borne, Lochak, and Stumpf (2001) and Stumpf and Borne (1994).

In order to find a dynamical equation for the functional states, we apply the Heisenberg formula to the antisymmetrized operator products [Eq. (1.5)]:

$$i\frac{\partial}{\partial t}A(\psi_{I_1}\ldots\psi_{I_n}) = \left[A(\psi_{I_1}\ldots\psi_{I_n})H\right]_- \quad n = 1\ldots\infty, \tag{1.9}$$

where H is the Hamiltonian of the system under consideration.

If $|0\rangle$ as well as $|a\rangle$ are assumed to be eigenstates of H, then between both states, the matrix elements of [Eq. (1.9)] can be formed, and subsequent evaluation of these expressions leads to the functional equation

$$\begin{aligned} E_0^a|\mathcal{A}(\mathbf{j}; a)\rangle = {} & \left[K_{I_1I_2}\mathbf{j}_{I_1}\mathbf{d}_{I_2} - W_{I_1I_2I_3I_4}\mathbf{j}_{I_1}\left(\mathbf{d}_{I_4}\mathbf{d}_{I_3}\mathbf{d}_{I_2} + A_{I_4J_1}A_{I_3J_2}\mathbf{j}_{J_1}\mathbf{j}_{J_2}\mathbf{d}_{I_2}\right)\right] \\ & \times |\mathcal{A}(\mathbf{j}; a)\rangle, \end{aligned} \tag{1.10}$$

with $E_0^a = E_a - E_0$. For details of the derivation, see Borne, Lochak, and Stumpf (2001) and Stumpf and Borne (1994).

The symbols used in Eq. (1.10) are defined by the following relations:

$$K_{I_1I_2} := iD^0_{I_1I}\left(D^k\partial_k - m\right)_{II_2}, \tag{1.11}$$

with

$$\begin{aligned} D^\mu_{I_1I_2} &:= i\gamma^\mu_{\alpha_1\alpha_2}\delta_{A_1A_2}\delta_{\Lambda_1\Lambda_2}\delta_{i_1i_2}\delta(\mathbf{r}_1 - \mathbf{r}_2) \text{ and} \\ m_{I_1I_2} &:= m_{i_1}\delta_{\alpha_1\alpha_2}\delta_{A_1A_2}\delta_{\Lambda_1\Lambda_2}\delta_{i_1i_2}\delta(\mathbf{r}_1 - \mathbf{r}_2); \end{aligned} \tag{1.12}$$

and $W_{I_1I_2I_3I_4} := -iD^0_{I_1I} V_{II_2I_3I_4}$, with

$$V_{I_1I_2I_3I_4} := \sum_{h=1}^{2} g\lambda_{i_1} B_{i_2i_3i_4} v^h_{\alpha_1\alpha_2} \delta_{A_1A_2}\delta_{\Lambda_1\Lambda_2}\left(v^h C\right)_{\alpha_3\alpha_4}\delta_{A_3A_4}\delta_{\Lambda_3 1}\delta_{\Lambda_4 2}\times$$
$$\delta(\mathbf{r}_1-\mathbf{r}_2)\delta(\mathbf{r}_1-\mathbf{r}_3)\delta(\mathbf{r}_1-\mathbf{r}_4),$$
$$\text{and } B_{i_2i_3i_4} := 1, \quad i_a = 1,2,3; \quad a = 2,3,4 \tag{1.13}$$

and the anticommutator matrices

$$A_{I_1I_2} := \lambda_{i_1}\left(C\gamma^0\right)_{\alpha_1\alpha_2}\delta_{A_1A_2}\sigma^1_{\Lambda_1\Lambda_2}\delta_{i_1i_2}\delta(\mathbf{r}_1-\mathbf{r}_2). \tag{1.14}$$

It should be emphasized that Eq. (1.10) holds for any algebraic representation and thus is in this way indeterminate. Therefore, it is necessary to select a special representation by choosing a special vacuum. This can be achieved by the introduction of normal ordered functionals, which are defined by

$$|\mathcal{F}(\mathbf{j};a)\rangle := \exp\left[\frac{1}{2}\mathbf{j}_{I_1}F_{I_1I_2}\mathbf{j}_{I_2}\right]|\mathcal{A}(\mathbf{j};a)\rangle =: \sum_{n=1}^{\infty}\frac{i^n}{n!}\varphi_n(I_1\ldots I_n|a)\mathbf{j}_{i_1}\ldots\mathbf{j}_{I_n}|0\rangle_f, \tag{1.15}$$

where the two-point function

$$F_{I_1I_2} := \langle 0|\mathcal{A}\{\psi_{Z_1}(\mathbf{r}_1,t)\psi_{Z_2}(\mathbf{r}_2,t)\}|0\rangle \tag{1.16}$$

contains information about the ground state and thus fixes the representation. The normal ordered functional equation then reads

$$E_0^a|\mathcal{F}(\mathbf{j};a)\rangle = \mathcal{H}_F(\mathbf{j};\mathbf{d})|\mathcal{F}(\mathbf{j};a)\rangle, \tag{1.17}$$

with

$$\begin{aligned}\mathcal{H}_F(\mathbf{j},\mathbf{d}) := {} & \mathbf{j}_{I_1}K_{I_1I_2}\mathbf{d}_{I_2} + W_{I_1I_2I_3I_4}\Big[\mathbf{j}_{I_1}\mathbf{d}_{I_4}\mathbf{d}_{I_3}\mathbf{d}_{I_2} - 3F_{I_4K}\mathbf{j}_{I_1}\mathbf{j}_K\mathbf{d}_{I_3}\mathbf{d}_{I_2} \\ & + 3\left(F_{I_4K_1}F_{I_3K_2} + \frac{1}{4}A_{I_4K_1}A_{I_3K_2}\right)\mathbf{j}_{I_1}\mathbf{j}_{K_1}\mathbf{j}_{K_2}\mathbf{d}_{I_2} \\ & - \left(F_{I_4K_1}F_{I_3K_2} + \frac{1}{4}A_{I_4K_1}A_{I_3K_2}\right)F_{I_2K_3}\mathbf{j}_{I_1}\mathbf{j}_{K_1}\mathbf{j}_{K_2}\mathbf{j}_{K_3}\Big]\end{aligned} \tag{1.18}$$

Eq. (1.17) is the algebraic Schroedinger representation of the spinor-field Lagrangian [Eq. (1.1)], written in functional space with a fixed definite algebraic state space.

1.2 NONPERTURBATIVE SELF-REGULARIZATION

Conventional quantum field theory is confronted with divergencies. This holds in particular for the nonrenormalizable, first-order, nonlinear spinor-field quantum theory (i.e., Fermi theory). However, if one follows the intellectual traces of de Broglie's and Heisenberg's ideas, spinor theories are of particular interest.

An analysis of attempts in this field shows that the real problem of a nonlinear spinor theory is *compatibility* with the demands of relativistic invariance, locality and causality, finiteness, and probability interpretation. With spinor theory models, these demands could not be satisfied in one or the other way in the past, and progress with respect to the compatibility of these demands was achieved in Stumpf (2000, 2004d).

In view of this history, the claim that a consistent nonlinear spinor field quantum theory exists is exceptional. Although for brevity no details of the proofs in Stumpf (2000, 2004d) are given here, it is helpful to know that in order to understand this claim, the so-called decomposition theorem plays a central role [cf. Stumpf (1982)]:

Proposition:

i) Let $\Psi \equiv \Psi(x)$ be a solution of the Nth-order equation

$$\prod_{i=1}^{N}(D - m_i)\Psi = V[\Psi], \tag{1.19}$$

where $V[\Psi]$ is a nonlinear interaction term; and let $\psi_i \equiv \psi_i(x), \mathrm{i} = 1, \ldots, \mathrm{N}$ be auxiliary fields that are defined by

$$\psi_\iota := \lambda_i \prod_{\substack{k=1 \\ k\neq 1}}^{N} (D - m_i)\Psi, \quad i = 1, \ldots, N \tag{1.20}$$

$$\lambda_i := \prod_{\substack{k=1 \\ k\neq i}}^{N} (m_i - m_k)^{-1}, \quad i = 1, \ldots, N \tag{1.21}$$

Then the auxiliary fields are solutions of the equations

$$(D - m_i)\psi_i = \lambda_i V\left[\sum_{j=1}^{N}\psi_j\right], \quad i = 1, \ldots, N \tag{1.22}$$

$$\Psi = \sum_{i=1}^{N} \psi_i, \tag{1.23}$$

with no summation over i on the left side in Eq. (1.22).

ii) If the fields $\psi_i(x) = 1,\ldots, N$, are solutions of Eq. (1.22) with λ_i of Eq. (1.21), and if $\Psi \equiv \Psi(x)$ is defined by Eq. (1.23), then the relation [Eq. (1.20)] holds and Ψ is a solution of Eq. (1.19) $\diamond$.

This theorem—first formulated and proved for $N = 3$ in Stumpf (1982) and later generalized to arbitrary N in Grosser (1983)—is described in extensive detail in Borne, Lochak, and Stumpf (2001, section 2.3) and Stumpf and Borne (1994, section 1.2).

One easily verifies that by means of this theorem for $N = 3$ and Eq. (1.21), the relations

$$\sum_{i=1}^{3} \lambda_i = 0, \sum_{i=1}^{3} \lambda_i m_i = 0 \tag{1.24}$$

can be derived. Eq. (1.24) is identical to the Pauli-Villars regularization conditions of perturbation theory and appears automatically in each operation of the algebraic theory. They are intrinsic properties of the theory, and the Lagrangian [Eq. (1.1)] itself is formulated in terms of auxiliary fields.

In contrast to the dim[3/2] of the spinor fields of the Fermi theory, the auxiliary fields of Eq. (1.1) have dim[1/2]; cf. Pfister (1990, p. 124). Thus, the dimension of the coupling constant evaluated for the fields of Eq. (1.1) [cf. Cheng and Li (1986, Eq. (2.135)], equals 2; i.e., in perturbation theory, the Lagrangian [Eq. (1.1)] defines a super-renormalizable model.

The regularization effect can be drastically seen if one derives the anticommutation rules for the original spinor field in Eq. (1.19). With Eqs. (1.2), (1.23), and (1.24), one obtains

$$\left[\Psi^{+}_{A\alpha}(\mathbf{r}, \mathrm{t})\Psi_{B\beta}(\mathbf{r}', t)\right]_{+} = 0. \tag{1.25}$$

The full set of anticommutation relations for this field is given in Borne, Lochak, and Stumpf [(2001, Eqs. (2.62) and (2.63)].

Thus, the canonical quantization of auxiliary fields is equivalent to a noncanonical quantization of the original field. The relation [Eq. (1.25)] was already assumed by Heisenberg in order to avoid singular terms in the course of field theoretic calculations; cf. Heisenberg (1966), but without the knowledge of the decomposition theorem. On the other hand, the regularizing effect of higher-order derivatives earlier discovered by Bopp

(1940); and independently by Podolsky (1942) implies an indefinite metric in state space, as can be concluded from Eqs. (1.2) and (1.21) or (1.24).

In gauge theories, an indefinite metric in state space appears as well, but it can be eliminated by an appropriate gauge fixing. In nonlinear spinor field quantum theory, a similar remedy does not work. However, without a solution to this problem, a spinor field model cannot be interpreted as a physical theory, which was the problem with Heisenberg's original approach.

At present, this difficulty can be overcome only by the decomposition theorem. The proofs of Stumpf (2000) demonstrate that as a consequence of this theorem, probability interpretation and regularization of the theory are compatible; i.e., one can treat the spinor field theory as a physical theory within the framework of a normal quantum theory. Of course, it is necessary to keep to the rules developed in Stumpf (2000). In particular, physical states arise only after the regularization has been done; cf. Stumpf (2000) and section 4.5 of this part for details.

1.3 SYMMETRIES AND SYMMETRY BREAKING

The algebraic representation is the basic formulation of the spinor field model. In order to avoid lengthy deductions, we refer to Borne, Lochak, and Stumpf (2001) and Stumpf and Borne (1994). The corresponding Lagrangian density is given by Eq. (1.1) and the algebra of the field operators is defined by anticommutators [Eq. (1.2)] originating from canonical quantization.

1.3.1 Conserved Symmetries

With respect to the Poincare group, this topic was treated in great detail in Borne, Lochak, and Stumpf (2001, sections 2.7 and 3.3). Concerning internal symmetry groups for the $SU(2) \otimes U(1)$ invariance of the initial spinor field equation or of the Lagrangian [Eq. (1.1)], respectively, the good quantum numbers result as eigenvalues of the symmetry generators in Borne, Lochak, and Stumpf (2001, Eqs. (6.80)−(6.83)], which in functional space were formulated in the form of Eqs. (6.84)−(6.89) from the same study.

Provided that the propagator is invariant under the action of these groups, the eigenvalue conditions of Borne, Lochak, and Stumpf (2001, Eqs. (6.84)−(6.89)] hold for normal transformed states as well.

1.3.2 Discrete Transformations

In the literature, corresponding operations and operators are explicitly constructed in Fock space with a Dirac vacuum; cf. for instance, Gross (1999) and Fonda and Ghirardi (1970). In Stumpf (2003a), these definitions were generalized to be valid in the algebraic formalism. In particular, it can be proved that the Lagragian [Eq. (1.1)] is invariant under parity operation; cf. Nachtmann (1986, pp. 62, 426) and Grimm (1994b, p. 49). That is, the Lagrangian [Eq. (1.1)] defines a left—right symmetric model. Furthermore, by construction of the model, the conditions for the application of the PCT- theorem [cf. Gross (1999)] are fulfilled. That is, the model is invariant under PCT-operation and the fermion propagator for conserved symmetries is PCT- and CP-invariant too [cf. Stumpf (2008)].

1.3.3 Antisymmetrization

For Fermi fields, time-ordered antisymmetric products of field operators were defined in covariant perturbation theory. The definition of such products was applied to the spinor fields of Eq. (1.1) and dynamical field equations were derived in functional space; cf. Borne, Lochak, and Stumpf (2001, Chapter 3). Against the use of this formalism beyond perturbation theory (i.e., for nonperturbation theoretical calculations), serious mathematical objections can be raised; cf. Borne, Lochak, and Stumpf (2001, section 3.7).

Beyond perturbation theory, these objections can be avoided only in the case of bound state calculations (cf. Chapter 2 of this part). An appropriate alternative treatment was developed by means of the *algebraic Schroedinger representation*, which is based on the Gelfand-Naimark-Segal (GNS) construction [cf. Borne, Lochak, and Stumpf (2001, section 4.4) and Stumpf and Borne (1994, section 3.5)]. It enables one to treat the composite many-particle problem and is exclusively reviewed later in this chapter, in section 1.4.

The GNS construction provides genuine representations of the basic field operator algebra, and the set of basis vectors is generated by the application of antisymmetric products of field operators on a spacelike hyperplane to the ground state (vacuum) $|0\rangle$ of the system. Furthermore, in order to obtain a simplified formulation of the states expressed by the GNS construction, the basis states can be integrated into generating functional states in functional space [cf. Eq. (1.7)]. Such functional states can then be chosen as eigenstates of energy [cf. Eq. (1.10)].

But as an infinite number of inequivalent vacuum states exists, it is impossible to draw any conclusions from the functional eigenvalue equation [Eq. (1.10)] without specifying the vacuum.

1.3.4 Symmetry Breaking and Parafermi States

The selection of a definite set of basis states can be done by the application of the normal ordering transformation [Eq. (1.15)]. A symmetry breaking can then be implemented by the assumption of an appropriately modified propagator in the normal ordering transformation, which replaces the free propagator for conserved symmetries. Stumpf (2008) demonstrated that a symmetry breaking is caused by a discharge in water. In water, one can distinguish the positive charges from the negative ones by their different damping levels. The latter leads to an irreversible time evolution; i.e., the time reversal invariance is violated. Owing to the conserved PCT invariance of the whole theory, the violation of time reversal has to be compensated by a CP symmetry breaking in order to maintain the PCT invariance of the theory. An alternative argument for this CP violation is given in Stumpf (2008).

From this discussion it follows: For particles without electric charge, like neutrinos, CP-violation decreases to P-violation. This fact has been incorporated in the present Standard Model that is an $SU(2)_L \otimes U(1)$ theory. In this theory maximal parity symmetry breaking is realized because only left-handed neutrino states but no right-handed ones are admitted, cf. Huang (1982, table 6.4, p. 114), whereas the charged particle states appear with both components. This construction is unsatisfactory as massive neutrinos without electric charge cannot be purely left-handed.

On the other hand in left-right symmetric models where charged and uncharged particles are treated on an equal basis, it must be explained how the discrimination of the right handed states for massive neutrinos comes about. Obviously an appropriate vacuum structure cf. Mohapatra (2003, sect. 6.1) must be assumed. This topic will not be treated in this article, because the calculations and conclusions in chapter 4 of this part are not affected by ignoring this problem. For its extensive treatment cf. Stumpf (2015).

The fact that by symmetry breaking various particle species can be identified prevents the use of antisymmetric states and leads to the application of **Parafermi** configurations discussed further in Chapter 2 of this part.

1.4 WEAK MAPPINGS IN FUNCTIONAL SPACE

In the spinor field model, it is assumed that the electroweak gauge bosons of the Standard Model possess a substructure consisting of two partons, while leptons and quarks are bound states of three partons. Starting with the spinor field dynamics [Eq. (1.18)], the mapping of the combined two-parton-three-parton many-particle system has to be studied and its effective dynamics derived, which represents that part of the theory related to experiments. In the following discussion, the latter effective dynamics is called the *extended Standard Model,* as it leads to modifications of the conventional theory.

In functional formulation, this proceeding can be expressed by mappings of Eq. (1.18) among appropriate functional spaces. The application of the functional formulation is indispensible, as otherwise without this aid the calculations would be too complicated.

Let the functional states of the corresponding effective field theory be defined by

$$|\mathcal{P}(\mathbf{b},\mathbf{f};a)\rangle := \sum_{n=2}^{\infty}\sum_{m=0}^{\infty}\sum_{r=0}^{\infty}\delta_{n,2m+3r}\frac{1}{(m!)(r!)} \times \tilde{\varrho}(k_1\ldots k_m,q_1\ldots q_r|a)\mathbf{b}_{k_1}\ldots\mathbf{b}_{k_m}\mathbf{f}_{q_1}\ldots\mathbf{f}_{q_r}|0\rangle_{BF}, \quad (1.26)$$

where $\mathbf{b}$, $\mathbf{d}^b$ and $\mathbf{f}$, $\mathbf{d}^f$ are the functional sources of the phenomenological bosons and fermions, respectively; and the $\tilde{\varrho}$-functions represent the matrix elements of the effective (phenomenological) boson-fermion theory. The latter functions are introduced in analogy to the definition of $\tau_n$in Eq. (1.6), but they are not completely antisymmetric. Furthermore, the effective theory is not normal-ordered by construction.

For conserved symmetries, faithful mapping theorems have been derived that are based on a complete antisymmetrization of the states [cf. Stumpf (2004d)]. However, these methods are rather complicated, and in the case of symmetry breaking, the corresponding mapping theorems are not applicable due to the associated loss of antisymmetry.

1.4.1 Chain Rule Mappings

Therefore, a simplified version of the weak mapping procedure should be applied, which is suitable to comprise symmetry breaking. To this end. one has to recall that the results of weak mappings are used for the derivation of effective theories. In general, exchange terms are neglected in effective theories. When these terms are disregarded, the necessity for the application of faithful mapping theorems is dropped and one can apply the so-called chain rule, which by its construction eliminates exchange terms [cf. Stumpf and Borne (1994, sections 4.7–4.9)].

The starting point of the chain rule is the assumption that the state [Eq. (1.26)] is only the representation of the original state [Eq. (1.15)] in another coordinate system, which leads to the vector invariance relation in functional space:

$$|\mathcal{F}(\mathbf{j}, a)\rangle \equiv |\mathcal{P}(\mathbf{b}, \mathbf{f}, a)\rangle. \tag{1.27}$$

Obviously, for conserved symmetries, this equivalence postulate violates the complete antisymmetry of the states [Eq. (1.15)] and requires the replacement of Eq. (1.27) by the faithful mapping formalism. However, for broken symmetry with the accompanying loss of antisymmetry, the equivalence [Eq. (1.27)] is no longer a contradiction and can thus be applied.

A corresponding transformed energy equation can be directly gained by substitution of Eq. (1.27) into Eq. (1.17); i.e.,

$$E_0^a|\mathcal{P}(\mathbf{b}, \mathbf{f}, a)\rangle = \mathcal{H}_F\left(\mathbf{j}, \mathbf{d}^j\right)|\mathcal{P}(\mathbf{b}, \mathbf{f}, a)\rangle. \tag{1.28}$$

In a first step, this equation can be transformed by the application of the chain rule:

$$\begin{aligned} \mathbf{d}_I^j|\mathcal{P}\rangle &= \left[\left(\mathbf{d}_I^j\mathbf{b}_k\right)\mathbf{d}_k^b + \left(\mathbf{d}_I^j\mathbf{f}_l\right)\mathbf{d}_l^f\right]|\mathcal{P}\rangle \\ &= \left[2C_{2,k}^{IK}\,\mathbf{j}_K\,\mathbf{d}_k^b + 3C_{3,l}^{IK_1K_2}\mathbf{j}_{K_1}\mathbf{j}_{K_2}\mathbf{d}_l^f\right]|\mathcal{P}\rangle, \end{aligned} \tag{1.29}$$

which can be repeatedly applied to Eq. (1.17); cf. Borne, Lochak, and Stumpf (2001, section 5.6) and Stumpf and Borne (1994, section 4.7).

In Eq. (1.29), the quantities $C_{II'}^l$ and $R_{II'}^n$ symbolize the boson states and their duals, while $C_{II'I''}^q$ and $R_{II'I''}^p$ are the fermion states and their duals, both of which are explicitly defined in functional space by

$$\mathbf{b}_{2,k} := C_{2,k}^{I_1I_2}\mathbf{j}_{I_1}\mathbf{j}_{I_2}, \quad \mathbf{f}_{3,l} := C_{3,l}^{I_1I_2I_3}\mathbf{j}_{I_1}\mathbf{j}_{I_2}\mathbf{j}_{I_3} \tag{1.30}$$

and

$$\sum_{I_1I_2} R_{I_1I_2}^{2,k}C_{2,k'}^{I_1I_2} = \delta_{k,k'} \quad \sum_{I_1I_2I_3} R_{I_1I_2I_3}^{3,l}C_{3,l'}^{I_1I_2I_3} = \delta_{l,l'}. \tag{1.31}$$

If the chain rule applications to Eq. (1.28) are combined with these duality relations, the operator $\mathcal{H}_f$ can be transformed into an operator $\mathcal{H}_F$ depending only on the new variables $\mathbf{b}$, $\mathbf{d}^b$, $\mathbf{f}$, and $\mathbf{d}^f$. The transformed energy operator $\mathcal{H}_C$ is explicitly given in Borne, Lochak, and Stumpf [(2001, Eq. (5.74)]. It contains leading terms and higher-order terms that are characterized by higher-order dual states stemming from dressed particle states; cf. Stumpf and Borne (1994, section 4.8) and Borne, Lochak, and Stumpf (2001, section 5.7).

The leading terms of the faithful and the chain rule method are nearly the same, although the calculation methods of the two mappings are different. In both cases, the resulting energy operator can be decomposed into

$$\mathcal{H} = \mathcal{H}_f + \mathcal{H}_b + \mathcal{H}_{bf}. \tag{1.32}$$

For a faithful weak mapping (i.e., for conserved symmetries), one obtains from Stumpf [(2004d, Eqs. (66) and (82)−(85)] for the leading terms the following expressions:

$$\begin{aligned}
\mathcal{H}_f &:= K^f_{qp}\mathbf{f}_q\mathbf{d}^f_q + M^f_{qp}\mathbf{f}_q\mathbf{d}^f_p \\
\mathcal{H}_b &:= K^b_{kl}\mathbf{b}_k\mathbf{d}^b_l + M^b_{kl}\mathbf{d}^b_l + W_1^{kl_1l_2}\mathbf{b}_k\mathbf{d}^b_{l_1}\mathbf{d}^b_{l_2} =: \mathcal{H}^1_b + \mathcal{H}^2_b + \mathcal{H}^3_b \\
\mathcal{H}_{bf} &:= W_2^{qlp}\mathbf{d}^b_l\mathbf{f}_q\mathbf{d}^f_p + W_4^{q_1q_2p_1p_2}R^k_{q_1q_2}\mathbf{b}_k\mathbf{d}^f_{p_1}\mathbf{d}^f_{p_2} =: \mathcal{H}^1_{b_f} + \mathcal{H}^2_{b_f}
\end{aligned} \tag{1.33}$$

with the coefficient functions

$$\begin{aligned}
K^f_{qp} &:= 3R^q_{II'I_1}K_{I_1I_2}C^p_{I_2II'} \\
M^f_{qp} &:= -9W_{I_1I_2I_3I_4}F_{I_4K}\left(R^q_{IKI_1}C^p_{I_2I_3I} - R^q_{I_1I'K}C^p_{I_2I_3I'}\right) \\
K^b_{kl} &:= 3R^k_{II_1}K_{I_1I_2}C^l_{I_2I} \\
M^b_{kl} &:= -6W_{I_1I_2I_3I_4}F_{I_4K}R^k_{KI_1}C^l_{I_2I_3} \\
W_1^{kl_1l_2} &:= 4W_{I_1I_2I_3I_4}R^k_{II_1}C^{l_1}_{I_4I}C^{l_2}_{I_2I_3} \\
W_2^{qlp} &:= 3W_{I_1I_2I_3I_4}R^q_{II'I_1}C^p_{I_4II'}C^l_{I_2I_3} \\
W_4^{q_1q_2p_1p_2} &:= -54W_{I_1I_2I_3I_4}F_{I_4K}R^{q_1}_{II'I_1}R^{q_2}_{K_1K_2K}C^{p_1}_{I'K_1K_2}C^{p_2}_{I_2I_3I},
\end{aligned} \tag{1.34}$$

which are equal to those of the chain rule in Borne, Lochak, and Stumpf [(2001, Eq. (5.74)]. apart from the W_4 term and an exchange term in W_2.

With respect to the monopole problem, the coupling of the composite fermions (leptons) to the composite vector bosons is of interest. In both mapping formulas, the corresponding term is of the lowest order and the chain rule expression differs from the expression of the exact mapping term only by an additional overlap term.

For the two mapping versions, the common term reads

$$\mathcal{H}^1_{bf} := W_2^{qlp}\mathbf{d}^b_l\mathbf{f}_q\mathbf{d}^f_p = 3W_{I_1I_2I_3I_4}R^q_{II'I_1}C^l_{I_2I_3}C^p_{I_4II'}\mathbf{d}^b_l\mathbf{f}_q\mathbf{d}^f_p \tag{1.35}$$

and can be taken from Eqs. (1.33) and (1.34), or equivalently, from Borne, Lochak, and Stumpf [(2001, Eq. (5.74)], where the overlap term here and in the following is disregarded.

In Eq. (1.35), the symmetry breaking manifests itself exclusively in the internal structures of the wave functions, which will be discussed further in Chapter 2.

To conclude this chapter, a comment should be made about the meaning of the higher-order terms. For faithful mappings, an evaluation and estimates of higher-order terms were carried out in Borne, Lochak, and Stumpf (2001, section 8.7). The results show that for high masses of the auxiliary fields (incorporated in the bound states under consideration), the higher terms are negligible compared with the leading terms. Also, for higher-order terms of the chain rule, estimates can be given. These terms are defined by higher-order, dressed particle duals. The corresponding formula reads [cf. Borne, Lochak, and Stumpf (2001, sect. 5.7)]:

$$R^{f,k}_{I_1\ldots I_{2n+f}} = 2^{2n+f}\sum_{k'}\left(g^{-1}\right)_{k,k'}\lambda^{-1}_{i_1}\ldots\lambda^{-1}_{i_{2n+f}}\left(C^{I_1\ldots I_{2n+f}}_{j,k'}\right)^*, \tag{1.36}$$

where detailed information about these duals is given in section 5.7 of the reference above and where λ_i is defined by Eq. (1.21). For probability conservation, λ_i^{-1} becomes infinitely small.

The essential point of this kind of estimate consists of the fact that in matrix elements between states and duals the original wave functions are finite, so their values cannot compensate for the convergence effects of the associated product factors in (1.36). This means that the higher-order terms in the mapping formulas can be disregarded, and only the genuine scalar products in these formulas remain.

CHAPTER 2

Composite Particle States Above the Ground State

2.1 RELATIVISTIC EQUATIONS FOR COMPOSITE BOSONS

In the covariant formulation of the spinor field dynamic, wave equations for single composite particles can be derived (Borne, Lochak, and Stumpf, 2001; Stumpf and Borne, 1994). The latter states are used for the construction of many-composite particle states that are required for the application of weak mapping theorems in order to obtain effective theories, such as the extended Standard Model.

However, before beginning a detailed discussion, it should be emphasized that in the framework of such a theory, the existence of composite photon states is not in contradiction to "no go" theorems [cf. Weinberg and Witten (1980)]. This is because these theorems do not take into account the consequences of regularization.

We start with the hard-core equations for composite photon states, where a highly symbolic notation is used in order to obtain clearly organized expressions. In particular, the following definitions are applied:

- $\mathbf{r} \in R^3$, $x \in M^4$, and $Z = (i, \kappa, \alpha)$, where κ means superspin-isospin index, α is the Dirac spinor index, and i is the auxiliary field index. The latter index characterizes the subfermion fields that are needed for the regularization procedure.

Let $\varphi_{Z_1 Z_2}(x_1, x_2)$ be the covariant, antisymmetric state amplitude of the composite particle (quantum). In the algebraic representation of quantum field theory, these amplitudes are expressed by time-ordered transition matrix elements:

$$\varphi_{Z_1 Z_2}(x_1, x_2) := \langle 0 | T\{\psi_{Z_1}(x_1)\psi_{Z_2}(x_2)\} | a \rangle, \tag{2.1}$$

where $\langle 0|$ is the ground state and $|a\rangle$ an excited state above the ground state. The state $\langle 0|$, generally known as a vacuum, is, even for conserved symmetries, a nontrivial state:

> *According to present ideas there is no vacuum in the ordinary sense of tranquil nothingness. There is instead a fluctuating quantum vacuum.*
>
> ***Milonni (1994)***

This *a fortiori* holds for the case of broken symmetry, where in particular the antisymmetry of the amplitudes [Eq. (2.1)] can no longer be maintained and where a material medium is present. We first discuss the case of conserved symmetries.

For the amplitudes [Eq. (2.1)], the following set of covariant photon equations can be derived:

$$\begin{aligned} &\left[D^{\mu}_{Z_1 X_1}\partial_\mu(x_1) - m_{Z_1 X_1}\right]\varphi_{X_1 Z_2}(x_1, x_2) \\ &\quad = 3U_{Z_1 X_2 X_3 X_4}F_{X_4 Z_2}(x_1 - x_2)\varphi_{X_2 X_3}(x_1, x_1) \end{aligned} \tag{2.2}$$

and owing to antisymmetry simultaneously with Eq. (2.1), the equation

$$\left[D^{\mu}_{Z_2X_2}\partial_\mu(x_2) - m_{Z_2X_2}\right]\varphi_{Z_1X_2}(x_1,x_2)$$
$$= -3U_{Z_2X_2X_3X_4}F_{X_4Z_1}(x_2 - x_1)\varphi_{X_2X_3}(x_2,x_2) \tag{2.3}$$

has to hold. In this representation, the following definitions are used:

$$D^{\mu}_{Z_1Z_2} := i\gamma^{\mu}_{\alpha_1\alpha_2}\delta_{\kappa_1\kappa_2}\delta_{i_1i_2} \tag{2.4}$$

and

$$m_{Z_1Z_2} := m_{i_1}\delta_{\alpha_1\alpha_2}\delta_{\kappa_1\kappa_2}\delta_{i_1i_2}, \tag{2.5}$$

and the propagator is referred to an experimental situation where no material medium is present and describes the effect of the vacuum in case of conserved symmetry It reads:

$$F_{Z_1Z_2}(x_1 - x_2) := -i\lambda_{i_1}\delta_{i_1i_2}\gamma^5_{\kappa_1\kappa_2}\left[\left(i\gamma^\mu\partial_\mu(x_1) + m_{i_1}\right)C\right]_{\alpha_1\alpha_2}\Delta(x_1 - x_2, m_i), \tag{2.6}$$

with $\Delta(x_1 - x_2, m_{i_1})$ as the scalar Feynman propagator.

The meaning of the index κ can be explained by decomposing it into two parts: $\kappa := (\Lambda, A)$, with $\Lambda = 1, 2$ as the index of spinors and charge-conjugated spinors; and A = 1,2 as the isospin index. This can be equivalently expressed by $\kappa = 1, 2, 3, 4$. cf. Stumpf and Borne (1994, eq. (5.7)).

The vertex terms in Eqs. (2.2) and (2.3) are fixed by the definition:

$$U_{Z_1Z_2Z_3Z_4} := \lambda_{\iota_1}B_{i_2i_3i_4}V^{\kappa_1\kappa_2\kappa_3\kappa_4}_{\alpha_1\alpha_2\alpha_3\alpha_4}, \tag{2.7}$$

where $B_{i_2i_3i_4}$ indicates the summation over the auxiliary field indices and where the vertex is given by a scalar and a pseudoscalar coupling of the subfermion fields:

$$V^{\kappa_1\kappa_2\kappa_3\kappa_4}_{\alpha_1\alpha_2\alpha_3\alpha_4} := \frac{g}{2}\left\{\left[\delta_{\alpha_1\alpha_1}C_{\alpha_3\alpha_4} - \gamma^5_{\alpha_1\alpha_2}\left(\gamma^5C\right)_{\alpha_3\alpha_4}\right]\right.$$
$$\left.\times\,\delta_{\kappa_1\kappa_2}\left[\gamma^5\left(1-\gamma^0\right)\right]_{\kappa_3\kappa_4}\right\}_{as[2,3,4]}. \tag{2.8}$$

The parameter λ_i originates from the regularization procedure and fulfills the conditions $\sum_i\lambda_i = 0$ and $\sum_i\lambda_\iota m_i = 0$, which guarantee the finiteness of the regularized expressions.

The properties of these equations can be summarized in the following points [see Borne, Lochak, and Stumpf (2001); Stumpf and Borne (1994); Stumpf (2002, 2003c, 2004b)]:

1. These equations are relativistically invariant, quantum mechanical many-body equations.
2. They contain nontrivial interactions among their constituents.

3. The interactions are mediated by causal functions.
4. The solutions of the secular equations for the mass values are finite by inherent self-regularization.
5. These equations admit a probability interpretation with finite norms for the corresponding wave functions.
6. For vanishing coupling constant $g = 0$, de Broglie's original fusion equations for local photons are obtained.
7. A Schrödinger representation for equal times $t_1 = t_2 = t$ exists exactly.
8. All the previous statements also hold for any n.

Therefore, these equations can be considered to be generalized de Broglie—Bargmann-Wigner equations. Although Bargmann and Wigner published their analysis of de Broglie's original equations 16 years after his first paper, in the Anglo-American literature, these equations are frequently named *Bargmann-Wigner equations*. So to do justice to their inventor and to facilitate identification, I chose the combinations used previously.

Eq. (2.1) admits exact solutions that are antisymmetric, and for a solution of the set [Eqs. (2.2) and (2.3)], only Eq. (2.2) has to be used if the wave functions are antisymmetric. We only give the result of such calculations; for more detail, see Borne, Lochak, and Stumpf (2001), Stumpf and Borne (1994), and Stumpf (2002, 2003c, 2004a)

For vector bosons, the exact solutions read in a general form, as follows:

$$\varphi_{Z_1 Z_2}(x_1, x_2) = T^a_{\kappa_1 \kappa_1} \exp\left[- i\frac{\kappa}{2}(x_1 + x_2)\right] A^\mu \chi^{i_1, i_2}_{\mu, \alpha_1 \alpha_2}(x_1 - x_2|\kappa) \tag{2.9}$$

If one excludes states with fermion number 2, which decouple from the effective dynamics, for electroweak bosons, the superspin-isospin part is given by a singlet and a triplet matrix representation and defined by the following sets of symmetric and antisymmetric matrices:

$$S^l := \begin{pmatrix} 0 & \sigma^l \\ (-1)^{l+1}\sigma^l & 0 \end{pmatrix}; \quad T^l := \begin{pmatrix} 0 & \sigma^l \\ (-1)^l \sigma^l & 0 \end{pmatrix}, \quad l = 1, 2, 3 \tag{2.10}$$

for the triplet and

$$S^0 := \begin{pmatrix} 0 & \sigma^0 \\ \sigma^0 & 0 \end{pmatrix}; \quad T^0 := \begin{pmatrix} 0 & \sigma^0 \\ -\sigma^0 & 0 \end{pmatrix} \tag{2.11}$$

for the singlet.

The quantum numbers of the bosons can be expressed by the eigenvalues of Q (charge operator) and F (fermion number operator), where Q and F are defined in superspin-isospin space in accordance with quantum field theoretic definitions. One obtains the following result [see Borne, Lochak, and Stumpf (2001); Stumpf and Borne (1994)]:

$$Q = \left(G_3 + \frac{1}{2} F \right) \tag{2.12}$$

with the isospin generator

$$G_3 := \frac{1}{2} \begin{pmatrix} \sigma^3 & 0 \\ 0 & -\sigma^3 \end{pmatrix}_{K_1 K_2} \tag{2.13}$$

and the fermion number generator

$$F := \frac{1}{3} \begin{pmatrix} 1 & 0 \\ 0 & -1 \end{pmatrix}_{K_1 K_2} \tag{2.14}$$

In the two-body system, an eigenstate of Eqs. (2.12) and (2.14) is represented by a superspin-isospin tensor $\boldsymbol{\Theta}$, which satisfies the conditions

$$Q_{K_1 K} \Theta_{K K_2} + \Theta_{K_1 K} Q_{K K_2} = q \Theta_{K_1 K_2} \tag{2.15}$$

and

$$F_{K_1 K} \Theta_{K K_2} + \Theta_{K_1 K} F_{K K_2} = f \Theta_{K_1 K_2} \tag{2.16}$$

According to Eq. (2.9), corresponding eigenstates can be formed by linear combinations of elements of Eq. (2.10) or (2.11), respectively. This gives for the antisymmetric T of Eq. (2.9) the following combinations for charge and fermion quantum numbers:

$$\begin{aligned} \frac{1}{2}\left(T^0 + T^3\right) &= \Theta_1^a(q = 0,\ f = 0) \\ \frac{1}{2}\left(T^0 - T^3\right) &= \Theta_2^a(q = 0,\ f = 0) \\ \frac{1}{2}\left(T^1 + T^2\right) &= \Theta_3^a(q = 1,\ f = 0) \\ \frac{1}{2}\left(T^1 - T^2\right) &= \Theta_4^a(q = -1,\ f = 0) \end{aligned} \tag{2.17}$$

Without additional symmetry breaking, the states [Eq. (2.17)] are associated with the quantum numbers of electroweak vector bosons, where the photon and the Z-boson correspond to linear combinations of Θ_1^a and

Θ_2^a that make after isospin symmetry breaking the Z-boson massive, while the photon remains massless. By the same mechanism, the charged bosons acquire masses. For a discussion of this mechanism, see Borne, Lochak, and Stumpf (2001).

2.2 PCT- AND CP-INVARIANT FERMION PROPAGATORS

As the propagator represents the "vacuum" (i.e., the ground state), its symmetry properties are basic for the following treatment and thus deserve special attention. First, one has to know the symmetry properties of the propagator for conserved symmetries, as no understanding of its change under symmetry breaking can be achieved otherwise.

This propagator is given by Eq. (2.6) and reads equivalently:

$$F_{Z_1Z_2}(x_1,x_2) = -i(2\pi)^{-4}\lambda_{i_1}\delta_{i_1i_2}\gamma^5_{\kappa_1\kappa_2} \times \int d^4p\left(\frac{\gamma^\mu p_\mu + m_{i_1}}{p^2 - m_{i_1}^2 + i\varepsilon}C\right)_{\alpha_1\alpha_2} e^{-ip(x_1-x_2)}. \tag{2.18}$$

The meaning of the indices was explained after Eq. (2.6); cf. Borne, Lochak, and Stumpf [2001, Eq. (6.7)].

The propagator is a free fermion propagator with superspin-isospin degrees and was used in the preceding calculations. But note that it is not a quantity for its own sake. This propagator is used to act as the initial condition for the start of nonperturbative calculations containing the unobservable subfermions as constituents of the observable physical particles. Calculations which for instance are done by solutions of the generalized de Broglie—Bargmann-Wigner (GBBW) equations and the weak mapping formalism. Justification of this procedure is ultimately the comparison of its results with experiments.

For discrete transformations P, C, and T, the tranformation properties of the general spinor fields were discussed in Stumpf (2003a). These transformation formulas represent an extension of the corresponding formulas of the conventional theory by the inclusion of the superspin-isospin indices of the spinor fields.

In this formalism, the action of a PCT-transformation on the spinor fields is defined by the antiunitary (or even more general) operator $\mathcal{A} := \mathcal{PCT}$ and leads to the following definition [Stumpf, 2003a, Eq. (35)]:

$$\psi'(x) := \mathcal{A}\psi_{\kappa\alpha i}(x)\mathcal{A}^{-1} = \left(\gamma^5\gamma^0\right)_{\kappa\kappa'}\left(\gamma^5\gamma^0 C\right)_{\alpha\alpha'}\psi_{\kappa'\alpha' i}(x') \tag{2.19}$$

with $x' = -x$. Eq. (2.19) holds for free fields χ as well, as the transformation laws for free and interacting fields are the same.

Furthermore, by combining the formulas [Stumpf (2003a, Eqs. (3) and (36)], one obtains for the unitary PC-transformation the formula

$$\psi'(x) = \mathcal{U}\psi_{\kappa\alpha i}(x)\mathcal{U}^{-1} = \left(\gamma^0\gamma^5\right)_{\kappa\kappa'}\gamma^0\psi_{\kappa'\alpha' i}(x'), \tag{2.20}$$

with $\mathcal{U} := \mathcal{CP}$ and $x' = (-\mathbf{r}, t)$.

If these formulas are decomposed and retranslated into the conventional representation of the spinor fields, they coincide with those of Gross (1999), Fonda and Ghirardi (1970), and Lurie (1968). The existence of such transformations implies relations between original and transformed matrix elements, as described next.

Let $\mathcal{O}$ be any element of the field operator algebra and $|a\rangle$ and $|b\rangle$ elements of a corresponding state space. Then, for antiunitary transformations, the relation

$$\langle a|\mathcal{O}|b\rangle = \langle a'|\mathcal{O}'|b'\rangle^* = \langle b'|(\mathcal{O}')^+|a'\rangle \tag{2.21}$$

can be derived [cf. Gross 1999, Eq. (8.84)], where the primed quantities are defined by $|b'\rangle = \mathcal{A}|b\rangle$ and $\mathcal{O}' = \mathcal{A}\mathcal{O}\mathcal{A}^{-1}$. The transformations [Eqs. (2.19) and (2.21)] are compatible; see Gross (1999, p. 225).

For unitary transformations, one obtains for matrix elements

$$\langle a|\mathcal{O}|b\rangle = \langle a'|\mathcal{O}'|b'\rangle, \tag{2.22}$$

with $|b'\rangle = \mathcal{U}|b\rangle$ and $\mathcal{O}' = \mathcal{U}\mathcal{O}\mathcal{U}^{-1}$.

The covariant propagator [Eq. (2.18)] can be expressed as a special matrix element

$$F_{Z_1 Z_2}(x_1, x_2) := \langle 0|\mathcal{O}_{Z_1 Z_2}(x_1, x_2)|0\rangle, \tag{2.23}$$

with

$$\mathcal{O} := \left[\Theta(t_1 - t_2)\chi_{Z_1}(x_1)\chi_{Z_2}(x_2) - \Theta(t_2 - t_1)\chi_{Z_2}(x_2)\chi_{Z_1}(x_1)\right], \tag{2.24}$$

where $|0\rangle$ is the physical ground state of the system and $\chi := \psi^f$ are the associated free spinor fields to the general fields ψ.

In Wigner's definition of the symmetry of a quantum system, the matrix relations [Eqs. (2.21) and (2.22)] can be used to show the invariance of the transition probability between states under such antiunitary or unitary tranformations.

But Wigner's definition of the symmetry of quantum systems is not suitable for the characterization of the symmetry properties of the propagator.

A better insight into these properties can be gained by the following definition, which corresponds to the second active point of view [see Fonda and Ghirardi (1970, p. 45)].

Consider the sets $\{|b\rangle\}$ and $\{\mathcal{A}|b\rangle \equiv |b'\rangle\}$ or $\{\mathcal{U}|b\rangle \equiv |b'\rangle\}$ as different representation spaces of the field operator algebra, defined by

$$F_{Z_1Z_2}(x_1,x_2)' := \langle 0'|\mathcal{O}_{Z_1Z_2}(x_1,x_2)|0'\rangle \tag{2.25}$$

as the corresponding different representations of the operator $\mathcal{O}_{Z_1Z_2}(x_1,x_2)$. Then the following theorem can be derived:

Proposition 1: Under the change of the representation spaces for PCT and CP transformations, the integral representation [Eq. (2.18)] of the free fermion propagator is form-invariant.

Proof: (i) According to the transformation laws for states and operators, the *identity*

$$\langle a'|\mathcal{O}'|b'\rangle^* = \langle a|\mathcal{A}^{-1}\mathcal{A}\mathcal{O}\mathcal{A}^{-1}\mathcal{A}|b\rangle^* \tag{2.26}$$

holds and is valid in particular for $\mathcal{O}$ as given by Eq. (2.24).

The operator $\mathcal{A}$ acts exclusively in the state space. Real c-number functions like Θ are not changed by its application. Therefore, Eq. (2.26) can be rewritten as

$$\begin{aligned}\langle 0'|\mathcal{O}'|0'\rangle^* = \Theta(t_1-t_2)\langle 0'|\mathcal{A}\chi_{Z_1}(x_1)\mathcal{A}^{-1}\mathcal{A}\chi_{Z_2}(x_2)\mathcal{A}^{-1}|0'\rangle^* \\ -\Theta(t_2-t_1)\langle 0'|\mathcal{A}\chi_{Z_2}(x_2)\mathcal{A}^{-1}\mathcal{A}\chi_{Z_1}(x_1)\mathcal{A}^{-1}|0'\rangle^*.\end{aligned} \tag{2.27}$$

If intermediate states are introduced with $\langle 0'|\chi'_{Z_1}(x')|b\rangle^* = \langle b|\chi'_Z(x')^+|0'\rangle$ in Eq. (2.27), one obtains from Eqs. (2.21) and (2.27) the expression

$$\begin{aligned}\langle 0|T[\chi_{Z_1}(x_1)\chi_{Z_2}(x_2)]|0\rangle = \Theta(t_1-t_2)\langle 0'|\chi'_{Z_2}(x_2)^+ + \chi'_{Z_1}(x_1)^+|0'\rangle \\ -\Theta(t_2-t_1)\langle 0'|\chi'_{Z_1}(x_1)^+ + \chi'_{Z_2}(x_2)^+|0'\rangle,\end{aligned} \tag{2.28}$$

where χ' is the $\mathcal{A}$-transform of χ and $|0'\rangle = \mathcal{A}|0\rangle$.

Because the right side of Eq. (2.28) is no proper time-ordered product, it must be rearranged to allow a reasonable interpretation. In superspinor-isospinor notation, the general relation

$$\psi_{\kappa\alpha i}(x)^+ := (\gamma^0 C)_{\alpha\alpha'}\gamma^5_{\kappa\kappa'}\psi_{\kappa'\alpha' i}(x) \tag{2.29}$$

can be derived. This relation holds for any spinor field and thus also for $\chi'(x)^+$. If Eq. (2.19) is also applied to $\chi'(x)$ in this case, the combination of both formulas yields

$$\chi'_{\kappa\alpha i}(x)^+ := \gamma^0_{\kappa\kappa'}\gamma^5_{\alpha\alpha'}\chi_{\kappa'\alpha' i}(x'). \tag{2.30}$$

Substitution of Eq. (2.30) into Eq. (2.28), and owing to Eq. (2.19) the replacement of t by $-t_j$, leads to the relation

$$F^{\kappa_1\kappa_2}_{\alpha_1\alpha_2}(x_1,x_2)_{i_1,i_2} = -\gamma^0_{\kappa_1\kappa'_1}\gamma^0_{\kappa_2\kappa'_2}\gamma^5_{\alpha_1\alpha'_1}\gamma^5_{\alpha_2\alpha'_2}\left\langle 0'\middle| T'\left[\chi_{\kappa'_1\alpha'_1 i_1}(x'_1)\chi_{\kappa'_2\alpha'_2 i_2}(x'_2)\right]\middle|0'\right\rangle. \tag{2.31}$$

The right side of Eq. (2.31) is the representation of the operator $\mathcal{O}$ defined in (2.24) at the points x'_1, x'_2 in the transformed state space with the transformed vacuum $|0'\rangle$. It is thus the propagator referred to in these transformed states.

In accordance with Eq. (2.25), we therefore define

$$F^{\kappa_1\kappa_2}_{\alpha_1\alpha_2}(x'_1,x'_2)'_{i_1,i_2} := \langle 0'|T'\left[X_{\kappa_1\alpha_1 i_1}(x'_1)X_{\kappa_2\alpha_2 i_2}(x'_2)\right]|0'\rangle, \tag{2.32}$$

where time-ordering T′ is referred to the primed coordinates. Then Eq. (2.31) can be resolved for F', which gives

$$F^{\kappa_1\kappa_2}_{\alpha_1\alpha_2}(x'_1,x'_2)'_{i_1,i_2} = -\gamma^0_{\kappa_1\kappa'_1}\gamma^0_{\kappa_2\kappa'_2}\gamma^5_{\alpha_1\alpha'_1}\gamma^5_{\alpha_2\alpha'_2}F^{\kappa'_1\kappa'_2}_{\alpha'_1\alpha'_2}(x_1,x_2)_{i_1,i_2}. \tag{2.33}$$

If on the right side of Eq. (2.33) the integral representation [Eq. (2.18)] of the propagator is inserted, it can be verified that F' has the same integral representation at the corresponding points x'_1, x'_2; i.e., F is forminvariant under PCT-transformation.

ii) Concerning the CP transformation, it is unitary in contrast to the anti-unitary PCT-transformation. In this case, no special transformation for the matrix elements is required other than the transformation of the field operators. Then, with Eq. (2.22), one gets the propagator relation

$$\langle 0|T\left[\chi_{Z_1}(x_1)\chi_{Z_2}(x_2)\right]|0\rangle = \langle 0'|T\left[\chi'_{Z_1}(x_1)\chi'_{Z_2}(x_2)\right]|0'\rangle \equiv F' \tag{2.34}$$

as $\mathcal{U}$ commutes with time ordering.

Owing to Eq. (2.20), Eq. (2.34) can equivalently be written

$$F^{\kappa_1\kappa_2}_{\alpha_1\alpha_2}(x_1,x_2)_{i_1,i_2} = (\gamma^0\gamma^5)_{\kappa_1\kappa'_1}(\gamma^0\gamma^5)_{\kappa_2\kappa'_2}\gamma^0_{\alpha_1\alpha'_1}\gamma^0_{\alpha_2\alpha'_2}F^{\kappa'_1\kappa'_2}_{\alpha'_1\alpha'_2}(x'_1,x'_2)'_{i_1,i_2}. \tag{2.35}$$

This relation can be resolved for F'. If in the resulting equation the integral representation [Eq. (2.18)] for F is inserted, the algebra can be directly evaluated. In the resulting integral, the transformation $p' = (-\mathbf{p}, p_0)$ can be performed, which eventually gives $F(x'_1, x'_2)$. Therefore, one obtains from this relation

$$F^{\kappa_1\kappa_2}_{\alpha_1\alpha_2}(x'_1, x'_2)'_{i_1,i_2} = F^{\kappa_1\kappa_2}_{\alpha_1\alpha_2}(x'_1, x'_2)_{i_1,i_2}, \tag{2.36}$$

which means forminvariance under CP transformation. ◇

For the next discussion, we employ the inhomogeneous Dirac equation, which is satisfied by the free fermion propagator [Eq. (2.18)]. In the notation of Borne, Lochak, and Stumpf [2001, Eq. (3.109)], it reads

$$\left[D^\mu \partial_\mu(x_1) - m\right]_{ZZ_1} F_{Z_1Z_2}(x_1, x_2) = D^0_{ZZ_1} A_{Z_1Z_2} \delta(x_1 - x_2). \tag{2.37}$$

For this equation, the following theorem holds:

Proposition 2: The propagator equation [Eq. (2.37)] is forminvariant under PCT and CP transformations.

Proof: With explicit indexing, Eq. (2.37) reads as

$$\begin{aligned}&\left[\iota\gamma^\mu_{\alpha\alpha_1}\partial_\mu(x_1) - m_i\delta_{\alpha\alpha_1}\right]\delta_{\kappa\kappa_1}\delta_{ii_1}F^{\kappa_1\kappa_2}_{\alpha_1\alpha_2}(x_1, x_2)_{i_1,i_2}\\ &\quad = -\iota C_{\alpha_1\alpha_2}\gamma^5_{\kappa\kappa_2}\lambda_i\delta_{ii_1}\delta(x_1 - x_2).\end{aligned} \tag{2.38}$$

i) For a PCT-transformation replace in Eqs. (2.31) on the right side by (2.32) and insert the resulting expression into Eq. (2.38). This yields

$$\begin{aligned}&\left[\iota\gamma^\mu_{\alpha\alpha_1}\partial_\mu(x_1) - m_i\delta_{\alpha\alpha_1}\right]\delta_{\kappa\kappa_1}\delta_{ii_1}\gamma^0_{\kappa_1\kappa'_1}\gamma^0_{\kappa_2\kappa'_2}\gamma^5_{\alpha_1\alpha'_1}\gamma^5_{\alpha_2\alpha'_2}F^{\kappa'_1\kappa'_2}_{\alpha'_1\alpha'_2}(x'_1, x'_2)'_{i_1,i_2}\\ &\quad = \iota C_{\alpha\alpha_2}\gamma^5_{\kappa\kappa_2}\lambda_i\delta_{ii_2}\delta(x'_1 - x'_2).\end{aligned} \tag{2.39}$$

Elimination of the four matrices (γ^0, γ^5) in Eq. (2.39) by multiplication of the whole equation by their dual matrices and with transformation $x = -x'$ in the partial derivative, this equation yields

$$\begin{aligned}&\left[\iota\gamma^\mu_{\beta_1\alpha'_1}\partial_\mu(x'_1) - m_i\delta_{\alpha\alpha'_1}\right]\delta_{ii_1}\delta_{\lambda_1\kappa'_1}F^{\kappa'_1\lambda_2}_{\alpha'_1\beta_2}(x'_1, x'_2)'_{i_1,i_2}\\ &\quad = -\iota C_{\beta_1\beta_2}\gamma^5_{\lambda_1\lambda_2}\lambda_i\delta_{ii_2}\delta(x'_1 - x'_2);\end{aligned} \tag{2.40}$$

comparison of (2.40) with (2.38) shows forminvariance of Eq. (2.30) under PCT transformation.

ii) Concerning the CP transformation, we substitute Eq. (2.35) into Eq. (2.38). This gives

$$\begin{aligned}&\left[\iota\gamma^{\mu}_{\alpha\alpha_1}\partial_\mu(x_1) - m_i\delta_{\alpha\alpha_1}\right]\delta_{\kappa\kappa_1}\delta_{ii_1}\left(\gamma^0\gamma^5\right)_{\kappa_1\kappa_1'}\left(\gamma^0\gamma^5\right)_{\kappa_2\kappa_2'}\gamma^0_{\alpha_1\alpha_1'}\gamma^0_{\alpha_2\alpha_2'}\\ &\times F^{\kappa_1'\kappa_2'}_{\alpha_1'\alpha_2'}\left(x_1', x_2'\right)'_{i_1,i_2} = \iota C_{\alpha\alpha_2}\gamma^5_{\kappa\kappa_2}\lambda_i\delta_{ii_2}\delta(x_1 - x_2),\end{aligned} \tag{2.41}$$

with $x' = (-\mathbf{r}, t)$. Eliminating the four transformation matrices $[(\gamma^0\gamma^5)$, etc.] by multiplication of the whole equation by their dual matrices and with $\mathbf{r} = \mathbf{r}'$ in the partial derivative and the δ-distribution, one finally obtains

$$\begin{aligned}&\left[\iota\gamma^{\mu}_{\alpha\alpha_1}\partial_\mu\left(x_1'\right) - m_i\delta_{\alpha\alpha_1}\right]\delta_{kk_1}\delta_{ii_1}F^{\kappa_1\kappa_2}_{\alpha_1\alpha_2}\left(x_1'x_2'\right)'_{i_1 i_2}\\ &= -\iota C_{\alpha\alpha_2}\gamma^5_{\kappa\kappa_2}\lambda_i\delta_{ii_2}\delta\left(x_1' - x_2'\right);\end{aligned} \tag{2.42}$$

i.e., equation (2.38) is forminvariant under CP transformation. ◇

Next, we consider the propagator and its equation under the influence of an electromagnetic field. The propagator is referred to superspin-isospin states of parton fields, which means that the coupling of partons to an electromagnetic field has to be derived. The effective SU(2) ⊗ U(1)−gauge theory of composite electroweak vector bosons in interaction with partons was extensively studied by weak mapping calculations by Grimm (1994b) and Pfister (1993). This effective theory is identical to the corresponding phenomenological gauge theory of the interaction of electroweak gauge bosons with fermions.

For the sake of brevity, we will skip the corresponding calculations which, among other things, lead to the covariant derivative for the coupling of partons to electromagnetic fields:

$$D_\mu := \partial_\mu - ieQ_{\mathrm{kk}'}A_\mu(x), \tag{2.43}$$

where the charge operator Q of the parton fields ψ is given in superspinor-isospinor representation by

$$Q_{\kappa\kappa'} := \frac{1}{3}\begin{pmatrix} 2 & 0 & 0 & 0\\ 0 & -1 & 0 & 0\\ 0 & 0 & -2 & 0\\ 0 & 0 & 0 & 1\end{pmatrix} \equiv \frac{1}{6}\left(\gamma^0 - 3\gamma^5\gamma^3\right)_{\kappa\kappa'}; \tag{2.44}$$

see Borne, Lochak, and Stumpf [2001, Eq. (6.103)] and the comments therein.

Thereforem due to Eqs. (2.43) and (2.44), in this case the propagator equation reads

$$\left\{\left[\iota\gamma^{\mu}_{\alpha\alpha_1}\partial_\mu(x_1) - m_i\delta_{\alpha\alpha_1}\right]\delta_{\kappa\kappa_1} + e\gamma^{\mu}_{\alpha\alpha_1}Q_{\kappa\kappa_1}A_\mu(x_1)\right\}\delta_{ii_1}F^{\kappa_1\kappa_2}_{\alpha_1\alpha_2}(x_1,x_2)_{i_1i_2}$$
$$= -\iota C_{\alpha\alpha_2}\gamma^5_{\kappa\kappa_2}\,\lambda_i\delta_{ii_2}\delta(x_1 - x_2). \tag{2.45}$$

Then for the PCT and CP transformations of Eq. (2.45) with external fields, the following transformation properties must be postulated for these fields in order to achieve a uniform transformation behavior of the entire equation under these operations.

Proposition 3: The propagator equation [Eq. (2.45)] is forminvariant under PCT and CP transformations provided that the relations

$$\mathcal{A}A_\mu(x)\mathcal{A}^{-1} = -A_\mu(-x), \quad UA_\mu(x)U^{-1} = -\eta^{\mu\mu'}A_{\mu'}(-\mathbf{r},t) \tag{2.46}$$

are satisfied.

Proof: The proof runs along the lines of Proposition 2. Hence, in this context, we only have to study the coupling term in Eq. (2.45).

i) If a PCT transformation [Eqs. (2.31) and (2.32)] is substituted into Eq. (2.45), the minus sign of Eq. (2.31) is removed from Eq. (2.45) by multiplication of the whole equation by (-1). Thus, one obtains for the transformed coupling term:

$$e\gamma^{\mu}_{\alpha_1\alpha'_1}Q_{\kappa\kappa_1}A_\mu(x_1)\gamma^0_{\kappa_1\kappa'_1}\gamma^0_{\kappa_2\kappa'_2}\gamma^5_{\alpha_1\alpha'_1}\gamma^5_{\alpha_2\alpha'_2}\delta_{ii_1}F^{\kappa'_1\kappa'_2}_{\alpha'_1\alpha'_2}(x'_1,x'_2)'_{i_1i_2}$$
$$= e\gamma^0_{\kappa\kappa_1}\gamma^0_{\kappa_2\kappa'_2}\gamma^5_{\alpha_1\alpha_1}\gamma^5_{\alpha_2\alpha'_2}\gamma^{\mu}_{\alpha_1\alpha'_1}Q_{\kappa_1\kappa'_1}A_\mu(x_1)\delta_{ii_1}F^{\kappa'_1\kappa'_2}_{\alpha'_1\alpha'_2}(x'_1,x'_2)'_{i_1i_2}. \tag{2.47}$$

In the first line, the minus sign of Eq. (2.31) drops out and $[Q,\gamma^0]_- = 0$. Furthermore, from Eq. (2.46), it follows with $A'_\mu(x) \equiv \mathcal{A}A_\mu(x)\mathcal{A}^{-1}$ that $-A'_\mu(-x) = A_\mu(x)$ or, equivalently, $A_\mu(x) = -A'_\mu(x')$. If the latter relation is substituted into the second line of Eq. (2.47), and if in analogy to Proposition 2 the transformation matrices are eliminated from the propagator equation, the following transformed equation results:

$$\left\{\left[\iota\gamma^{\mu}_{\beta\alpha'_1}\partial_\mu(x'_1) - m_i\delta_{\beta\alpha'}\right]\delta_{\lambda\kappa_1} + e\gamma^{\mu}_{\beta\alpha'_1}Q_{\lambda_1\kappa'_1}A'_\mu(x'_1)\right\}\delta_{ii_1}F^{\kappa'_1\lambda_2}_{\alpha'_1\beta_2}(x'_1,x'_2)_{i_1i_2}$$
$$= -\iota C_{\beta_1\beta_2}\gamma^5_{\lambda_1\lambda_2}\,\lambda_i\delta_{ii_2}\delta(x'_1 - x'_2); \tag{2.48}$$

i.e., Eq. (2.45) is forminvariant under PCT transformations.

ii) In a similar way, the forminvariance under PC transformations of Eq. (2.45) can be verified. ◇

Addendum: The transformation relations [Eq. (2.46)] coincide with those of the phenomenological theory; see Fonda and Ghirardi [1970, Eqs. (6.276),(6.149), and (6.164a)].

2.3 SPINOR FIELD VERSION OF LOCHAK'S PHOTON THEORY

Although de Broglie's fusion theory and the spinor field theory are based on the same physical idea, their theoretical treatments are completely different. To get confidence in the weak mapping method, which is essential for the spinor field theory, the results of the extended de Broglie-Lochak photon theory will be compared with those of a simplified spinor field test model with pointlike fermions and composite bosons.

The photon fusion theory of de Broglie and Lochak refers to conserved symmetries and to the coupling of composite photons to observable pointlike fermions (leptons, quarks, etc.). In order to compare the results of this theory with the test model, the latter model has been evaluated under the same assumptions as those that are valid for the fusion theory.

Owing to this postulated structural equivalence of both models, a comparison is of special interest because in contrast to conventional belief, Lochak's photon theory is *not* a theory of spin 1 particles but a theory of *maximum* spin 1 particles, which also automatically includes spin 0 parts (i.e., scalar boson fields). For the interpretation, we refer to Lochak's contribution in Part 1 of this book.

For conserved symmetries, the propagator of the pointlike fermions in Eqs. (2.2) and (2.3) must be identified with Eq. (2.18). In this case, the boson equations can be exactly solved and their solutions are thus available for weak mapping calculations. For such mappings, the chain rule can be used, which is considerably easier to handle than the exact mapping theorems. Although the chain rule mapping is closely related to symmetry breaking (cf. section 1.4) of this part, it can also be applied in the case of conserved symmetries, if only the derivation of effective theories is intended

Stumpf (1986b) first introduced the chain rule for the derivation of an effective SU(2) vector boson theory. This approach was followed by several papers treating the abovementioned simplified spinor field model by the same method, but with increasing accuracy.

In addition to the thesis of Pfister (1990), which was acknowledged in Borne, Lochak, and Stumpf (2001), the findings of Grimm (1994a, 1994b) and of Fuss (2004) must be mentioned. Grimm treated the group theoretical implications of an $SU(2) \otimes U(1)$ effective gauge theory with composite bosons but pointlike fermions, while Fuss treated an $SU(3) \otimes U(1)$ effective gauge theory with composite bosons (gluons) and pointlike fermions (quarks) by extensive state calculations.

In particular, the results of Grimm are suitable for comparison with the extended de Broglie–Lochak photon theory. A prerequisite to understand Grimm's results is to understand the formation of the elements of the state space by the boson functions (for pointlike fermions, the state space is trivial—i.e., free Dirac waves). With respect to the boson functions, one must remember that according to Eqs. (2.2) and (2.3), the boson solutions of the GBBW equations are already determined by their local amplitudes $\varphi(x,x)$. This pecularity holds only for $n = 2$.

For comparison, the numbering of the formulas of Grimm's original paper is used and marked by the label *Gr.*, Grimm (1994b)

One gets for the local boson amplitude

$$\begin{aligned}\varphi(x,x) = {} & A^{\mu}(x)\gamma_{\mu}C + \frac{1}{2}F^{\mu\nu}(x)\Sigma_{\mu\nu}C \\ & + G^{\mu}(x)\gamma_{\mu}\gamma_5 C + i\varphi(x)\gamma_5 C + Z(x)C.\end{aligned} \qquad \text{(Gr3.97)}$$

Then one can define the basis functions of the state space by splitting the exact solutions into parts characterized by the electric vector potential $A^{\mu}(x)$, the magnetic axial vector potential $G^{\mu}(x)$, and the common field strength tensor $F^{\mu\nu}(x)$, as well as the scalar function Z(x) and the pseudoscalar φ (x).

The dependence of these quantities on x is justified by the influence of the surroundings on the free vector boson states if the latter are embedded in a medium and interact with it. This interaction can be expressed by variable coefficients (to be identified later) with the effective field variables, which appear as a result of the chain rule method.

To compare the de Broglie–Lochak photon theory with Grimm's results, one can use the corresponding equations, which can be found in Borne, Lochak, and Stumpf (2001, section 1.4.2). These equations are referred to a vacuum without polarization and magnetization.

Furthermore the de Broglie–Lochak photon theory is formulated without isospin. In order to obtain comparable results with Grimm's simplified $SU(2) \otimes U(1)$ spinor field model, one has to consider isospin singlet states in this theory. In the following formulas, the absence of any term referring to the $SU(2)$ structure is because of this fact, and it is important to keep in mind that Grimm's results are derived by application of the chain rule mapping but without any linear approximation.

Electric Photon spin 1 (M)—cf. Borne, Lochak, and Stumpf [2001, (1.23)] and Grimm [1994b, (Gr 5.175)]

i) dynamical field equations

$$\text{(a)}\quad -\partial_t \mathbf{H} = rot\mathbf{E} \quad \text{(b)}\quad \partial_t \mathbf{E} = rot\mathbf{H} - k_0^2 \mathbf{A}$$

ii) field definition by potentials

$$\text{(c)}\quad \mathbf{H} = rot\mathbf{A} \quad \text{(d)}\quad \mathbf{E} = -gradA^0 - \partial_t \mathbf{A}$$

iii) constraints

$$\text{(e)}\quad \partial_t A^0 + \partial_k A^k = 0 \quad \text{(f)}\quad div\mathbf{E} = \mu^2 A^0 \quad \text{(g)}\quad div\mathbf{H} = 0,$$

where V of (1.23) was renamed by A^0.

While the spin 1 part (M) is the famous result of de Broglie's fusion theory and Grimm's calculations lead to no additional insights, the spin 0 part of de Broglie's ansatz has led to his vanishing "antifields" and has only recently been correctly interpreted by Lochak.

The corresponding equations are unknown and thus will be directly compared with Grimm's group theoretical results, which in this case makes for better understanding.

Electric photon spin 0 (NM)—cf. Borne, Lochak, and Stumpf (2001, (1.24)] and Grimm (1994b, (Gr 5.176)]

This case is characterized by $I_1 = 0$, which, in accordance with Grimm's group theoretical analysis, corresponds to $Z(x) = 0$. Furthermore, for comparison, the pseudo-quadripotential (W, $\mathbf{B}$) must be renamed as (G^0, $\mathbf{G}$) and $I_2 = i\varphi_5$ as $\boldsymbol{\Phi}$. Then, putting c = 1 for simplification, the following equivalences can be established:

$$\begin{aligned}
&(a') \quad -\partial_t I_2 = k_0 W \equiv -\partial_t \boldsymbol{\Phi} = G^0 \\
&(b') \quad \partial_t W + div\mathbf{B} = k_0 I_2 \equiv \partial_t G^0 + div\mathbf{G} = k_0 \boldsymbol{\Phi} \\
&(c') \quad gradI_2 = k_0 \mathbf{B} \equiv grad\boldsymbol{\Phi} = k_0 \mathbf{G} \\
&(d') \quad gradW + \partial_t \mathbf{B} = 0 \equiv gradG^0 + \partial_t \mathbf{G} = 0.
\end{aligned}$$

Finally, the equivalence $rot\mathbf{B} = rot\mathbf{G} = 0$ follows from (c′).

In spite of these equivalences, there is a difference between the photon fusion equations and the weak mapping equations: while the fusion equations are exact, these weak mapping equations are only leading terms that are followed by higher-order terms and quantization terms. In these latter terms, the substructure of the bosons manifests itself in contrast to the local construction of the wave functions in fusion theory. For sufficiently large masses of the auxiliary fields, these latter terms are tiny and can be disregarded.

Furthermore, in the Standard Model, the photon arises as a mixture of SU(2) and U(1) parts, and in the full nonlinear spinor theory, the fermionic substructure of leptons and quarks has to be taken into account because only in this way can photons be related to weak interactions, which is essential for the formation of magnetic monopoles and physical interpretation of the corresponding processes. Thus, the simplified model of the nonlinear spinor field discussed in this section can only be a test version.

Finally, it should be mentioned that Grimm and Fuss partly published their results in journals, but these papers cannot replace their original theses if you wish to achieve a thorough understanding.

2.4 PROPAGATOR FOR SYMMETRY BREAKING EXPERIMENTS

In accordance with the statements in the introduction to this article, we assume that the experiments of Urutskoiev *et al.* should be interpreted as manifestations of a symmetry breaking of the vacuum, leading to a new inequivalent vacuum state, which has already been covered in previous studies (e.g., Stumpf, 2004b, 2005, 2006).

As a suitable canditate for this symmetry breaking, a violation of the commonly assumed CP invariance has been considered. Hence, in the following discussion, we try to show how this symmetry breaking comes about and how the theoretical counterpart of the experimental arrangement has to be formulated.

First, we give a further statement about the CP-invariant superspin-isospin propagator.

Proposition 4: For free spinor fields χ_Z, the associated superspin-isospin propagator $F_{Z_1 Z_2}(x_1 - x_2)$ contains the conventional fermion and the conventional antifermion propagators at any time ($t := \mathrm{t}_1 - \mathrm{t}_2$).

Proof: For the proof, we refer to definitions (2.22) and (2.23) with Z : = (A, Λ, α, i) the indices A and i are spectator indices and will be suppressed. This yields

$$\begin{aligned} F^{\alpha_1\alpha_2}_{\Lambda_1\Lambda_2}(x_1 - x_2) := \; & \Theta(t_1 - t_2)\langle 0|\chi_{\Lambda_1\alpha_1}(x_1)\chi_{\Lambda_2\alpha_2}(x_2)|0\rangle \\ & -\Theta(t_2 - t_1)\langle 0|\chi_{\Lambda_2\alpha_2}(x_2)\chi_{\Lambda_1\alpha_1}(x_1)|0\rangle \end{aligned} \tag{2.49}$$

Without loss of generality, we assume $t_1 - t_2 > 0$. Then Eq. (49) reads

$$F^{\alpha_1\alpha_2}_{\Lambda_1\Lambda_2}(x_1 - x_2) = \langle 0|\chi_{\Lambda_1\alpha_1}(x_1)\chi_{\Lambda_2\alpha_2}(x_2)|0\rangle$$
$$= \delta_{\Lambda_1 1}\delta_{\Lambda_2 2}\langle 0|\chi_{1\alpha_1}(x_1)\chi_{2\alpha_2}(x_2)|0\rangle$$
$$+ \delta_{\Lambda_1 2}\delta_{\Lambda_2 1}\langle 0|\chi_{2\alpha_1}(x_1)\chi_{1\alpha_2}(x_2)|0\rangle \quad (2.50)$$

Hence eq. (2.50) can be written

$$F^{\alpha_1\alpha_2}_{\Lambda_1\Lambda_2} = \begin{pmatrix} 0 & \langle 0|\chi_{\alpha_1}(x_1)\chi^C_{\alpha_2}(x_2)|0\rangle \\ \langle 0|\chi^C_{\alpha_1}(x_1)\chi_{\alpha_2}(x_2)|0\rangle & 0 \end{pmatrix}_{\Lambda_1\Lambda_2} \quad (2.51)$$

and with $\chi^C = C\overline{\chi}^T = -C\overline{\chi}$ one gets for the matrix elements of (2.51).

$$\langle 0|\chi_{\alpha_1}(x_1)\chi_{\alpha_2}(x_2)|0\rangle = C_{\alpha_2\alpha'_2}\langle 0|\chi_{\alpha_1}(x_1)\overline{\chi}_{\alpha'_2}(x_2)|0\rangle$$
$$= -iC_{\alpha_2\alpha'_2}S^+_{\alpha_1\alpha'_2}(x_1 - x_2). \quad (2.52)$$

In the same way, it follows that

$$\langle 0|\chi^c_{\alpha_1}(x_1)\chi_{\alpha_2}(x_2)|0\rangle = -iC_{\alpha_1\alpha'_1}S^-_{\alpha_2\alpha'_1}(x_2 - x_1), \quad (2.53)$$

where S^+ is the conventional particle propagator and S^- is the conventional antiparticle propagator for Dirac fermions.

Based on this theorem, one can analyze the modifications that are necessary to adapt the theoretical description of the vacuum to the experimental arrangements. Two properties are characteristic of these experiments:

i) The reactions are confined to the interior of a closed vessel.

ii) The reactions by discharges proceed within a fluid medium.

I concentrate on the theoretical description of (ii) because an essential change of the properties of the vacuum has to be expected. This is because for discharges in fluid media, the motion of the charge carriers is damped (i.e., accompanied by energy losses). One of the standard fluids used in these experiments is water.

The stopping power of matter for fast particles has been extensively discussed in Heitler (1949, section 23). For electrons in water, the following ranges have been calculated in multiples of the primary energy $m_ec^2 = 0.510$ MeV:

Primary energy	0.1	1	10	100	1000	m_ec^2
H_2O	0.47×10^{-2}	0.19	2.6	19	78	cm

For low energies, the magnitudes of these ranges fit into the dimensions of the vessels in these experiments.

The fact that positrons can be annihilated somewhere in their paths diminishes the average ranges of positrons in comparison with those of electrons. Corresponding formulas describing these differences are given in Heitler [1949, section 23, Eqs. (21) and (22)]. Apart from the importance of numerical values, the principal effect of these differences consists in the signal of a symmetry breaking. Charged particles and their antiparticles behave differently in a medium, which leads to C- or CP-symmetry breaking, respectively.

Discharges are triggered by electrons that on their way in the fluid ionize molecules, generate secondary electrons, and other similar actions. But to avoid theoretical difficulties in the description of the rather complicated processes of a discharge, we simplify the theoretical treatment by considering only an average damping effect of electrons.

In such discharges, no positrons occur and their real presence is not necessary because the inclusion of positrons is a theoretical concept to study the behavior of the system under charge conjugation. The absence of positrons in experiments, therefore, is not an argument against positrons in the theoretical treatment.

In this model, electrons and positrons are assumed to have a fermionic substructure. Therefore, the question is: do these considerations apply to their fermionic constituents as well?

The group theoretical representation of lepton states with respect to superspin-isospin combinations is exact. In particular, for the superspin part, one obtains [see Stumpf (2005, section 4)]

$$e^{+} \rightarrow \delta_{1\Lambda_1}\delta_{1\Lambda_2}\delta_{1\Lambda_3}; \quad e^{-} \rightarrow \delta_{2\Lambda_1}\delta_{2\Lambda_1}\delta_{2\Lambda_1}, \tag{2.54}$$

where Λ is the superspinor index. With respect to this index, the fermion number f is defined. With these fermion numbers, one gets for e^+ the configuration (1/3, 1/3, 1/3), while e^- leads to the set (−1/3, −1/3, −1/3). Fermion numbers are used to discriminate particles from antiparticles by changing f into $-f$ by convention.

Thus, from these sets of fermion numbers, it follows that electrons consist only of particles, and positrons only of antiparticles. In the complete representation of the wave functions, this property is not changed. Therefore, the partons of the electrons and the antipartons of the positrons share their behavior with that of electrons or positrons, respectively, and we can base our arguments on the spinor field propagator

[Eq. (2.18)] instead of the phenomenological electron-positron Feynman propagator.

In the next step, we consider the influence of damping on the motion of partons and antipartons. Their motion is described by the propagator, and if damping is effective, their motion ceases in a finite time interval. This fact can be expressed by a damping factor in the integral representation [Eq. (2.18)] of the propagator. According to Richard Feynman, this integral can be evaluated by giving the mass an infinitesimal negative imaginary part; i.e., $m \to m - i\delta$, $\delta > 0$. If δ is allowed to have a finite value, this leads for $t_1 - t_2 > 0$ to a damping factor $\exp[-\delta'(t_1 - t_2)]$ with $\delta' := m\delta$, while for $t_1 - t_2 < 0$, one obtains the damping factor $\exp[\delta'(t_1 - t_2)]$ in the space-time representation of the propagator.

The damping factor introduced in this way is independent of superspin-isospin states and it does not allow a different behavior of partons and antipartons. But according to Proposition 4, the propagator [Eq. (2.18)] can be decomposed into a pure particle and a pure antiparticle propagator. Thus, these propagators can be treated separately with different damping factors, which leads to the experimentally observed different behavior of particles and antiparticles in the fluid. Formally, this can be described by adding in the propagation equation Eq. (2.37) to the mass a superspin-dependent damping factor that respects the above decomposition.

According to Heitler (1949), the damping factors of particles and antiparticles differ only weakly compared with their average absolute values. We therefore use the following formulation of this fact:

$$\delta_{ii'}\delta_{KK'}\delta_{\alpha\alpha'}m \to \left(m^*\delta_{KK'} - i\delta\gamma^0_{KK'}\right)\delta_{\alpha\alpha'}\delta_{ii'}, \tag{2.55}$$

with $\gamma^0_{KK'} := \delta^3_{\Lambda\Lambda'} \otimes \delta_{AA'}$, and with $m^* := m - i\,\delta^*$, where δ^* is the average damping factor and δ means the small difference between the damping factors of particles and antiparticles.

Proposition 5: In the propagator equation [Eq. (2.38)], the mass term [Eq. (2.55)] violates the CP forminvariance of the propagator and of the equation.

Proof: According to Propositions 1 and 2, the propagator equation [Eq. (2.38)] for mass m and the propagator itself are forminvariant under CP transformation. Therefore, if in Eq. (2.38) the mass m is replaced by Eq. (2.55), a violation of CP invariance can only be caused by this modified mass term.

The CP transformation of the propagator equation [Eq. (2.38)] with the mass term [Eq. (2.55)] is carried out by substitution of Eq. (2.35) into Eq. (2.38) and the subsequent elimination of the transformation matrices by multiplication of the whole equation by duals of these matrices. This gives for the modified mass term the following expression:

$$\begin{aligned}&\left(\gamma^0\gamma^5\right)^+_{\lambda\kappa}\left(\gamma^0\gamma^5\right)^+_{\lambda_2\kappa_2}\gamma^0_{\beta\alpha}\gamma^0_{\beta_2\alpha_2}\left(m^*\delta_{\kappa\kappa_1}-\mathrm{i}\delta\gamma^0_{\kappa\kappa_1}\right)\delta_{\alpha\alpha_1}\delta_{ii_1}\times\\&\left(\gamma^0\gamma^5\right)_{\kappa_1\kappa_1'}\left(\gamma^0\gamma^5\right)_{\kappa_2\kappa_2'}\gamma^0_{\alpha_1\alpha_1'}\gamma^0_{\alpha_2\alpha_2'}F^{\kappa_1'\kappa_2'}_{\alpha_1'\alpha_2'}\left(x_1',x_2'\right)'_{i_1i_2}\\&\equiv\left(m^*\delta_{\lambda\kappa_1'}+\mathrm{i}\delta\gamma^0_{\lambda\kappa_1'}\right)\delta_{\beta\alpha_1'}\delta_{ii_1}F^{\kappa_1'\lambda_2'}_{\alpha_1'\beta_2'}\left(x_1',x_2'\right)'_{i_1i_2}.\end{aligned}\tag{2.56}$$

If the whole equation is transformed in this way, one obtains

$$\begin{aligned}&\left[\mathrm{i}\gamma^\mu_{\beta\alpha_1'}\partial_\mu(x_1')\delta_{\lambda\kappa_1'}-\left(m^*\delta_{\lambda\kappa_1'}+\mathrm{i}\delta\gamma^0_{\lambda\kappa_1'}\right)\delta_{\beta\alpha_1'}\right]\delta_{ii_1}F^{\kappa_1'\lambda_2}_{\alpha_1'\beta_2}\left(x_1',x_2'\right)'_{i_1i_2}\\&=-\mathrm{i}C_{\beta\beta_2}\gamma^5_{\lambda\lambda_2}\lambda_i\delta_{ii_2}\delta(x_1'-x_2').\end{aligned}\tag{2.57}$$

By an appropriate change of indexing, one can reestablish the original denotation of Eq. (2.38). Then a comparison between Eq. (2.38) with Eqs. (2.55) and (2.57) shows that the damping term has changed its sign under CP transformation (i.e., CP invariance is violated). This then holds for the solution of Eq. (2.38) (i.e., the propagator) too. ◇

Apart from discrete symmetry operations, the propagator equation [Eq. (2.38)] admits the application of the continuous SU(2)-isospin group and the abelian U(1)-fermion number group. We are particularly interested in the isospin group.

Proposition 6: The propagator equation [Eq. (2.38)] is forminvariant under global isospin transformations.

Proof: We apply the decomposition of the superspin-isospin index κ into the pair of indices (Λ, A), where the index A denumerates isospin states.

To verify this statement, the charge-conjugated spinor fields χ^C have to be replaced by G-parity spinors χ^G, which are defined by

$$\chi^G_A := c^{-1}_{AA'}\chi^c_{A'},\tag{2.58}$$

with $c := -\mathrm{i}\,\sigma_2$. In contrast to χ^c, the G-parity spinor [Eq. (2.58)] transforms cogrediently to χ for isospin transformations. This guarantees a homogenous isospin transformation law for superspinor-isospinor fields.

Denoting the pair (χ_A, χ^c_A) by χ_κ and the pair (χ_A, χ^G_A) by $\tilde{\chi}_\kappa$, $\kappa = 1, 2, 3, 4$, one can express the relation between the superspinor-isospinor fields χ_κ and $\tilde{\chi}_\kappa$ by

$$\chi_\kappa = G_{\kappa\kappa'}\tilde{\chi}_{\kappa_1'};\quad G := \begin{pmatrix}1 & 0\\ 0 & c\end{pmatrix}.\tag{2.59}$$

If Eq. (2.59) is inserted in the definition of the propagator [Eqs. (2.23) and (2.24)], this leads to the transformation

$$F^{\kappa_1\kappa_2}_{\alpha_1\alpha_2}(x_1,x_2)_{i_1i_2} = G_{\kappa_1\kappa'_1} G_{\kappa_2\kappa'_2} \tilde{F}^{\kappa'_1\kappa'_2}_{\alpha_1\alpha_2}(x_1,x_2)_{i_1i_2}, \tag{2.60}$$

where $\tilde{F}$ refers to superspinor–isospinor fields $\tilde{\chi}$.

The insertion of Eq. (2.60) into Eq. (2.38) and elimination of the transformation matrices G yield

$$\begin{aligned}&\left[\iota\gamma^{\mu}_{\alpha\alpha_1}\partial_{\mu}(x_1) - m_i\delta_{\alpha\alpha_1}\right]\delta_{\lambda\kappa'_1}\delta_{ii_1}\tilde{F}^{\kappa'_1\lambda_2}_{\alpha_1\alpha_2}(x_1,x_2)_{i_1i_2}\\&\quad = -\gamma^2_{\lambda\lambda_2}C_{\alpha\alpha_2}\lambda_i\delta_{ii_2}\delta(x_1-x_2);\end{aligned} \tag{2.61}$$

i.e., the propagator equation in the field representation $\tilde{\chi}$. Owing to the homogenous isospin transformation of $\tilde{\chi}$, the isospin transformation properties of Eq. (2.61) can be analyzed. If one returns to the double indexing (Λ, A), and if the index value $\Lambda = 2$ is now related to χ^{G}, Eq. (2.61) can be written as

$$\begin{aligned}&\left[\iota\gamma^{\mu}_{\alpha\alpha_1}\partial_{\mu}(x_1) - m_i\delta_{\alpha\alpha_1}\right]\delta_{\Lambda\Lambda_1}\delta_{AA_1}\delta_{ii_1}\tilde{F}^{\Lambda_1A_1\Lambda_2A_2}_{\alpha_1\alpha_2}(x_1,x_2)_{i_1i_2}\\&\quad = -\iota\sigma^2_{\Lambda\Lambda_2}\sigma^2_{AA_2}C_{\alpha\alpha_2}\lambda_i\delta_{ii_2}\delta(x_1-x_2).\end{aligned} \tag{2.62}$$

It is sufficient to study the invariance properties of Eq. (2.62) under infinitesimal transformations, which are given by

$$dU := 1 - \mathrm{i}\frac{\varepsilon^a}{n}\sigma_a \quad dU^{-1} := 1 + \mathrm{i}\frac{\varepsilon^a}{n}\sigma_a \quad n \to \infty. \tag{2.63}$$

Then one obtains the following relation between $\tilde{F}$ and its isospin transform $\tilde{F}'$:

$$\tilde{F}^{\Lambda_1A_1\Lambda_2A_2}_{\alpha_1\alpha_2}(x_1,x_2)_{i_1i_2} = dU_{A_1A'_1}dU_{A_2A'_2}\tilde{F}^{\Lambda_1A'_1\Lambda_2A'_2}_{\alpha_1\alpha_2}(x_1,x_2)'_{i_1i_2}. \tag{2.64}$$

Insertion of Eq. (2.64) into Eq. (2.62) and elimination of the transformation matrices yield

$$dU^{-1}_{BA}dU^{-1}_{B_2A_2}\sigma^2_{AA_2} = \sigma^2_{BB_2}\mathrm{O}\left(\left(\frac{\varepsilon}{n}\right)^2\right), \tag{2.65}$$

and the equation

$$\begin{aligned}&\left[i\gamma^{\mu}_{\alpha\alpha_1}\partial_{\mu}(x_1) - m_i\delta_{\alpha\alpha_1}\right]\delta_{\Lambda\Lambda_1}\delta_{BB_1}\delta_{ii_1}\tilde{F}^{\Lambda_1 B_1 \Lambda_2 B_2}_{\alpha_1\alpha_2}(x_1, x_2)'_{i_1 i_2}\\&\quad = -i\sigma^2_{\Lambda\Lambda_2}\sigma^2_{BB_2}C_{\alpha\alpha_2}\lambda_i\delta_{ii_2}\delta(x_1 - x_2).\end{aligned} \tag{2.66}$$

In the limit $n \to \infty$, the last term in Eq. (2.65) vanishes. This means for infinitesimal isospin transformations, Eq. (2.66) is identical to Eq. (2.62). On account of the group properties of infinitesimal transformations, this result holds for finite transformations too—i.e., Eq. (2.62) is forminvariant under these transformations. ◇

In the same manner, one can show the following:

Proposition 7: The mass term [Eq. (2.55)] with additional CP symmetry breaking damping term is forminvariant under global isospin transformations.

However, invariance under local and global electroweak isospin transformations is a theoretical concept that is not realized in physical reality.

Although isospin symmetry breaking manifests itself only on the phenomenological level, the proper treatment of the phenomenological theory as an effective theory requires isospin symmetry breaking already on the parton level; i.e., in the parton propagator [Eq. (2.18)].

The modification of this propagator was extensively treated in Borne, Lochak, and Stumpf (2001, section 8.3). We adopt this symmetry breaking mass correction term from this research and introduce it directly into the propagator equation. The essential point is that the CP-symmetry violating term as well as the isospin symmetry violating term are addictive corrections of the mass m^* in (2.55) which referred to the representation with charge-conjugated spinors, leads to the propagation equation

$$\begin{aligned}&\left[i\gamma^{\mu}_{\alpha\alpha_1}\partial_{\kappa\kappa_1}(x_1) - m_i^*\delta_{\alpha\alpha_1}\delta_{\kappa\kappa_1} - i\theta\delta_{\alpha\alpha_1}\gamma^0_{\kappa\kappa_1} + \theta'\delta_{\alpha\alpha_1}\left(\gamma^0\gamma^3\right)_{\kappa\kappa_1}\right]\\&\quad\times\delta_{ii_1}\tilde{F}^{\kappa_1\kappa_2}_{\alpha_1\alpha_2}(x_1, x_2)_{i_1 i_2} = -iC_{\alpha\alpha_2}\gamma^5_{\kappa\kappa_2}\lambda_i\delta_{ii_2}\delta(x_1 - x_2).\end{aligned} \tag{2.67}$$

Applying the method of the proofs of Propositions 5 and 6, one gets.

Proposition 8: Eq. (2.67) breaks CP invariance as well as global isospin invariance. If the corresponding propagator is used for the derivation of an effective electroweak theory, it also causes the violation of the local isospin invariance.

In ordinary quantum theory and quantum field theory, symmetry breakings are frequently treated by perturbation theory. But it must be emphasized: The treatment of symmetry breaking by discharges in water must be done nonperturbatively, or else no meaningful results can be obtained.

2.5 SUMMARY

The finding of this section—namely, that discharges in water cause CP symmetry breaking—is amazing and surprising. To get greater confidence in these results, one should recall that the CP symmetry violation of the K-meson decays is described by means of complex mass-matrices; i.e., an assumption that is equivalent to Eq. (2.55). The difference between both approaches is that in K-meson physics, the complex mass-matrices were formally introduced, while for water, the physical reactions leading to Eq. (2.55) can be explained in great detail if necessary.

With respect to the consequences for the case under consideration, one should also recall that the basic Lagrangian is left-right symmetric, and that in the algebraic theory, the basic physical laws are independent of special representations. Therefore, any parity symmetry violation can only arise from the vacuum, which is the basis of a corresponding state space construction. If such a situation occurs, an immediate consequence is that there must be right-handed neutrinos, the neutrinos must be massive [cf. Mohapatra (2003, p. 130)], and that for composite neutrinos, the antisymmetry among the constituents is removed, which is the most important effect for the actual calculations.

2.6 PARAFERMI ELECTRIC AND MAGNETIC BOSON STATES

In this section, we first treat the effect of the CP symmetry breaking on the composite boson states as only by the coupling of the bosons to magnetic monopoles in the effective theory the latter can be identified. The calculation will be nonperturbative. As the starting point for a detailed calculation of the boson states, we use the integral form of the GBBW-equations [Eq. (2.2)]. cf. Stumpf and Borne (1994, Sect. 52) where the Green function G is the inverse of the Dirac operator in eq. (2.2)

$$\varphi_{Z_1 Z_2}(x_1, x_2) = 3 \int d^4x G_{Z_1 X_1}(x_1 - x) U_{X_1 X_2 X_3 X_4} F_{X_2 Z_2}(x - x_2) \varphi_{X_3 X_4}(x, x). \tag{2.68}$$

If in the propagator [Eq. (2.18)] the CP symmetry breaking mass term [Eq. (2.55)] is inserted, and if this propagator is applied in Eq. (2.68), the CP-invariance violation of the propagator induces a violation of the permutation invariance; i.e., a violation of the antisymmetry of the wave functions in Eq. (2.68).

It should be noted that the symmetry properties of the vertex U in Eq. (2.68) are not affected by CP symmetry breaking because U is related to the dynamical law of the underlying quantum field, which does not depend on the special representation induced by the associated special vacuum expectation values.

Symmetry breaking means that the invariance of the effective Hamiltonian under a certain symmetry group is violated. As a result, the degeneracy of the energy eigenvalues is removed with respect to the members of group representations; i.e., the members of a multiplet aquire different energy eigenvalues and can thus individually identified (meaning that they are no longer indistinguishable).

It is obvious that this fact must find its expression in the properties of the representation of the field operator algebra in the corresponding state space. Thus, the anticommutation relations in the case of the unbroken symmetry group must be partly replaced by commutation relations in the case of the broken symmetry. This necessity is simply illustrated by the fact that in the hydrogen atom, the wave function is not antisymmetrized between proton and electron.

In such a case, according to Green (1953), a construction with mixed commutators and anticommutators of fermion field operators should be called *parafermi statistics*. Ohnuki and Kamefuchi (1982) give further details of this parafermi algebra for field operators. But here, we directly discuss the effect of this modified algebra on representations.

For conserved symmetries, the exact boson eigenstates can be written in the general form of Eq. (2.9). With the superspin-isospin index $\kappa = \Lambda, A$ the set of antisymmetric T-matrices in Eqs. (2.10) and (2.11) can be expressed in the following form:

$$
\begin{aligned}
&T^{0}_{\kappa_1\kappa_2} \equiv i\sigma^{2}_{\Lambda_1\Lambda_2} \otimes \sigma^{0}_{A_1A_2} \\
&T^{1}_{\kappa_1\kappa_2} \equiv i\sigma^{2}_{\Lambda_1\Lambda_2} \otimes \sigma^{1}_{A_1A_2}, \quad T^{2}_{\kappa_1\kappa_2} \equiv \sigma^{1}_{\Lambda_1\Lambda_2} \otimes \sigma^{2}_{A_1A_2}, \quad T^{3}_{\kappa_1\kappa_2} \equiv i\sigma^{2}_{\Lambda_1\Lambda_2} \otimes \sigma^{3}_{A_1A_2}.
\end{aligned}
\tag{2.69}
$$

The whole wave function [Eq. (2.9)] is antisymmetric. To establish this, the matrices T^{l} must have the exact form of Eq. (2.69). The matrices $\sigma^{a}_{\Lambda\Lambda'}$

characterize the content of spinor fields and charge-conjugated spinor fields in the corresponding boson state. As in σ^2 and in σ^1, only the elements $\sigma^2_{12}, \sigma^2_{21}$ and $\sigma^1_{12}, \sigma^1_{21}$ are unequal to zero. This means that in all T-matrices, either symmetrized or antisymmetrized products of one spinor field and one charge-conjugated spinor field occur.

If CP symmetry is broken and spinors and charge-conjugated spinors can be discriminated accordingly, the permutation representations for these fields are destroyed. For comparison, we discuss the treatment of conserved and of broken CP symmetry in parallel terms.

The symmetry-conserving case was extensively treated in Stumpf (2002), cf. also Borne, Lochak, and Stumpf (2001, section 6.2). Therefore, only a short review will be given here.

In general form, the exact wave functions are defined by Eq. (2.9) for $a =$ 0, 1, 2, 3.

On the right side of the integral equation [Eq. (2.68)] appears the local wave function, where $\widehat{\varphi}$ means summation over i_1 and i_2

$$\widehat{\varphi}^{K_1K_2}_{\alpha_1\alpha_2}(x,x) = \exp(-ikx)\widehat{\chi}_{\alpha_1\alpha_2}(0|k)T^a_{K_1K_2}. \tag{2.70}$$

Owing to the antisymmetry of the wave functions, which must also hold for Eq. (2.70), the spin-tensor $\widehat{\chi}$ can be expanded on a Dirac spinor basis with exclusively symmetric elements:

$$\widehat{\chi}_{\alpha_1\alpha_2}(0|k) := \left[A^0_\mu(k)(\gamma^\mu C)_{\alpha_1\alpha_2} + F^0_{\mu\nu}(k)(\Sigma^{\mu\nu}C)_{\alpha_1\alpha_2}\right]. \tag{2.71}$$

The green's function of eq. (2.68) is defined by the corresponding Feynman propagator cf. Schweber, Bethe, Hoffman (1956, Vol. I, chpt. 8, eq. (37)) in integral representation

$$\begin{aligned} &G_{Z_1Z_2}(x_1 - x_2) \\ &= \delta_{i_1i_2}\delta_{K_1K_2}(2\pi)^{-4}\int d^4p(\gamma^u p_u + m_{i1})_{\alpha_1\alpha_2}\left(p^2 - m^2_{i_1}\right)^{-1} e^{-ip(x_1-x_2)}. \end{aligned} \tag{2.72}$$

Then, with application of the symmetry-conserving propagator [Eq. (2.18)] in Eq. (2.68) after inserting Eqs. (2.9), (2.70), and (2.72) into Eq. (2.68), the center of mass motion can be eliminated and only an equation for χ results. With $k' := \frac{1}{2}k$, this equation reads as

$$\chi^{\kappa_1\kappa_2}_{\alpha_1\alpha_2}(x_1-x_2|\kappa)_{i_1,i_2}=g'\int d^4pe^{-ip(x_1-x_2)}\lambda_{i_1}f^+_{i_1}\left[\left(p_\mu+\kappa'_\mu\right)\gamma^\mu+m_{i_1}\right]_{\alpha_1\beta_1}\delta_{\kappa_1\nu_1}$$
$$\times\sum_h\left[v^h_{\beta_1\beta_2}\delta_{\nu_1\nu_2}\left(v^hC\right)_{\beta_3\beta_4}\gamma^5_{\nu_3\nu_4}-v^h_{\beta_1\beta_3}\delta_{\nu_1\nu_3}\left(v^hC\right)_{\beta_2\beta_4}\gamma^5_{\nu_2\nu_4}-v^h_{\beta_1\beta_4}\delta_{\nu_1\nu_4}\left(v^hC\right)_{\beta_3\beta_2}\gamma^5_{\nu_3\nu_2}\right]$$
$$\times(-i)\lambda_{i_2}\left\{f^-_{i_2}\left[(p_\lambda-\kappa'_\lambda)\gamma^\lambda C+m_{i_2}C\right]_{\beta_2\alpha_2}\gamma^5_{\nu_2\kappa_2}\right\}\widehat{\chi}^{\nu_3\nu_4}_{\beta_3\beta_4}(0|\kappa) \tag{2.73}$$

where the definitions

$$f^+_i:=\left[(p+\kappa')^2-m_i^2\right]^{-1},\quad f^-_i:=\left[(p-\kappa')^2-m_i^2\right]^{-1} \tag{2.74}$$

have been introduced. In addition in g' all numerical constants are enclosed which is not relevant to this review.

After some algebraic rearrangements the superspin-isospin matrices T^a can be completely eliminated and one obtains from (2.73) with insertion of (2.71) into the right side

$$\chi^{i_1i_2}_{\alpha_1\alpha_2}(x_1-x_2|\kappa)=g'\int d^4pe^{-ip(x_1-x_2)}\lambda_{i_1}f^+_{i_1}\left[\left(p_\mu+\kappa'_\mu\right)\gamma^\mu+m_{i_1}C\right]_{\alpha_1\beta_1}$$
$$A^0_\eta(\kappa)\gamma^\eta_{\beta_1\beta_2}f^-_{i_2}\left[(p_\lambda-\kappa'_\lambda)\gamma^\lambda C+m_{i_2}C\right]_{\beta_2\alpha_2} \tag{2.75}$$

Then if on both sides of Eq. (2.75) the limit $x_1\to x_2$ and the sum over i_1,i_2 are performed, with insertion of Eq. (2.71) into the left side of (2.75) one gets a set of equations for the calculation of $A^a_\mu(k)$ and $F^a_{\mu\nu}(k)$, together with the mass eigenvalue $k^2=m_B^2$.

However note: The elimination of the superspin-isospin part in eq. (2.73) indicates that owing to the isospin invariance of the hard-core equations the mass-spectrum for the states a = 0, 1, 2, 3 is degenerate.

As a result of this degeneracy the effective field variables $A^a_\mu(k)$ and $F^a_{\mu\nu}(k)$ are independent of a and the eigenvalue equations for a = 0 of the photon can be studied regardless of the necessary Weinberg-transformation. From Stumpf (2002, eqs (80), (81)) one gets for this case the formulas

$$A'_\mu=\frac{2ig}{(2\pi)^4}\left\{\left[J_0+J_1k^2+2J_2\right]A'_\mu-2J_1k_\mu k^\nu A'_\nu\right\}\quad F_{\mu\nu}=-i\left\{A'_\mu k_\nu\right\}_{as(\mu\nu)} \tag{2.76}$$

where the coefficient functions in these formulas are given in the above reference. By self-regularization all their values are finite. If the Lorentz-condition $k^{\mu}A'_{\mu} = 0$ is assumed an eigenvalue equation for k^2 can be derived, cf. Stumpf (2002, sect. 5).

The case of broken CP symmetry, leads to parafermi states and was fully treated by Stumpf (2004b). With reference to this paper only a review is given. In this case, the set of symmetric matrices S^a in Eqs. (2.10) and (2.11) comes into play. For this set, the following representation holds:

$$\begin{aligned} &S^0_{K_1K_2} \equiv \sigma^1_{\Lambda_1\Lambda_2} \otimes \sigma^0_{A_1A_2} \\ &S^1_{K_1K_2} \equiv \sigma^1_{\Lambda_1\Lambda_2} \otimes \sigma^1_{A_1A_2}, \quad S^2_{K_1K_2} \equiv i\sigma^2_{\Lambda_1\Lambda_2} \otimes \sigma^2_{A_1A_2}, \quad S^3_{K_1K_2} \equiv \sigma^1_{\Lambda_1\Lambda_2} \otimes \sigma^3_{A_1A_2}. \end{aligned} \tag{2.77}$$

Then, for Eqs. (2.69) and (2.77), the relations

$$\left(T^a\gamma^0\right) \equiv S^a; \quad \left(S^a\gamma^0\right) \equiv T^a \quad a = 0, 1, 2, 3 \tag{2.78}$$

hold, which are essential for the solution procedure. From Eq. (2.78), it follows that

$$\left(T^a + S^a\right)\gamma^0 = \left(S^a + T^a\right) \quad a = 0, 1, 2, 3, \tag{2.79}$$

and for these combinations, one obtains the representation

$$\left(T^{\alpha} + S^a\right)_{K_1K_2} = \left(\sigma^1 + i\sigma^2\right)_{\Lambda_1\Lambda_2} \otimes \sigma^a_{\Lambda_1\Lambda_2} \quad a = 0, 1, 2, 3. \tag{2.80}$$

In comparison with the superspin parts of the conserved symmetry, this matrix is neither symmetric nor antisymmetric. But if one defines Θ^s_b by replacing T^b in Eq. (2.17) by S^b, the elements of the set $\{(\Theta^a_b + \Theta^s_b)$, b = 0, 1, 2, 3$\}$ are eigenstates of the operators [Eqs. (2.12) and (2.14)] too; i.e., the set [Eq. (2.80)] allows the same state classification as for conserved symmetry.

In addition to the set [Eq. (2.80)], one can form the set

$$\begin{aligned} &\left(T^0 - S^0\right)_{K_1K_2} = -\left(\sigma^1 - i\sigma^2\right)_{\Lambda_1\Lambda_2} \otimes \sigma^0_{A_1A_2} \\ &\left(T^a - S^a\right)_{K_1K_2} = (-1)^a\left(\sigma^1 - i\sigma^2\right)_{\Lambda_1\Lambda_2} \otimes \sigma^0_{A_1A_2} \quad a = 1, 2, 3. \end{aligned} \tag{2.81}$$

Therefore, one should apply the ansatz

$$\varphi_{Z_1 Z_2}(x_1, x_2|k, a) = (T^a + S^a)_{\kappa_1\kappa_2} \exp\left[-i\frac{k}{2}(x_1 + x_2)\right] \chi^{i_1 i_2}_{\alpha_1\alpha_2}(x_1 - x_2|k, a)$$

$$+ (T^a - S^a)_{\kappa_1\kappa_2} \exp\left[-i\frac{k}{2}(x_1 + x_2)\right] \tilde{\chi}^{i_1 i_2}_{\alpha_1\alpha_2}(x_1 - x_2|k, a). \tag{2.82}$$

for the left side of Eq. (2.68), which for its right side yields

$$\begin{aligned}\widehat{\varphi}^{\kappa_1\kappa_2}_{\alpha_1\alpha_2}(x, x|k, a) &= \exp(-ikx)\times \\ &= \Big\{\Big[A^a_u(k)(\gamma^u C)_{\alpha_1\alpha_2} + G^a_u(k)\big(\gamma^5\gamma^u C\big)_{\alpha_1\alpha_2} \\ &\quad + F^a_{\mu\nu}(k)(\Sigma^{\mu\nu}C)_{\alpha_1\alpha_2}\Big](T^a + S^a)_{\kappa_1\kappa_2} \\ &\quad + \Big[\tilde{A}^a_u(k)(\gamma^u C)_{\alpha_1\alpha_2} + \tilde{G}^a_\mu(k)\big(\gamma^5\gamma^u C\big)_{\alpha_1\alpha_2} \\ &\quad + F^a_{\mu\nu}(k)(\Sigma^{\mu\nu}C)_{\alpha_1\alpha_2}\Big](T^a - S^a)_{\kappa_1\kappa_2}\Big\}\end{aligned} \tag{2.83}$$

and which replaces the expansion [Eqs. (2.70) and (2.71)].

The set $\{(T^0 - S^0), (-1)^{a+1}(T^a - S^a), a = 1, 2, 3\}$ is the CP transform of Eq. (2.80). This can be verified by the application of the CP-transformation formula (2.35) for the propagation to the second term of (2.83).

As by the local term (2.83) the solutions of the boson equation are fixed is follows that the part (2.81) in (2.83) is redundant. Therefore we omit the additional contribution of the CP transform (2.81) and apply only the ansatz:

$$\begin{aligned}&\widehat{\chi}^{\kappa_1\kappa_2}_{\alpha_1\alpha_2}(0|k, a) \\ &= \Big[A^a_\mu(k)(\gamma^u C)_{\alpha_1\alpha_2} + G^a_\mu(k)\big(\gamma^5\gamma^\mu C\big)_{\alpha_1\alpha_2} \\ &\quad + F^a_{\mu\nu}(k)(\Sigma^{\mu\nu}C)_{\alpha_1\alpha_2}\Big](T^a + S^a)_{\kappa_1\kappa_2}.\end{aligned} \tag{2.84}$$

If in Eq. (2.68) the symmetry-breaking propagator—i.e., the propagator with the mass term [Eq. (2.55)]—is substituted, then the following integral equation results:

$$
\begin{aligned}
\chi^{\kappa_1\kappa_2}_{\alpha_1\alpha_2}(x_1-x_2|k)_{i_1i_2} = g' \int d^4pe^{-ip(x_1-x_2)}\lambda_{i_1}f^{+}_{i_1}\left[\left(p_u+k'_\mu\right)\gamma^\mu+m_{i_1}\right]_{\alpha_1\beta_1}\delta_{\kappa_1\nu_1} \\
\times \sum_h \Big[v^h_{\beta_1\beta_2}\delta_{\nu_1\nu_2}\left(v^hC\right)_{\beta_3\beta_4}\gamma^5_{\nu_3\nu_4} \\
- v^h_{\beta_1\beta_3}\delta_{\nu_1\nu_3}\left(v^hC\right)_{\beta_2\beta_4}\gamma^5_{\nu_2\nu_4} \\
- v^h_{\beta_1\beta_4}\delta_{\nu_1\nu_4}\left(v^hC\right)_{\beta_3\beta_2}\gamma^5_{\nu_3\nu_2}\Big] \\
\times(-i)\lambda_{i_2}\Big\{f^-_{i_2}\left[(p_\lambda-k'_\lambda)\gamma^\lambda C+m_{i2}C\right]_{\beta_2\alpha_2}\gamma^5_{\nu_2\kappa_2} \\
+f^-_{i_2}(\delta m)C_{\beta_2\alpha_2}\left(\gamma^5\gamma^0\right)_{\nu_2\kappa_2}\Big\}\widehat{\chi}^{\nu_3\nu_4}_{\beta_3\beta_4}(0|k),
\end{aligned}
\tag{2.85}
$$

Equation (2.85) corresponds to the equation (31) in Stumpf (2004b); but note that in that paper Lochak's notation **B** of the axial vector potential is used instead of the notation **G** in this article.

The above equation can be solved by putting $x_1 - x_2 = 0$ on both sides. Furthermore by violating only the CP-invariance the isopspin invariance is not affected. This means that the isospin state spectrum is degenerate.

It is interesting to relate this degeneracy of the isospin states to the physical particle spectrum that is defined in terms of $SU(2) \otimes U(1)$ representations. In superspinor-isospinor space the quantum numbers for the electroweak vector bosons are given in eq. (2.17) for the antisymmetric algebra elements Θ^a_n, $n = 0, 1, 2, 3$. An analogous definition holds for the symmetric algebra elements Θ^s_n, $n = 0, 1, 2, 3$. Then we replace the ansatz (2.84) by

$$
\begin{aligned}
&\widehat{\chi}^{\kappa_1\kappa_2}_{\alpha_1\alpha_2}(0|\kappa,a) = \\
&\left[A^a_\mu(\kappa)(\gamma^\mu C)_{\alpha_1\alpha_2} + G^a_\mu(\kappa)\left(\gamma^5\gamma^\mu C\right)_{\alpha_1\alpha_2} + F^a_{\mu\nu}(\kappa)(\Sigma^{\mu\nu}C)_{\alpha_1\alpha_2}\right]\times \\
&\left(\Theta^a_n+\Theta^a_n\right)_{\kappa_1\kappa_2}
\end{aligned}
\tag{2.86}
$$

where owing to fermion number 0 the charge quantum numbers are proportional to the isospin values, i.e. all four states are degenerate. Substitution of one of these states yields equation (2.87) which corresponds to the equation (41) in the original paper.

$$\chi^{i_1 i_2}_{\alpha_1\alpha_2}(x_1 - x_2|k) = g' \int d^4p e^{-ip(x_1-x_2)} \lambda_{i_1} f^{+}_{i_1} \left[\left(p_\mu + k'_\mu\right)\gamma^\mu + m_{i1}\right]_{\alpha_1\beta_1}$$
$$\times\left\{\left[A^0_\eta(k)\gamma^\eta_{\beta_1\beta_2} + G^0_\eta(k)\left(\gamma^5\gamma^\eta\right)_{\beta_1\beta_2}\right] i\lambda_{i_2}\right.$$
$$f^{-}_{i_2}\left[\left(p_\lambda - k'_\lambda\right)\left(\gamma^\lambda C\right) + m_{i2}C\right]_{\beta_2\alpha_2}$$
$$\left.+\left[A^0_\eta(k)\gamma^\eta_{\beta_1\beta_2} + G^0_\eta(k)\left(\gamma^5\gamma^\eta\right)_{\beta_1\beta_2}\right] i\lambda_{i_2} f^{-}_{i_2}(\delta m) C_{\beta_2\alpha_2}\right\}, \tag{2.87}$$

Analogous equations can be derived for $a = 1, 2, 3$. If on the left side the limit $x_1 \to x_2$ and the sum over i_1, i_2 are performed, with insertion of Eq. (2.86) in the left side of the resulting equation, one gets the set of equations for the calculation of $A^0_\mu(k)$, $G^0_\mu(k)$ and $F^0_{\mu\nu}(k)$, as well as of the mass eigenvalue $k^2 = m^2_B$. For this evaluation, see Stumpf (2004b). In all these calculations, due to self-regularization, no divergencies occur; i.e., for CP-invariant as well as for broken CP symmetry, exact finite boson eigenstates can be derived.

2.7 PHYSICAL EFFECT OF SYMMETRY BREAKING

This effect results from the comparison of Eq. (2.71) for conserved symmetries with Eq. (2.86) for broken CP symmetry. The consequence is that in the case of conserved symmetry with Eq. (2.71), only the electric (conventional) vector bosons couple to fermions, whereas for CP symmetry breaking with Eq. (2.84), electric and magnetic vector bosons couple to the fermions.

This can be demonstrated by means of Eq. (1.30) of chapter 1, which is part of the effective theory and which independently holds of the symmetry character. Apart from several other research in conserved symmetries, this was done with unsurpassed lucidness in Fuss (2004, section 7.2.2), where the part

$$W_{I_1 I_2 I_3 I_4} C^{l}_{I_2 I_3} \tag{2.88}$$

was evaluated. If the same term is evaluated for broken CP symmetry, then in contrast to conserved symmetry, two coupling terms appear that represent the coupling to electric and magnetic vector bosons. A comprehensive evaluation for broken symmetry will be given later in this part, in section 4.3 of Chapter 4, and the meaning of the term *magnetic vector boson* will be explained there.

2.8 RELATIVISTIC EQUATIONS FOR COMPOSITE LEPTONS

The question whether excited neutrinos act as magnetic monopoles cannot be answered without general information about the observable fermion states in the extended Standard Model, which are all composites. As in the case of composite electroweak vector bosons, the first step in such an investigation is the treatment of states for conserved symmetries.

We start with the field theoretic version of the theory for hard-core states, which can be expressed by a single (covariant) functional equation [cf. Borne, Lochak, and Stumpf (2001, Chapter 6) and Stumpf and Borne (1994, Chapter 6)]. Then in symbolic notation of section 1.1, of this part this hard-core functional equation reads as follows (using the summation convention):

$$K_{I_1 I}\mathbf{d}_I|F_n\rangle = U_{I_1 I_2 I_3 I_4}[F_{I_2 I}\mathbf{j}_I\mathbf{d}_{I_4}\mathbf{d}_{I_3} + F_{I_3 I}\mathbf{j}_I\mathbf{d}_{I_2}\mathbf{d}_{I_4} + F_{I_4 I}\mathbf{j}_I\mathbf{d}_{I_3}\mathbf{d}_{I_2}]|F_n\rangle. \tag{2.89}$$

At first, we explain the states $|F_n\rangle$, which are defined by

$$|F_n(\mathbf{j})\rangle = \varphi_n(I_1 \ldots I_n)\mathbf{j}_{I_1} \cdots \mathbf{j}_{I_n}|0\rangle, \tag{2.90}$$

where φ_n is a formally normal-ordered matrix element of the parton dynamics for hard-core states,

With regard to the application of Eq. (2.89) to the case $n = 3$, we choose in Eq. (2.90) the corresponding states and project [Eq. (2.89)] from the left side with $\langle 0|\mathbf{d}_{N_1}\mathbf{d}_{N2}$. This yields

$$\begin{aligned}&\sum_N K_{N_3 N}\mathcal{A}_{N_1 N_2 N}\varphi_{N_1 N_2 N} = \\ &\sum_{I_2 I_3 I_4} U_{N_3 I_2 I_3 I_4}\left[-3F_{I_2 N_2}\mathcal{A}_{N_1 I_3 I_4}\varphi_{N_1 I_3 I_4} + 3F_{I_2 N_1}\mathcal{A}_{N_2 I_3 I_4}\varphi_{N_2 I_3 I_4}\right],\end{aligned} \tag{2.91}$$

where the symbol $\mathcal{A}$ means antisymmetrization in the corresponding indices. In all the following calculations, we omit the $\mathcal{A}$ symbols for brevity, but keep in mind that they are always present. In particular, we define $\mathbf{r} \in R^3$, $x \in M^4$, and $Z = (i, \kappa, \alpha)$, where κ means the superspin-isospin index, α is the Dirac spinor index, i is the auxiliary field index. The latter characterizes the subfermion fields, which are needed for the regulatization procedure.

Let $\varphi_{Z_1 Z_2 Z_3}(x_1, x_2, x_3)$ be the covariant, antisymmetric state amplitude for the case $n = 3$. Then from Eq. (2.91) for this state, the following equation can be derived:

$$\begin{aligned}&\left[D^{\mu}_{Z_3 X_3}\partial_\mu(x_3) - m_{Z_3 X_3}\right]\varphi_{Z_1 Z_2 Z_3}(x_1, x_2, x_3) = \\ &3U_{Z_3 X_2 X_3 X_4}\left[-F_{X_2 Z_2}(x_3 - x_2)\varphi_{Z_1 X_3 X_4}(x_1, x_3, x_3)\right. \\ &\quad \left. + F_{X_2 Z_1}(x_3 - x_1)\varphi_{Z_2 X_3 X_4}(x_2, x_3, x_3)\right]\end{aligned} \tag{2.92}$$

Owing to the antisymmetrization in Eq. (2.91), one obtains two additional equations if the Dirac operator on the left side of Eq. (2.92) is applied to the coordinates x_1 and x_2.

For brevity, these two equations are not explicitly given because apart from one exception (namely, the derivation of the energy representation), these two equations are not needed if antisymmetrization is secured in every step of the calculation.

For vanishing coupling constant $g = 0$, de Broglie's original fusion equations for the local three fermion states are obtained, and for a solution of the whole set of equations, only Eq. (2.92) has to be used. As for antisymmetric wave functions, the remaining equations can be derived from Eq. (2.92) by interchange of indices.

The physical interpretation of the wave functions is closely related to the role of the auxiliary fields (indices), which appear in the corresponding equations and their solutions. First, we refer to the role of auxiliary fields in regularization, leading to the definition of the physical wave functions. We consider the wave functions of Eq. (2.92) with full dependence on the auxiliary fields as being unobservable (i.e., unphysical). In order to obtain the physical, singularity-free wave functions in the case of three-parton states, we decompose the index $Z := (\alpha, \kappa, i)$ into $Z := (z, i)$ and sum over i_1, i_2, i_3. This gives

$$\widehat{\varphi}_{z_1 z_2 z_3}(x_1, x_2, x_3) := \sum_{i_1 i_2 i_3} \varphi_{Z_1 Z_2 Z_3}(x_1, x_2, x_3). \tag{2.93}$$

These functions are by definition physical states. It can be easily verified that the physical wave function $\widehat{\varphi}$ has the same transformation properties as the original wave function φ. In order to derive a probability interpretation for the physical parton wave functions, the single time formulation of Eq. (2.93) has to be used [cf. Borne, Lochak, and Stumpf (2001) and Stumpf and Borne (1994)]. In addition, the single time energy equation has to be derived from Eq. (2.92) [cf. Stumpf (2000) and Stumpf and Borne (1994)]. Then, with the single time density,

$$\widehat{\varphi}^{\dagger}\widehat{\varphi} := \sum_{z_1 z_2 z_3} \widehat{\varphi}_{z_1 z_2 z_3}(\mathbf{r}_1, \mathbf{r}_2, \mathbf{r}_3, t)^{*} \widehat{\varphi}_{z_1 z_2 z_3}(\mathbf{r}_1, \mathbf{r}_2, \mathbf{r}_3, t), \tag{2.94}$$

for a general time-dependent solution of the energy equation, one obtains from this equation with $m_i = m + \delta m_i$, in the limit $\delta m_i \to 0$ current conservation, cf. Stumpf (2000, section 4) for details and also section 4.5 of this part where an example is treated.

This limit can be performed in the regularized wave functions with no difficulty after all calculations have been carried out. Owing to current conservation, the densities [Eq. (2.94)] are conserved positive quantities; i.e., the physical state amplitudes $\widehat{\varphi}$ are elements of a corresponding Hilbert space with the norm expression (in the case under consideration)

$$\langle\widehat{\varphi}|\widehat{\varphi}\rangle = \int d^3r_1 d^3r_2 d^3r_3 \widehat{\varphi}_{z_1z_2z_3}(\mathbf{r}_1,\mathbf{r}_2,\mathbf{r}_3,t)^* \widehat{\varphi}_{z_1z_2z_3}(\mathbf{r}_1,\mathbf{r}_2,\mathbf{r}_3,t), \quad (2.95)$$

and they describe the states of the system with interaction. Hence, one is able to extract all quantum mechanical information about this system from its state space.

Finally, it should be noted that in the latter limit, the coupling constants λig in the vertex [Eqs. (2.7) and (2.8)] of the three-parton equation diverge. But the essential point is that the regularized solutions of these equations remain finite in the whole range $(0, \infty)$ of λig. Hence, as the auxiliary fields are unobservable and the whole physics depends on the regularized solutions, this behavior of the coupling constants has no observable consequences.

2.9 EIGENSTATES OF ENERGY AND ANGULAR MOMENTUM

By inversion of the Dirac operator in Eq. (2.92), this equation can be replaced for bound states by the equivalent integral equation

$$\varphi_{Z_1Z_2Z_3}(x_1,x_2,x_3) = U_{X_1X_2X_3X_4}\int d^4x G_{Z_3X_1}(x_3-x)\times$$
$$3\big[-F_{X_2Z_2}(x-x_2)\varphi_{Z_1X_3X_4}(x_1,x,x) + F_{X_2Z_1}(x-x_1)\varphi_{Z_2X_3X_4}(x_2,x,x)\big] \quad (2.96)$$

In order to secure the antisymmetry of the solutions in the course of calculations, Eq. (2.96) can explicitly be antisymmetrized. This gives

$$\varphi_{Z_1Z_2Z_3}(x_1,x_2,x_3)_{as} = U_{X_1X_2X_3X_4}\sum_{h_1h_2h_3}(-1)^P\int d^4x G_{Z_{h_1}X_1}(x_{h_1}-x)$$
$$\times F_{X_2Z_{h_2}}(x-x_{h_2})\varphi_{Z_{h_3}X_3X_4}(x_{h_3},x,x). \quad (2.97)$$

Using the decomposition $Z_l = (\alpha_l, \kappa_l, i_l)$, $l = 1, 2, 3$, and $X_k = (\beta_k\ \varrho_k\ j_k)$, $k = 1, 2, 3$ in Eq. (2.97), it can be summed over i_1, i_2, and i_3. With Eqs. (2.7) and (2.93), this leads to

$$\widehat{\varphi}^{\kappa_1\kappa_2\kappa_3}_{\alpha_1\alpha_2\alpha_3}(x_1, x_2, x_3)a_s = V^{\varrho_1\varrho_2\varrho_3\varrho_4}_{\beta_1\beta_2\beta_3\beta_4} \sum_{h_1h_2h_3} (-1)^P \int d^4x \widehat{G}^{\kappa_{h_1}}_{\alpha_{h_1}}(x_{h_1} - x) \times \widehat{F}^{\varrho_2\kappa_{h_2}}_{\beta_2\alpha_{h_2}}(x - x_{h_2})\widehat{\varphi}^{\kappa_{h_3}\varrho_3\varrho_4}_{\alpha_{h_3}\beta_2\beta_3}(x_{h_3}, x, x). \quad (2.98)$$

Due to the translational invariance of Eq. (2.98), their solutions admit a representation where the total four-momentum is diagonalized. Therefore, the ansatz

$$\widehat{\varphi}^{\kappa_1\kappa_2\kappa_3}_{\alpha_1\alpha_2\alpha_3}(x_1, x_2, x_3) = \exp\left[-ik\frac{1}{3}(x_1 + x_2 + x_3)\right]\widehat{\chi}^{\kappa_1\kappa_2\kappa_3}_{\alpha_1\alpha_2\alpha_3}(x_2 - x_1, x_3 - x_2) \quad (2.99)$$

can be used, where k is the center of mass four-momentum. With the transformation

$$z = \frac{1}{3}(x_1 + x_2 + x_3); \quad u = (x_2 - x_1); \quad v = (x_3 - x_2) \quad (2.100)$$

and its inversion

$$x_1 = z - \frac{1}{3}(2u + v); \quad x_2 = z + \frac{1}{3}(u - v); \quad x_3 = z + \frac{1}{3}(u + 2v), \quad (2.101)$$

one obtains (by suppressing the algebraic indices)

$$\widehat{\varphi}(x_1, x_2, x_3) = \exp(-ikz)\widehat{\chi}(u, v). \quad (2.102)$$

If $x_2 = x_3 = x$ is assumed, this results in $v = 0$ and

$$\begin{aligned}\widehat{\varphi}(x_1, x, x) &= \exp\left[-ik\frac{1}{3}(x_1 + 2x)\right]\widehat{\chi}(x - x_1, 0) \\ &= \exp\left[-ik\frac{1}{3}\left(z - \frac{2}{3}u + 2x\right)\right]\widehat{\chi}\left(x - z + \frac{2}{3}u, 0\right).\end{aligned} \quad (2.103)$$

Transformation of the corresponding term in Eq. (2.98) by $x = x' + z - \frac{2}{3}u$ gives under the integral

$$\widehat{\varphi}(x_1, x, x) = \exp\left[-ikz - ik\frac{1}{3}(2x' - 2u)\right]\widehat{\chi}(x', 0). \quad (2.104)$$

Similarly, $\widehat{\varphi}(x_2, x, x)$ and $\widehat{\varphi}(x_3, x, x)$ can be expressed if one observes that for any set of coordinates x_a,x_b,x_c, the corresponding x is given by definition by $\widehat{\chi}(x_b - x_a, x_c - x_b)$.

Transforming in an analogous way all other terms of Eq. (2.98), the center of mass part exp($-ikz$) can be eliminated. The resulting equation is independent of z and reads as

$$
\begin{aligned}
\widehat{\chi}^{\kappa_1\kappa_2\kappa_3}_{\alpha_1\alpha_2\alpha_3}(u,0) = &-V^{\varrho_1\varrho_2\varrho_3\varrho_4}_{\beta_1\beta_2\beta_3\beta_4}\int d^4x'\times\\
&\Big\{\widehat{G}^{\kappa_1\varrho_1}_{\alpha_1\beta_1}(-u-x')\exp\left[-ik\frac{1}{3}(u+2x')\right]\widehat{F}^{\varrho_2\kappa_3}_{\beta_2\alpha_3}(x')\widehat{\chi}^{\kappa_2\varrho_3\varrho_4}_{\alpha_2\beta_3\beta_4}(x',0)\\
&-\widehat{G}^{\kappa_1\varrho_1}_{\alpha_1\beta_1}(-u-x')\exp\left[-ik\frac{1}{3}(u+2x')\right]\widehat{F}^{\varrho_2\kappa_2}_{\beta_2\alpha_2}(x')\widehat{\chi}^{\kappa_3\varrho_3\varrho_4}_{\alpha_3\beta_3\beta_4}(x',0)\\
&+\widehat{G}^{\kappa_2\varrho_1}_{\alpha_2\beta_1}(-x')\exp\left[-ik\frac{1}{3}(u+2x')\right]\widehat{F}^{\varrho_2\kappa_2}_{\beta_2\alpha_1}(x'+u)\widehat{\chi}^{\kappa_3\varrho_3\varrho_4}_{\alpha_3\beta_3\beta_4}(x',0)\\
&-\widehat{G}^{\kappa_2\varrho_1}_{\alpha_2\beta_1}(u-x')\exp\left[-ik\frac{1}{3}(-2u+2x')\right]\widehat{F}^{\varrho_2\kappa_3}_{\beta_2\alpha_3}(x'-u)\widehat{\chi}^{\kappa_1\varrho_3\varrho_4}_{\alpha_1\beta_3\beta_4}(x',0)\\
&+\widehat{G}^{\kappa_3\varrho_1}_{\alpha_3\beta_1}(u-x')\exp\left[-ik\frac{1}{3}(-2u+2x')\right]\widehat{F}^{\varrho_2\kappa_2}_{\beta_2\alpha_2}(x'-u)\widehat{\chi}^{\kappa_1\varrho_3\varrho_4}_{\alpha_1\beta_3\beta_4}(x',0)\\
&-\widehat{G}^{\kappa_3\varrho_1}_{\alpha_3\beta_1}(-x')\exp\left[-ik\frac{1}{3}(u+2x')\right]\widehat{F}^{\varrho_2\kappa_1}_{\beta_2\alpha_1}(u+x')\widehat{\chi}^{\kappa_2\varrho_3\varrho_4}_{\alpha_2\beta_3\beta_4}(x',0)\Big\}
\end{aligned}
\tag{2.105}
$$

The derivation of Eq. (2.105) was first given in Stumpf (2003b), with restriction to the rest system. For arbitrary motion, the left side has to be completed by the factor exp(-ikz), while on the right side, the factor $\exp\left(-ik\frac{1}{3}z\right)$ has to be included. This proceeding was correctly formulated in Stumpf [2004c]. In contrast to the above cited papers in (2.105) the super-spin-isospin part of the wave function has not been eliminated.

Concerning the angular momentum, one can study this problem in the rest system without loss of generality. As Eqs. (2.92) and (2.98) are relativistically invariant equations, their solutions can be relativistically covariant transformed into any other convenient system.

Relativistically, the angular momentum can be defined by the Pauli-Lubanski spin vector. In the rest system, this vector is reduced to the non-relativistic angular momentum operators J_i, $i = 1,2,3$, which are defined by

$$J_i = L_i + S_i = \sum_{\alpha=1,2,3} i\varepsilon_{ijk}\left[x_j^\alpha \partial_k^\alpha - x_k^\alpha \partial_j^\alpha\right]$$
$$\frac{1}{2}\left[\Sigma^i_{\alpha_1\beta_1}\delta_{\alpha_2\beta_2}\delta_{\alpha_3\beta_3} + \delta_{\alpha_1\beta_1}\Sigma^i_{\alpha_2\beta_2}\delta_{\alpha_3\beta_3} + \delta_{\alpha_1\beta_1}\delta_{\alpha_2\beta_2}\Sigma^i_{\alpha_3\beta_3}\right. \tag{2.106}$$

and which are the generators of little group transformations in the rest system. For further information about this formula cf. Borne, Lochak, Stumpf (2001, sec. 6.4).

In Stumpf (2003b), it was shown that J_3 commutes with the integral kernel of Eq. (2.98). But the proof is so general that for any J_i, the same conclusion holds. Thus, $\mathbf{J}^2$ and J_3 commute with the integral kernel of Eq. (2.98). In the rest system, the center of mass part does nothing to contribute to the angular momentum quantum numbers, and therefore, $\mathbf{J}^2$ and J_3 also commute with the integral kernel of Eq. (2.105).

The transformation of the angular momentum operators into the new coordinates u, v, x' and their compatibilty with antisymmetrization is treated in detail in Stumpf (2004c, section 5). According to this property of the integral kernel, it follows that for any given eigenstate of the integral equation, the application of a little group operation to this solution generates a state that is an eigenstate of this integral equation as well. Thus, these group operations applied to eigenstates generate a set of eigenstates that define a representation space of the little group. But nothing can be said about the dimensions of such representations.

In quantum mechanics, Hamiltonians that are invariant under a symmetry group are generally projected on the state spaces of representations of this group in order to obtain information about the participation of various representations in the spectrum of the Hamiltonian. But in this way, Eq. (2.105) cannot be treated because its integral kernel is not symmetric. In addition, it contains only the reduced function $\widehat{\chi}(x', 0)$.

2.10 GROUP THEORY OF FERMIONS FOR FULL SYMMETRY

The three-parton states under consideration are assumed to be solutions of generalized de Broglie—Bargmann-Wigner equations for $n = 3$

(Stumpf, 2003b, 2004c). For conserved symmetries, their quantum numbers can be studied, which result from group theoretical constraints being imposed on these solutions owing to the specific symmetry properties of the theory. In particular, the symmetry groups are defined by the continuous superspin-isospin group $SU(2) \otimes U(1)$, the finite permutation group S_3, and the relativistic invariance against space-time transformations. Then the algebraic symmetry constraints are defined by the diagonalization condition of the Casimir operators of the corresponding Lie algebras for the continuous groups.

Owing to their construction, all solutions depend on the auxiliary field indices i_1, i_2, and i_3, which are needed for regularization. But the symmetries to be discussed are not correlated to these auxiliary field indices (Stumpf, 2003b, 2004c). Hence, one can study the corresponding physical (i.e., regularized) wave functions $\widehat{\varphi}$, which arise from the original wave functions φ by summations over i_1, i_2, and i_3.

Let $\widehat{\varphi}_{I_1 I_2 I_3}$ be such a solution with formal coordinates I_1, I_2, and I_3, which, in order to simplify matters in contrast to the definition in Eq. (1.5) of this part are defined by $I = x, \alpha, \kappa$ and $z = x, \alpha$ in this section. Then this function can be written with full indexing:

$$\widehat{\varphi}_{I_1 I_2 I_3} := \widehat{\varphi}^{\alpha_1 \alpha_2 \alpha_3}_{\kappa_1 \kappa_2 \kappa_3}(x_1, x_2, x_3) =: \widehat{\varphi}_{\kappa_1 \kappa_2 \kappa_3}(z_1, z_2, z_3). \tag{2.107}$$

While in Borne, Lochak, and Stumpf (2001), the algebraic group theoretical constraints were introduced by reference to the field theoretic background of the wave functions, in Stumpf (2003b, 2004c), the invariance of generalized de Broglie—Bargmann-Wigner equations under superspin-isospin transformations was directly proven. This invariance is valid without any loss of generality and thus leads to the same algebraic constraints that were adopted from the field theoretical formalism.

These constraints are expressed by the following conditions on $\widehat{\varphi}$, which result from diagonalizing G^2, G_3, F, and Q [see Borne, Lochak, and Stumpf (2001); Stumpf (2003b, 2004c)]:

$$\frac{9}{4}\widehat{\varphi}_{I_1 I_2 I_3} + 2\Big[G^k_{I_1 I} G^k_{I_2 K} \widehat{\varphi}_{IKI_3} + G^k_{I_1 I} G^k_{I_3 K} \widehat{\varphi}_{II_2K} + G^k_{I_2 I} G^k_{I_3 K} \widehat{\varphi}_{I_1 IK}$$
$$= t(t+1)\widehat{\varphi}_{I_1 I_2 I_3} \tag{2.108}$$

$$G^3_{I_1 I}\widehat{\varphi}_{II_2 I_3} + G^3_{I_2 I}\widehat{\varphi}_{I_1 II_3} + G^3_{I_3 I}\widehat{\varphi}_{I_1 I_2 I} = t_3 \widehat{\varphi}_{I_1 I_2 I_3} \tag{2.109}$$

$$F_{I_1 I}\widehat{\varphi}_{I I_2 I_3} + F_{I_2 I}\widehat{\varphi}_{I_1 I I_3} + F_{I_3 I}\widehat{\varphi}_{I_1 I_2 I} = f\widehat{\varphi}_{I_1 I_2 I_3} \tag{2.110}$$

$$Q_{I_1 I}\widehat{\varphi}_{I I_2 I_3} + Q_{I_2 I}\widehat{\varphi}_{I_1 I I_3} + Q_{I_3 I}\widehat{\varphi}_{I_1 I_2 I} = q\widehat{\varphi}_{I_1 I_2 I_3}, \tag{2.111}$$

where the operators G^k, $k = 1, 2, 3$ are defined by

$$G^k := \frac{1}{2}\begin{pmatrix} \sigma^k & 0 \\ 0 & (-1)^k \sigma^k \end{pmatrix}_{\kappa_1 \kappa_2} \delta_{z_1 z_2} = \frac{1}{2}\begin{pmatrix} \sigma^k & 0 \\ 0 & \left(\sigma^k\right)^* \end{pmatrix}_{\kappa_1 \kappa_2} \delta_{z_1 z_2}, \tag{2.112}$$

while F and Q are given by

$$F := \frac{1}{3}\begin{pmatrix} 1 & 0 \\ 0 & -1 \end{pmatrix}_{\kappa_1 \kappa_2} \delta_{z_1 z_2}; \quad Q = G^3 + \frac{1}{2}F. \tag{2.113}$$

These constraints have to be satisfied for exact and for approximate solutions; otherwise, it would be impossible to give such solutions a physical meaning. Obviously, the condition [Eq. (2.111)] for the eigenstates of Eqs. (2.109) and (2.110) is reduced to the c-number condition $q = t^3 + \frac{1}{2}f$. Hence, only Eqs. (2.108)−(2.110) are nontrivial.

Furthermore, the symmetry operations of the permutation group S_3 commute with the algebraic operators G^2, G_3, F, and Q, which means that any representation of S_3 can be generated by the eigenfunctions of Eqs. (2.108)−(2.111).

As any transposition P_j commutes with G^2, G_3, F, and Q, and as any permutation can be generated by transpositions, antisymmetrization and the constraints [Eqs. (2.108)−(2.111)] are compatible conditions; i.e., the solutions of the constraints can be studied without regard to antisymmetry operations.

In Stumpf (2003b), it was shown that Eqs. (2.108)−(2.111) must hold for any solution of Eq. (2.92). In order to study these constraints separately, we assume that without loss of generality, a general solution can be written in the following form:

$$\widehat{\varphi}_{I_1 I_2 I_3} = \sum_k \Theta^k_{\kappa_1 \kappa_2 \kappa_3} \phi^k(z_1, z_2, z_3) \tag{2.114}$$

To satisfy all constraints, we proceed in several steps, the last of which, of course, should be the calculation of the mass eigenvalue.

In the first step, we try to construct eigenstates of Eqs. (2.108)−(2.111) by considering the tensors Θ^k separately for any k and by postponing the treatment of the antisymmetry condition. Any tensor Θ^k can be generated by products of single-particle states, given in the algebraic case by a set of four independent vectors in superspin-isospin space $\{\varphi^j_{m.\kappa}\}$ with $m = 1,\ldots, 4$ and $k = 1,\ldots, 4$ and leading to the following expansion:

$$\Theta^k_{\kappa_1\kappa_2\kappa_3} = \sum_{j_1 m_1 j_2 m_2 j_3 m_3} C^k_{j_1 m_1 j_2 m_2 j_3 m_3} \varphi^{j_1}_{m_1\kappa_1} \varphi^{j_2}_{m_2\kappa_2} \varphi^{j_3}_{m_3\kappa_3}, \tag{2.115}$$

where j is the number of the particle and m characterizes the superspin-isospin state of this particle.

Before studying a representation of Eq. (2.114) or (2.115), respectively, we adopt the quantum numbers of the single-particle states with respect to G^2, G_3, F, and Q from Greiner (1979), leading to the following table:

	$m = 1$	$m = 2$	$m = 3$	$m = 4$
t	1/2	1/2	1/2	1/2
t_3	1/2	−1/2	−1/2	1/2
f	1/3	1/3	−1/3	−1/3
q	2/3	−1/3	−2/3	1/3

In the next step, we turn to the construction of basis states that simultaneously satisfy Eqs. (2.108)−(2.111).

It is convenient to explicitly separate and distinguish isospin indices from superspin indices by applying the decomposition $\kappa = (\Lambda, A)$, with Λ being the superspin index and A the isospin index. This decomposition is defined by the following correspondence:

$$\begin{aligned} \kappa = 1,2 \quad &\rightarrow \quad \Lambda = 1,\ \ \mathrm{A} = 1,2 \\ \kappa = 3,4 \quad &\rightarrow \quad \Lambda = 2,\ \ \mathrm{A} = 1,2. \end{aligned} \tag{2.116}$$

In this indexing, one obtains for the generators [Eqs. (2.112) and (2.113)]

$$\begin{aligned} G^k_{\kappa_1\kappa_2} &\equiv G^k_{\Lambda_1 A_1,\Lambda_2 A_2} = \frac{1}{2}\left[\delta_{\Lambda_1 1}\delta_{\Lambda_2 1}\sigma^k_{A_1 A_2} - \delta_{\Lambda_1 2}\delta_{\Lambda_2 2}\left(\sigma^k_{A_1 A_2}\right)^*\right] \\ F_{\kappa_1\kappa_2} &\equiv F_{\Lambda_1 A_1,\Lambda_2 A_2} = \frac{1}{3}\sigma^3_{\Lambda_1\Lambda_2}\delta_{A_1 A_2}, \end{aligned} \tag{2.117}$$

while the single-particle states of Eq. (2.115) are to be represented by the following relation:

$$\widehat{\varphi}^j_{m\kappa} \quad \rightarrow \quad \widehat{\varphi}^j_{(m_\Lambda m_A)\Lambda A}. \tag{2.118}$$

Likewise, Eq. (2.92) can be transcribed into this indexing by observing that with $\nu^1 := 1$ and $\nu^2 := \gamma^5$, the antisymmetrized vertex [Eq. (2.8)] has the following form:

$$V^{\kappa_1\kappa_2\kappa_3\kappa_4}_{\alpha_1\alpha_2\alpha_3\alpha_4} :=$$

$$\sum_h \Big[\nu^h_{\alpha_1\alpha_2}\delta_{\kappa_1\kappa_2}\left(\nu^h C\right)_{\alpha_3\alpha_4}\Upsilon^5_{\kappa_3\kappa_4} - \nu^h_{\alpha_1\alpha_3}\delta_{\kappa_1\kappa_3}\left(\nu^h C\right)_{\alpha_2\alpha_4}\Upsilon^5_{\kappa_2\kappa_4} - \nu^h_{\alpha_1\alpha_4}\delta_{\kappa_1\kappa_4}\left(\nu^h C\right)_{\alpha_3\alpha_2}\Upsilon^5_{\kappa_3\kappa_2}\Big]. \tag{2.119}$$

Then this transcription can be performed by the use of

$$\delta_{KK'} = \delta_{\Lambda\Lambda'}\delta_{AA'} \quad \Upsilon^5_{KK'} = \sigma^1_{\Lambda\Lambda'}\delta_{AA'} \tag{2.120}$$

and Eqs. (2.108)–(2.110) read as follows for one-particle states with $\Lambda = 1$:

$$\begin{aligned} \frac{1}{4}\left(\sigma^k\sigma^k\right)_{A_1A_2}\varphi^j_{(m_\Lambda m_A)1A_2} &= t(t+1)\varphi^j_{(m_\Lambda m_A)1A_1} \\ \frac{1}{2}\sigma^3_{A_1A_2}\varphi^j_{(m_\Lambda m_A)1A_2} &= t^3\varphi^j_{(m_\Lambda m_A)1A_1} \\ \frac{1}{3}\delta_{A_1A_2}\varphi^j_{(m_\Lambda m_A)1A_2} &= f\varphi^j_{(m_\Lambda m_A)1A_1}, \end{aligned} \tag{2.121}$$

and with $\Lambda = 2$:

$$\begin{aligned} \frac{1}{4}\left(\sigma^{k*}\sigma^{k*}\right)_{A_1A_2}\varphi^j_{(m_\Lambda m_A)2A_2} &= t(t+1)\varphi^j_{(m_\Lambda m_A)2A_1} \\ -\frac{1}{2}\sigma^{3*}_{A_1A_2}\varphi^j_{(m_\Lambda m_A)2A_2} &= t^3\varphi^j_{(m_\Lambda m_A)2A_1} \\ -\frac{1}{3}\delta_{A_1A_2}\varphi^j_{(m_\Lambda m_A)2A_2} &= f\varphi^j_{(m_\Lambda m_A)2A_1}. \end{aligned} \tag{2.122}$$

From these relations, it follows that for the calculation of isospin states, one needs only the algebra $\{\sigma^k\}$ because the complex conjugate algebra $\{\sigma^{k*}\}$ is isomorphic to the original one. This fact can be expressed in a more convenient way by representations with G-conjugated fields.

Considering isospin transformations only, the spinor fields transform in isospace according to

$$\psi' = \exp\left(-i\varepsilon_k\sigma^k\right)\psi \tag{2.123}$$

while the charge-conjugated spinor fields transform in isospace according to

$$\psi^{c'} = \exp\left(i\varepsilon_k\sigma^*_k\right)\psi^c, \tag{2.124}$$

which is expressed by the special form of the generators [Eqs. (2.112) and (2.113)].

Owing to the equivalence of the σ-algebra with the σ^*-algebra, it is reasonable to assume that for these transformations, the system contains a hidden symmetry. Indeed, this symmetry can be realized by replacing the charge-conjugated spinor fields by G-conjugated spinor fields. The latter are defined by the relation cf. [Stancu (1996, section 8.5)]:

$$\psi^G_{A\alpha i}(x) = c^{-1}_{AA'}\psi^c_{A'\alpha i}(x), \tag{2.125}$$

with c $:= -i\sigma_2$. One easily verifies that under isospin transformations, ψ^G transforms according to

$$\psi^{G'} = \exp\left(-i\varepsilon_k\sigma^k\right)\psi^G. \tag{2.126}$$

Therefore, with respect to isospin transformations, the fields ψ and ψ^G cannot be distinguished. Furthermore, it can be shown that for Lorentz transformations, the fields ψ and ψ^c, as well as $-\psi^G$, transform according to the same transformation law [see Roman (1960, p. 285) and Stumpf (2003c, Eq. 15)].

Thus, if the superspinor transformation is referred to ψ and ψ^G, it is impossible to distinguish these fields by means of their transformation properties. This fact allows for the description of bound states by mixtures of these fields without destroying homogenous transformation properties, which are required for an appropriate physical interpretation of these states.

Suppose now that the spinor field theory has been transformed into the $\psi - \psi^G$ the representation. Then the one-particle states [Eq. (2.118)] must be referred to this representation, where $\varphi_{(1)A}$ means either an isospinor state $m_A = 1$ with quantum number $t_3 = 1/2$ for the ψ-field or for the ψ^G-field, respectively; and the same holds for the isospinor state $\varphi_{(2)A}$ with $m_A = 2$ and the quantum number $t_3 = -1/2$. Consequently, both these fields satisfy the following constraints:

$$\begin{aligned} \frac{1}{4}\left(\sigma^k\sigma^k\right)_{A_1A_2}\varphi_{(u)A_2} &= t(t+1)\varphi_{(u)A_1} \\ \frac{1}{2}\sigma^3_{A_1A_2}\varphi_{(u)A_2} &= t_3\varphi_{(u)A_1} \end{aligned} \tag{2.127}$$

for $u = 1, 2$ and the corresponding t_3 values. Therefore, a product of three one-particle superspinor-isospinor states can be written in the following way:

$$\begin{aligned} &\varphi\left(m^1_\Lambda m^1_A\right)_{\Lambda_1A_1}\varphi\left(m^2_\Lambda m^2_A\right)_{\Lambda_2A_2}\varphi\left(m^3_\Lambda m^3_A\right)_{\Lambda_3A_3} = \\ &\varphi\left(m^1_A\right)_{A_1}\varphi\left(m^2_A\right)_{A_2}\varphi\left(m^3_A\right)_{A_3}\varphi\left(m^1_\Lambda\right)_{\Lambda_1}\varphi\left(m^2_\Lambda\right)_{\Lambda_2}\varphi\left(m^3_\Lambda\right)_{\Lambda_3}. \end{aligned} \tag{2.128}$$

As in the new representation, the spinor fields ψ and ψ^G are completely undiscernible under all relevant field transformations. It is impossible to assign any definite superspinor quantum number to the corresponding states in isospace, as well as in spin and coordinate space. This means that the ordering of the superspinors in Eq. (2.128) is irrelevant; and this fact leads to an additional selection rule concerning the admitted states.

This irrelevance of the order can be expressed either by separate symmetrization or by separate antisymmetrization of the superspin part. As will be justified in the three-parton case discussed later in this chapter, only symmetrization is compatible; i.e., the physical superspin-isospin basis vectors for three-parton states are given by

$$\begin{aligned}&\varphi\left(m_A^1\right)_{A_1}\varphi\left(m_A^2\right)_{A_2}\varphi\left(m_A^3\right)_{A_3}\left\{\varphi\left(m_\Lambda^1\right)_{\Lambda_1}\varphi\left(m_\Lambda^2\right)_{\Lambda_2}\varphi\left(m_\Lambda^3\right)_{\Lambda_3}\right\}_{Sym} = \\ &\varphi\left(m_A^1\right)\varphi\left(m_A^2\right)\varphi\left(m_A^3\right)\otimes\left\{\varphi\left(m_\Lambda^1\right)\varphi\left(m_\Lambda^2\right)\varphi\left(m_A^3\right)\right\}_{Sym}.\end{aligned} \tag{2.129}$$

To simplify the notation, we define for superspin states for $m_\Lambda = 1{,}2$:

$$\varphi_{\left(m_\Lambda^n=1\right)\Lambda_n} = \in(n); \quad \varphi_{\left(m_\Lambda^n=1\right)\Lambda_n} = \delta(n) \quad n = 1, 2, 3 \tag{2.130}$$

and for later use for isospin states $m_A = 1{,}2$:

$$\varphi_{\left(m_A^n=1\right)A_n} = \xi(n); \quad \varphi_{\left(m_A^n=2\right)A_n} = \eta(n) \quad n = 1, 2, 3, \tag{2.131}$$

which allows one to return to detailed indexing if necessary.

As far as the superspin part is concerned, the four combinations $m_\Lambda^1, m_\Lambda^2, m_\Lambda^3 = (1,1,1), (1,1,2), (1,2,2), (2,2,2)$ can be used. The corresponding symmetrized wave functions read as follows:

$$\begin{aligned}(1,1,1)_{\Lambda_1\Lambda_2\Lambda_3} &:= \in(1)\in(2)\in(3) \\ (1,1,1)_{\Lambda_1\Lambda_2\Lambda_3} &:= \left(\frac{1}{3}\right)^{\frac{1}{2}}[\in(1)\in(2)\delta(3) + \in(1)\delta(2)\in(3) + \delta(1)\in(2)\in(3)] \\ (1,1,1)_{\Lambda_1\Lambda_2\Lambda_3} &:= \left(\frac{1}{3}\right)^{\frac{1}{2}}[\delta(1)\delta(2)\in(3) + \in(1)\delta(2)\delta(3) + \delta(1)\in(2)\delta(3)] \\ (2,2,2)_{\Lambda_1\Lambda_2\Lambda_3} &:= \delta(1)\delta(2)\delta(3).\end{aligned} \tag{2.132}$$

The quantum numbers of these states are defined by the eigenvalues of F in the representation [Eq. (2.117)]; i.e., they are eigenvalues of $\sigma^3_{\Lambda\Lambda'}$.

This leads to the following table:

	(1,1,1)	(1,1,2)	(1,2,2)	(2,2,2)
f	1	1/3	−1/3	−1

For conserved symmetries, the permutation group representations play an essential role in the construction of appropriate wave functions. So long as no exact solutions of the three parton equations [Eq. (2.92)] are available, test functions have to be used. The group theoretical analysis of the three-parton system must guarantee that the resulting test functions possess quantum numbers that coincide with those of the fermions in phenomenological theory. But attention must be paid to the following fact:

The spinor field model is a left-right symmetric model by construction cf. Stumpf [(2003a), sect. 6]. For such a kind of model it holds:

> *"The basic premise of the left-right symmetric models is that the fundamental weak interaction Lagrangian (here (1.1)) is invariant under parity symmetry. … An immediate consequence of this hypothesis is that there must be right-handed neutrinos in nature and as a consequence neutrinos must be massive"*
>
> ***Mohapatra [(2003), chpt. 6, p. 130].***

However one should add: another consequence is the necessity of symmetry breaking for being in accordance with experiments. As long as this symmetry breaking has not been introduced, in a preliminary way it follows: For conserved symmetries the quantum numbers of the spinor field theory for both kinds of fermions (left-handed and right-handed) must coincide with the quantum numbers of the left handed fermions of the Standard Model. Hence, in the spinor field model for test functions of fermion states, these quantum numbers must be obtained. This can be achieved only by the use of mixed representations of the permutation group. Such mixed representations are generated by the application of Young operators C_{ik} (Pfister, 1995a; Kramer, John, & Schenzle, 1981).

For two-dimensional representations, these operators are defined by the relations

$$
\begin{aligned}
C_{11}^{[21]} &= \frac{1}{2}(1 - P_{12})\frac{1}{3}(2 + P_{13} + P_{23}) \\
C_{22}^{[21]} &= \frac{1}{2}(1 + P_{12})\frac{1}{3}(2 - P_{13} - P_{23}) \\
C_{12}^{[21]} &= \frac{1}{2}(1 - P_{12})\frac{\sqrt{3}}{3}(P_{23} - P_{13}) \\
C_{21}^{[21]} &= \frac{1}{2}(1 + P_{12})\frac{\sqrt{3}}{3}(P_{23} - P_{13}),
\end{aligned}
\tag{2.133}
$$

where P_{ik} means transposition, which interchanges arguments with indexes i and k. These operators are to be applied to superspin-isospin states as well as to spin-orbit states, and the definitions [Eq. (2.133)] of Pfister are equivalent to those of Kramer, John, and Schenzle (1981)

The use of the Young operators allows one to start with products of test wave functions that are not antisymmetrized from the beginning. For fermions, these products have to be formed by superspin-isospin test functions Θ^j and spin and orbit test functions $\Omega \otimes \psi$.

The superspin-isospin test functions are responsible for the definition of the phenomenological quantum numbers for isospin, charge, and fermion number, while the spin-orbit test functions should lead to the spin-1/2 of the fermions, to generation numbers, and to internal excitation levels. The latter two possibilities will be partly pursued in chapter 4.

After rearrangements, the general expression of Young combinations leads to the following antisymmetric test functions for fermions (Pfister, 1995b):

$$\begin{aligned} C^{\kappa_1\kappa_2\kappa_3}_{\alpha_1\alpha_2\alpha_3}(\mathbf{r}_1,\mathbf{r}_2,\mathbf{r}_3) &= \exp\left[-ik\frac{1}{3}(\mathbf{r}_1+\mathbf{r}_2+\mathbf{r}_3)\right]\times \\ &\left\{\left(C_{11}\Theta^j_{\kappa_1\kappa_2\kappa_3}\right)\times C_{22}\left[\Omega^n_{\alpha_1\alpha_2\alpha_3}\psi(\mathbf{r}_2-\mathbf{r}_1,\mathbf{r}_3-\mathbf{r}_2)\right]\right. \\ &\left.-\left(C_{21}\Theta^j_{\kappa_1\kappa_2\kappa_3}\right)\times C_{12}\left[\Omega^n_{\alpha_1\alpha_2\alpha_3}\psi(\mathbf{r}_2-\mathbf{r}_1,\mathbf{r}_3-\mathbf{r}_2)\right]\right\} \end{aligned} \tag{2.134}$$

A structurally transparent representation can be derived if charge-conjugated spinors are replaced by G-conjugated spinors and if this transformation is applied to the states in Borne, Lochak, and Stumpf [2001, Eqs. (6.108)]. If these states are used as initial information, one gets product representations between superspinor states [Eq. (2.132)] and isospinor states where the symmetric superspinor states commute with the Young operators. In isospinor space, one gets doublets and one can introduce indices r_1 and r_2 to characterize them, which read as

$$\begin{aligned} \chi^{1/2}_{1/2}(r_1)_{A_1A_2A_3} &= -\left(\frac{2}{3}\right)^{\frac{1}{2}}\xi(1)\xi(2)\eta(3) + \left(\frac{1}{6}\right)^{\frac{1}{2}}[\eta(1)\xi(2)+\xi(1)\eta(2)]\xi(3) \\ \chi^{1/2}_{1/2}(r_2)_{A_1A_2A_3} &= \left(\frac{1}{2}\right)^{\frac{1}{2}}[\xi(1)\eta(2)-\eta(1)\xi(2)]\xi(3) \end{aligned} \tag{2.135}$$

and

$$\chi^{1/2}_{-1/2}(r_1)_{A_1A_2A_3} = \left(\frac{2}{3}\right)^{\frac{1}{2}} \eta(1)\eta(2)\xi(3) - \left(\frac{1}{6}\right)^{\frac{1}{2}} [\eta(1)\xi(2)+\xi(1)\eta(2)]\eta(3)$$

$$\chi^{1/2}_{-1/2}(r_2)_{A_1A_2A_3} = \left(\frac{1}{2}\right)^{\frac{1}{2}} [\xi(1)\eta(2) - \eta(1)\xi(2)]\eta(3) \tag{2.136}$$

where the superscripts mean the t-value and the subscripts the t_3 values. The degeneracy parameters r_1 and r_2 allow an interpretation of Eqs. (2.135) and (2.136) by comparing them with mixed representations of the permutation group, where the states [Eq. (2.135)] correspond to the Young tableau (1, 2)(3), and the states [Eq. (2.136)] to the Young tableau (1, 3)(2).

Hence, by using these representations, one succeeds in constructing wave functions with full antisymmetry immediately. The corresponding formula is given by [cf. Hamermesh 1962, Eq. (7.211)]

$$\left[(-1)^3, 1\right] = \left(\frac{1}{2}\right)^{\frac{1}{2}} \sum_{i=1}^{2} \left[\lambda, r_i\right] \left[\tilde{\lambda}, r_i\right], \tag{2.137}$$

where $[(-1)^3,1]$ is the full, antisymmetric, one-dimensional representation of S_3, while $[\lambda, r_i]$ and $[\lambda, r_i]$ are the two-dimensional representations that arise from the Young tableau given previously.

While the states $[\lambda, r_i]$ are given by Eqs. (2.135) and (2.136), the states $[\tilde{\lambda}, r_i]$ should be constructed by means of spin-orbit states of the system under consideration. Here, we define them formally by

$$\left[\tilde{\lambda}, r_i\right] := \psi^{r_i}_{\alpha_1\alpha_2\alpha_3}(x_1, x_2, x_3). \tag{2.138}$$

Explicit construction of Eq. (2.138) was given in "Kramer and Stumpf (nd)" but in the connection at issue only formal use is made of it. Using the product representation of the superspinor states and the isospinor states, one can introduce a symbolic notation for the characterization of the complete wave function $\widehat{\psi}$, where $\widehat{\psi}$ refers to spinors and G-conjugated spinors. Then one gets

$$
\begin{aligned}
\widehat{\psi}^1 &:= (1,1,1)\otimes\sum_i \chi^{1/2}_{1/2}(r_i)\otimes\psi^{r_i}\\
\widehat{\psi}^2 &:= (1,1,1)\otimes\sum_i \chi^{1/2}_{-1/2}(r_i)\otimes\psi^{r_i}\\
\widehat{\psi}^3 &:= (2,2,2)\otimes\sum_i \chi^{1/2}_{1/2}(r_i)\otimes\psi^{r_i}\\
\widehat{\psi}^4 &:= (2,2,2)\otimes\sum_i \chi^{1/2}_{-1/2}(r_i)\otimes\psi^{r_i}\\
\widehat{\psi}^5 &:= (1,1,2)\otimes\sum_i \chi^{1/2}_{1/2}(r_i)\otimes\varphi^{r_i}\\
\widehat{\psi}^6 &:= (1,1,2)\otimes\sum_i \chi^{1/2}_{-1/2}(r_i)\otimes\varphi^{r_i}\\
\widehat{\psi}^7 &:= (1,2,2)\otimes\sum_i \chi^{1/2}_{1/2}(r_i)\otimes\varphi^{r_i}\\
\widehat{\psi}^8 &:= (1,2,2)\otimes\sum_i \chi^{1/2}_{-1/2}(r_i)\otimes\varphi^{r_i}.
\end{aligned}
\tag{2.139}
$$

In this set of states, the spin-orbit parts may be different for different algebraic quantum numbers. This is symbolically indicated by using two letters, ψ and φ, for the spin-orbit parts of leptons and quarks. The algebraic quantum numbers of the various components of these functions are completely fixed, and it should be emphasized that the three-parton equations [Eq. (2.92)] must be compatible with this scheme. The corresponding table of these algebraic quantum numbers reads as follows:

	$\widehat{\psi}^1$	$\widehat{\psi}^2$	$\widehat{\psi}^3$	$\widehat{\psi}^4$	$\widehat{\psi}^5$	$\widehat{\psi}^6$	$\widehat{\psi}^7$	$\widehat{\psi}^8$
t	1/2	1/2	1/2	1/2	1/2	1/2	1/2	1/2
t_3	1/2	−1/2	1/2	−1/2	1/2	−1/2	1/2	−1/2
f	1	1	−1	−1	1/3	1/3	−1/3	−1/3
q	1	0	0	−1	2/3	−1/3	1/3	−2/3
	$\bar{e}$	$\bar{\nu}$	ν	e	u	d	$\bar{d}$	$\bar{u}$

In this table, the isospin quantum numbers are to be identified with the quantum numbers of the unbroken electroweak SU(2) gauge group, and no distinction is made between left-handed and right-handed particles. The latter difference comes about by symmetry breaking effects, which have been excluded by definition. The fermion number in this scheme is connected with the weak hypercharge by $f = 2y$. Using this relation, the values of the table are in complete agreement with the phenomenological values of the left-handed particles; see, for instance, Nachtmann (1986, Table 22.2).

2.11 PARAFERMI BOSON AND LEPTON STATES

For carrying out the weak mapping procedure, the use and selection of states is closely related to the question of which observables are required for the theoretical description of intended or already-realized experiments.

Macroscopic observables are manifestations of the micro-world that can be registered by measuring devices in the macroworld. It depends on the state of art that can be measured. In any case, in order to gain macroscopic results, the observables have to be decoupled to a certain degree from the permanent motions in the microscopic world.

Effective theories are designed to derive macro-observables from micro-physics. Owing to the many potentialities to define such observables, the theory must offer sufficient flexibility for adaption to specific situations. This flexibility is provided by weak mappings apart from thermo-field theory and the algebraic methods of statistical mechanics; the latter two methods, however, are of only minor interest in the context of the problem being discussed here.

In the case under consideration, the effects of symmetry breaking at the classical level are of special interest, as the magnetic monopoles can be observed by the occurrence of isolated events in the macroworld (apart from indirect effects in the microworld). For the macroscopic observations, a physical interpretation of the associated functional energy equation [Eq. (1.28)] of Chapter 1 can be achieved by considering the classical limit of this equation. In this classical limit, the system is described by its effective classical equations of motion. The latter can be derived from Eq. (1.28) if mutual correlations in the matrix elements are suppressed. The transition from the functional version of the theory to its classical formulation can be performed by the ansatz [cf. Stumpf and Borne (1994, pp. 141,143) and Borne, Lochak, and Stumpf (2001, p. 242)]:

$$|P(\mathbf{b},\mathbf{f})\rangle = \exp[\mathcal{Z}_0(\mathbf{b},\mathbf{f})]0\rangle_V, \tag{2.140}$$

with

$$\mathcal{Z}_0(\mathbf{b},\mathbf{f}) = \int d^3r\Big[\sum_X \sum_{ka} \Theta_{ka}^{X,b}(\mathbf{r},t)\mathbf{b}_{ka}^X(\mathbf{r},t) - \sum_Y \sum_{k'a'} \Theta_{k'a'}^{Y,f}(\mathbf{r},t)\mathbf{f}_{k'a'}^Y(\mathbf{r},t)\Big], \tag{2.141}$$

where $\Theta_{ka}^{X,b}$ and $\Theta_{k'a'}^{Y,f}$ are to be the classical variables of the boson and fermion dynamics.

By comparing the ansatz [Eq. (2.140)] with the original functional state [Eq. (1.26)] of the weak mapping calculation, one verifies that the effect of this ansatz is the transition from one-particle matrix elements to classical functions by definition and a factorization of higher-order correlation functions in terms of these one-particle functions (i.e., a suppression of higher-order correlations).

This procedure is illustrated at some length for composite gravitons in Borne, Lochak, and Stumpf (2001, sections 9.3, 9.4, and 9.5). From this example, as well as from other calculations, one learns that possibly the choice of macro-observables might require the splitting of the complete bound state wave functions into parts and to perform the weak mapping by means of those parts. For further information, see Borne, Lochak, and Stumpf (2001).

To understand the physical meaning of the next discussion, it is useful to compare the wave functions for conserved symmetry and for broken symmetry. According to Eq. (2.68), the vector boson solutions are determined by the local state value $\varphi_{Z_1 Z_2}(x,x)$ on the right side of the equation.

For conserved symmetry, one gets from Eqs. (2.70) and (2.71) the regularized value

$$\widehat{\varphi}^{\kappa_1\kappa_2}_{\alpha_1\alpha_2}(x,x) = \exp(-ikx)T^a_{\kappa_1\kappa_2}\left[A^a_\mu(\mathbf{k})(\gamma^u C)_{\alpha_1\alpha_2} + F^a_{\mu\nu}(\mathbf{k})\left(\Sigma^{\mu\nu}C\right)_{\alpha_1\alpha_2}\right], \tag{2.142}$$

while for broken symmetry, one obtains from Eq. (2.83) the regularized value

$$\widehat{\varphi}^{\kappa_1\kappa_2}_{\alpha_1\alpha_2}(x,x) = \exp(-ikx)(T^a + S^a)_{\kappa_1\kappa_2}\left[A^a_\mu(\mathbf{k})(\gamma^u C)_{\alpha_1\alpha_2}\right.$$
$$\left. + G^a_\mu(\mathbf{k})\left(\gamma^5\gamma^u C\right)_{\alpha_1\alpha_2} + F^a_{\mu\nu}(\mathbf{k})\left(\Sigma^{\mu\nu}C\right)_{\alpha_1\alpha_2}\right], \tag{2.143}$$

where in accordance with Eq. (2.84), the terms with the redundant superspin-isospin part [Eq. (2.81)] are suppressed and the regularization procedure is not explicitly indicated in order to concentrate upon the important aspect of these expansions.

The denomination of the coefficients of the Dirac algebra in Eqs. (2.142) and (2.143) already shows that the map with such boson states should lead to effective electromagnetic or effective electroweak field equations with vector potentials and field-strength tensors.

Hence for the symmetry conserving case, Eq. (2.141) has to be defined by

$$\mathcal{Z}_0(\mathbf{b},\mathbf{f}) = \int d^3k \left[\sum_{\mathbf{k},la} \mathbf{b}^A_{l,a}(\mathbf{k}) A^l_a(\mathbf{k}) + \sum_{\mathbf{k},l,\nu a} \mathbf{b}^F_{l,\nu,a}(\mathbf{k}) F^{l\nu}_a(\mathbf{k}) \right]; \tag{2.144}$$

[cf. Borne, Lochak, and Stumpf 2001, Eqs. (7.114) and (7.115)] or Stumpf and Borne 1994, Eqs. (6.63) and (6.64). Whereas for the broken symmetry, one has to define

$$\mathcal{Z}_0(\mathbf{b},\mathbf{f}) = \int d^3k \left[\sum_{\mathbf{k},la} \mathbf{b}^A_{l,a}(\mathbf{k}) A^l_a(\mathbf{k}) + \sum_{\mathbf{k},la} \mathbf{b}^G_{l,a}(\mathbf{k}) G^l_a(\mathbf{k}) + \sum_{\mathbf{k},l,\nu a} \mathbf{b}^F_{l,\nu,a}(\mathbf{k}) F^{l\nu}_a(\mathbf{k}) \right]. \tag{2.145}$$

with "a" number of the electroweak symmetry group elements and "l" vector components.

For the symmetry-conserving case, the parts of the boson function have been calculated in Stumpf and Borne [1994, Eqs. (6.25),(6.26), and (6.27)].

In the case of CP symmetry breaking, exact solutions of the GBBW equations were derived for vector boson states in Stumpf (2004b). Hence, if one uses approximations to simplify the calculation for corresponding test functions, all group theoretical properties can be adopted from the exact solutions.

Furthermore, for the evaluation of the effective theory, single time wave functions are needed. To perform the transition to equal times, see Borne, Lochak, and Stumpf (2001, p. 153). Owing to the translational invariance of the system, we use the limit $t_1 = t_2 = 0$ without loss of generality. With $Z := (i, \alpha, \kappa)$ in this limit, the wave functions of the vector bosons contain the superspin-isospin factor $(S^a + T^a)^D = (S^a + T^a)^S$ in this special case, where the S-label denotes ordinary superspin-isospin representation (which is referred to by ψ and ψ^c), and the D-label denotes states that are referred to by ψ and ψ^G.

Then one can define an appropriate set of test functions by

$$\begin{aligned}
C^A_{Z_1Z_2}(\mathbf{r}_1,\mathbf{r}_2|\mathbf{k}a\mu) &:= (S^a+T^a)^S_{\kappa_1\kappa_2} e^{\left[-i\mathbf{k}\frac{1}{2}(\mathbf{r}_1+\mathbf{r}_2)\right]} (\gamma^u C)_{\alpha_1\alpha_2} f^A(\mathbf{r}_1-\mathbf{r}_2|\mathbf{k}a\mu)_{i_1i_2}\\
C^G_{Z_1Z_2}(\mathbf{r}_1,\mathbf{r}_2|\mathbf{k}a\mu) &:= (S^a+T^a)^S_{\kappa_1\kappa_2} e^{\left[-i\mathbf{k}\frac{1}{2}(\mathbf{r}_1+\mathbf{r}_2)\right]} (\gamma^5\gamma^u C)_{\alpha_1\alpha_2} f^G(\mathbf{r}_1-\mathbf{r}_2|\mathbf{k}a\mu)_{i_1i_2}\\
C^F_{Z_1Z_2}(\mathbf{r}_1,\mathbf{r}_2|\mathbf{k}a\mu,\nu) &:= (S^a+T^a)^S_{\kappa_1\kappa_2} e^{\left[-i\mathbf{k}\frac{1}{2}(\mathbf{r}_1+\mathbf{r}_2)\right]} (\Sigma^{\mu\nu} C)_{\alpha_1\alpha_2} f^F(\mathbf{r}_1-\mathbf{r}_2|\mathbf{k}a\mu\nu)_{i_1i_2}
\end{aligned} \tag{2.146}$$

if the exact vector boson solutions are split into vector potential parts and field parts, which later are connected to the effective phenomenological field variables by means of Z_0.

Following Lochak's hypothesis of excited neutrinos, it is imperative to assume that leptons must be composites. In order to gain confidence in the formalism, the relativistically invariant GBBW equations for three subfermions were treated for conserved symmetries (the associated group theory was discussed in sections 2.9 and in 2.10). The calculated quantum numbers of these states coincide with experimental data of the conventional Standard Model. Owing to conserved symmetry and relativistic invariance in both sections, the postulate of antisymmetry of the relativistic wave functions must be imposed.

The question of whether the antisymmetry of the relativistic wave functions for leptons must be abandoned for broken symmetries cannot be answered because calculations comparable to the boson case in section 2.6 have not yet been done. But actually such calculations are unnecessary because in the algebraic Schrödinger representation, only the associated GBBW-energy equations (which are single time equations) have to be treated.

Nevertheless, no exact solutions are available for this case. Therefore, an assumption of adequate test functions is necessary. For the latter test functions it is necessary to be compatible with CP-symmetry breaking, i.e. the lepton states must contain parafermi elements in analogy to the parafermi boson states and for the constructions of such lepton states it is prerequisite: Any superspin-isospin symmetry breaking states must still allow to identify neutrinos or the other leptons of the first generation.

In contrast to the Young construction in section 2.10, one can try to construct lepton states in the form of single products of superspin-isospin states and spin-orbit states. According to Borne, Lochak, and Stumpf [2001, Eq. (6.93)], such states read as

$$C^{\kappa_1\kappa_2\kappa_3}_{\alpha_1\alpha_2\alpha_3}(\mathbf{r}_1,\mathbf{r}_2,\mathbf{r}_3|\mathbf{k},j,n) = \exp\left[-i\mathbf{k}\frac{1}{3}(\mathbf{r}_1+\mathbf{r}_2+\mathbf{r}_3)\right] \times \left[\Theta^j_{\kappa_1\kappa_2\kappa_3}\Omega^n_{\alpha_1\alpha_2\alpha_3}\psi(\mathbf{r}_2-\mathbf{r}_1,\mathbf{r}_3-\mathbf{r}_2|\mathbf{k})\right]. \tag{2.147}$$

It was shown in Stumpf (2003b, 2004c) that for superspin-isospin states with permutation symmetry [such as Eq. (2.149)], this ansatz satisfies both symmetrical and asymmetric GBBW equations. Whether such a construction is useful in the case of symmetry breaking warrants further discussion, especially since no exact solutions can be obtained, and the ansatz can be considered only a general starting point for the elaboration of a test function.

For the spin tensor Ω^n, we apply products of elementary spin combinations that are *not* eigenstates of the Dirac operator for the definite k-vector. Furthermore, as the leptons are assumed to occupy the ground states of the three-parton system, the spin tensor and orbit functions must show the highest possible invariance under symmetry operations, which for these parts of the wave functions are the little group operations with all discrete transformations. This leads to the spin tensor and its adjoint tensor:

$$\Omega^n_{\alpha_1\alpha_2\alpha_3} = C_{\alpha_1\alpha_2}\xi^n_{\alpha_3}, \quad \overline{\Omega}^n_{\alpha_1\alpha_2\alpha_3} = C^+_{\alpha_1\alpha_2}\overline{\xi}^n_{\alpha_3}, \tag{2.148}$$

where ξ^n_α are the four unit spinors $\delta_{\alpha n}$, n = 1,2,3,4, while C is invariant under rotations and the discrete operation PC [cf. Grimm (1994b, p. 110)].

The C-matrix is an odd operator; i.e., in the conventional treatment of the Dirac equation, it mixes positive and negative energy states. But in the spinor field model, this statement is meaningless because no physical single particle subfermion states exist and in the bound states, no on-shell single particle states can be identified.

In Borne, Lochak, and Stumpf (2001, p. 171), a set of superspin-isospin states was given, which respects the identification of leptons by quantum numbers but violates the complete antisymmetrization of the composite lepton state. Hence, it has to be analyzed whether these states are suitable canditates for parafermionic representations.

For neutrinos, the corresponding superspin-isospin tensor is explicitly given in Borne, Lochak, and Stumpf [2001, Eq. (6.108)], by

$$\Theta^2_{\kappa_1\kappa_2\kappa_3} = 3^{-1/2}\left[\delta_{4,\kappa_1}\delta_{4,\kappa_2}\delta_{3,\kappa_3} + \delta_{4,\kappa_1}\delta_{3,\kappa_2}\delta_{4,\kappa_3} + \delta_{3,\kappa_1}\delta_{4,\kappa_2}\delta_{4,\kappa_3}\right], \tag{2.149}$$

where the index 2 serves for the identification of the neutrino state in Borne, Lochak, and Stumpf (2001). The complete list of superspin-isospin states in this source shows that abandoning the Young construction leads to higher isospin states and higher charge states, which at present have not been observed so far.

Although the Young construction is avoided by Eq. (2.147), the state [Eq. (2.149)] cannot be the correct description of the CP symmetry breaking because in this case, the exact superspin-isospin boson states are neither symmetric nor antisymmetric. Therefore, in analogy to the boson states, for CP symmetry breaking, the superspin-isospin lepton states should not have a permutation symmetry like Eq. (2.149).

Concerning the orbital states for broken symmetry, we consider only the lepton ground states of the first generation, which should have an s-wave character in the rest frame. Apart from the center of mass wave for this case, a suitable test function reads after self-regularization as

$$\psi(\mathbf{r}_2-\mathbf{r}_1,\mathbf{r}_3-\mathbf{r}_2|\mathrm{k}=0)=\exp\left[-\frac{1}{2}a(\mathbf{r}_2-\mathbf{r}_1)^2\right]\times\exp\left[-\frac{1}{2}a(\mathbf{r}_3-\mathbf{r}_2)^2\right]. \tag{2.150}$$

For practical applications, Eq. (2.150) needs further elaboration, which will be given in the following sections. The specific neutrino functions will be treated elsewhere.

2.12 COMPOSITE PARTICLE STATES FOR HIGH VELOCITIES

Electroweak processes partly proceed with particles with very small masses and the corresponding high velocities near the speed of light. In such cases in phenomenological theory, these particles are described by solutions of relativistically invariant wave equations. Indeed, this is possible even if the particles have a fermionic substructure. For composite particles, the generalized GBBW equations are relavistically invariant [cf. Borne, Lochak, and Stumpf (2001) and Stumpf and Borne (1994)] and can be used to describe the bound states of these particles.

However, for the interactions of composite particles due to the difficulties of relativistic calculations beyond perturbation theory [cf. Borne, Lochak, and Stumpf (2001, section 3.7.)], the algebraic Schrödinger representation was developed. In this representation, single time wave functions are exclusively used, and this raises the question: How can single time wave functions correctly describe covariant particle states?

For the two-fermion case, an answer was given in Stumpf [(1985a) p. 23] which subsequently was further evaluated.

Proposition 9: The covariant multi-time solution $\varphi^{(2)}_{z_1z_2}(x_1,x_2|a,k)$ of the two-body GBBW equation is completely determined by the single time solution $\varphi^{(2)}_{z_1z_2}(r_1,r_2|a,k)$ of the associated algebraic Schrödinger equation.

For the proof of this proposition, see Borne, Lochak, and Stumpf (2001, pp. 153, 154) and Stumpf and Borne (1994, pp. 110–112).

For $n > 2$ the treatment runs along similar lines but becomes rather complicated and hence these equations are of little practical value. However, independently of such calculations a more general statement can be given: In the rest frame of a single composite particle or a multi-composite particle state the single time algebraic Schroedinger representation is a limit state of the multitime covariant state function, cf. Borne, Lochak, Stumpf (2001, sect. 4.9) and Stumpf, Borne (1994, sect. 3.9).

As due to the self-regularization of the spinor field in the algebraic Schroedinger representation the energy eigenvalues can (in principle) be calculated, the fully covariant state description ought not to deviate from this result, i.e. the algebraic Schroedinger representation must be sufficient for the theoretical description of the composite particle system. This fact was illustrated in the statement by Milonni (1994, p. 449): *"The result of the old-fashioned (noncovariant) and of the covariant perturbation theories in quantum electrodynamics is the same."*

We leave aside electric phenomena and concentrate on the magnetic ones. These phenomena are connected with neutrinos, which in older theories were considered as massless. But owing to recent findings, the neutrinos possess small masses (Lindner & Weinheimer, 2011). This difference is essential because in that case, massive neutrinos move with velocities below the speed of light. Thus, one should find a rest frame for them and treat them as true particles. These remarks hold not only for the neutrino wave functions, but also for those of the other first-generation leptons. The existence of such rest frames is crucial for the subsequent conclusions. However, it remains the problem to provide suitable test functions without getting involved in much calculation.

Using test functions, we define them for the rest frame and then transform them into a moving frame in order to get high-velocity states. First we consider a composite particle in a relativistically invariant description by a hypothetical solution of the GBBW equations in the rest frame of this particle.

In any case, owing to the relativistic invariance, the associated solution depends on four-dimensional relative coordinates like $(x^1 - x^2)$, where $x^i := (\mathbf{x}^i, t^i)$, $i = 1, 2$, while the relativistic treatment of the center of mass part of the wave function gives an ordinary plane wave for the center of mass coordinates, which is trivial.

If one applies a relativistic transformation from the rest frame $\mathcal{K}$ with four dimensional coordinates of the center of mass x to a moving system $\overline{\mathcal{K}}$ with coordinates, $\overline{x}$ we adopt Eqs. (11.17)–(11.19) of Jackson (1999), which describe this transformation. With the notation $x_0 = ct$, $\boldsymbol{x} = x_1, x_2, x_3$ and the symbols

$$\vec{\beta} = \frac{\mathbf{v}}{c}, \quad \beta = |\vec{\beta}|, \quad \gamma = \left(1 - \beta^2\right)^{-1/2}, \tag{2.151}$$

the Lorentz transformation P from $\mathcal{K}$ to $\overline{\mathcal{K}}$ in an arbitrary direction $\mathbf{v}$ reads according to Eq. (11.19) of Jackson (1999):

$$\begin{aligned} \overline{x}_0 &= \gamma\left(x_0 - \vec{\beta}\cdot\mathbf{x}\right) \\ \overline{\mathbf{x}} &= \mathbf{x} + \frac{(\gamma-1)}{\beta^2}\left(\vec{\beta}\cdot\mathbf{x}\right)\vec{\beta} - \gamma\vec{\beta}x_0. \end{aligned} \tag{2.152}$$

As a first step, the transformation of Eq. (2.150) from the rest frame to a high-velocity state will be treated. Although no exact states are available, in principle, these states in $\mathcal{K}$ and $\overline{\mathcal{K}}$ must be solutions of the GBBW equations.

Given a solution ψ of these equations by application of the transformation $\mathcal{P}$, one gets a new state $\mathcal{P}\psi$, which is also a solution of the same GBBW equations due to their relativistic invariance; i.e., apart from the spinor transformation, $\mathcal{P}\psi$ is a state that arises from ψ by a coordinate transformation connected with $\mathcal{P}$.

In the algebraic Schrödinger representation, single time solutions of the associated GBBW energy equations are used for an adequate state definition in $\mathcal{K}$ and in $\overline{\mathcal{K}}$. Independently of that energy representation, it holds

The central condition for a relativistic transformation is the invariance relation

$$\left(\overline{x}^1_\mu - \overline{x}^2_\mu\right)\cdot\left(\overline{x}^{1,\mu} - \overline{x}^{2,\mu}\right) = \left(x^1_\mu - x^2_\mu\right)\cdot\left(x^{1,\mu} - x^{2,\mu}\right), \tag{2.153}$$

which already has been formulated in relative coordinates as the deformation of the internal structure of the particle is of interest. For these coordinates, the transformation $\mathcal{P}$ reads

$$\begin{aligned} \overline{x}^1_0 - \overline{x}^2_0 &= \gamma\left(x^1_0 - x^2_0 - \vec{\beta}\cdot\left(\mathbf{x}^1 - \mathbf{x}^2\right)\right) \\ \overline{\mathbf{x}}^1 - \overline{\mathbf{x}}^2 &= \mathbf{x}^1 - \mathbf{x}^2 + \frac{(\gamma-1)}{\beta^2}\left(\vec{\beta}\cdot\left(\mathbf{x}^1 - \mathbf{x}^2\right)\right)\vec{\beta} - \gamma\vec{\beta}\left(x^1_0 - x^2_0\right), \end{aligned} \tag{2.154}$$

where the indices 1, 2 characterize in a general way two relative coordinates of the system.

We now consider the transition from the rest frame to a rapidly moving frame, which takes the original particle with it. To start the calculation the composite particle is assumed to be at rest with respect to a single laboratory time $x_0^1 = x_0^2 = ct = x_0$ which corresponds to the algebraic Schroedinger representation referred to $\mathcal{K}$. Then one gets from (2.154)

$$\overline{x}_0^1 - \overline{x}_0^2 = -\gamma \overrightarrow{\beta} \cdot \left(\mathbf{x}^1 - \mathbf{x}^2\right) \text{ for } x_0^1 - x_0^2 = 0, \tag{2.155}$$

and the invariance relation [Eq. (2.153)] goes over into

$$\left(\overline{x}_\mu^1 - \overline{x}_\mu^2\right) \cdot \left(\overline{x}^{1\mu} - \overline{x}^{2\mu}\right) = \left(\mathbf{x}^1 - \mathbf{x}^2\right)^2. \tag{2.156}$$

On the other hand for the above particle which was at rest in $\mathcal{K}$ and which is now assumed to be in motion the frame $\overline{\mathcal{K}}$ must be the new rest frame where the algebraic Schroedinger representation has to be applied for calculations under the condition of vanishing time differences.

Then in $\mathcal{K}$ for the algebraic Schrödinger representation, the first equation in [Eq. (2.154)] yields

$$x_0^1 - x_0^2 = \overrightarrow{\beta} \cdot \left(\mathbf{x}^1 - \mathbf{x}^2\right) \text{ for } \overline{x}_0^1 - \overline{x}_0^2 = 0. \tag{2.157}$$

Therefore, with the second set of equations [Eq. (2.154)] in $\overline{\mathcal{K}}$, one gets with Eq. (2.157)

$$\overline{\mathbf{x}}^1 - \overline{\mathbf{x}}^2 = \left(\mathbf{x}^1 - \mathbf{x}^2\right) + \frac{(\gamma - 1)}{\beta^2}\left[\overrightarrow{\beta} \cdot \left(\mathbf{x}^1 - \mathbf{x}^2\right)\right]\overrightarrow{\beta} - \gamma \overrightarrow{\beta}\left[\overrightarrow{\beta} \cdot \left(\mathbf{x}^1 - \mathbf{x}^2\right)\right]. \tag{2.158}$$

Scalar multiplication of Eq. (2.158) with $\overline{\beta}$ yields

$$\overrightarrow{\beta} \cdot \left(\overline{\mathbf{x}}^1 - \overline{\mathbf{x}}^2\right) = \gamma\left(1 - \beta^2\right)\left[\overrightarrow{\beta} \cdot \left(\mathbf{x}^1 - \mathbf{x}^2\right)\right] \tag{2.159}$$

after some rearrangements, or, equivalently,

$$\gamma\left[\overrightarrow{\beta} \cdot \left(\overline{\mathbf{x}}^1 - \overline{\mathbf{x}}^2\right)\right] = \overrightarrow{\beta} \cdot \left(\mathbf{x}^1 - \mathbf{x}^2\right). \tag{2.160}$$

In the algebraic Schrödinger representation in $\overline{\mathcal{K}}$, it is $\overline{x}_0^1 - \overline{x}_0^2 = 0$ and $x_0^1 - x_0^2 \neq 0$. Hence, with Eq. (2.153), one gets

$$\left(\overline{\mathbf{x}}^1 - \overline{\mathbf{x}}^2\right) = \left(\mathbf{x}^1 - \mathbf{x}^2\right)^2 - \left(x_0^1 - x_0^2\right)^2, \tag{2.161}$$

which with Eqs. (2.157) and (2.160) ends in the final transformation formula:

$$\left(\overline{\mathbf{x}}^1 - \overline{\mathbf{x}}^2\right)^2 + \gamma^4 \left[\overrightarrow{\beta} \cdot \left(\overline{\mathbf{x}}^1 - \overline{\mathbf{x}}^2\right)\right]^2 = \left(\mathbf{x}^1 - \mathbf{x}^2\right)^2. \tag{2.162}$$

This formula can be applied immediately to the orbit part of Eq. (2.150) and can be interpreted as a transformation to a state of a moving particle in $\mathcal{K}$, as Eq. (2.162) is invariant under ν inversion. For the particle at rest in $\mathcal{K}$, apart from superspin-isospin and spin parts, the normalized orbital test function reads as

$$\psi(\mathbf{u}, \mathbf{w}|\mathbf{k} = 0) = \exp\left[-\frac{1}{2}a\mathbf{u}^2\right] \times \exp\left[-\frac{1}{2}a\mathbf{w}^2\right]\left(\frac{a}{\pi}\right)^{\frac{3}{2}} \exp(ik_0 z_0)\Big/_{z_0=0} \tag{2.163}$$

in the algebraic Schrödinger representation with $\mathbf{u} = \mathbf{r}_2 - \mathbf{r}_1$, $\mathbf{w} = \mathbf{r}_3 - \mathbf{r}_2$ and the center of mass coordinate $\mathbf{z} = \frac{1}{3}(\mathbf{r}_1 + \mathbf{r}_2 + \mathbf{r}_3)$, which is used to describe the motion of the composite particle as a whole by a plane wave in $\mathbf{R}^3$:

$$\psi(\mathbf{z}) = (2\pi)^{-3/2} \exp\left[-i\mathbf{kz}\right] \tag{2.164}$$

with the orthogonality property

$$(2\pi)^{-3} \int_{R^3} d^3z \exp\left[i(\mathbf{k} - \mathbf{k}')z\right] = \delta(\mathbf{k} - \mathbf{k}') \tag{2.165}$$

and the Fourier transformations

$$\begin{aligned} f(\mathbf{z}) &= (2\pi)^{-3/2} \int_{\mathbf{R}^3} d^3k A(\mathbf{k}) \exp(i\mathbf{k}z) \\ A(\mathrm{k}) &= (2\pi)^{-3/2} \int_{\mathbf{R}^3} d^3z f(\mathbf{k}) \exp(-i\mathbf{kz}) \end{aligned} \tag{2.166}$$

In the following (which also is discussed in section 4.3 of this part), the dimensionless normalization constants will be suppressed in order to concentrate on what is important. Only contributions to the normalization constant with nontrivial dimensions will be explicitly registered.

Because a nonvanishing wave vector k already indicates a moving system, the z-coordinate is not especially distinguished in contrast to the internal u- and w-coordinates.

Then, by the transformation of Eq. (2.163) into the moving frame $\overline{K}$, one gets for the mixed density with normalized wave functions from Eq. (4.31) of Chapter 4 of this part

$$\Upsilon(\overline{\mathbf{u}}, \overline{\mathbf{w}}) = |\psi(\overline{\mathbf{u}}, \overline{\mathbf{w}}|\mathbf{k}', \mathbf{k}'')|^2$$

$$= \exp\left[-\sum_{l=1}^{3} \overline{a}'_l \overline{\mathbf{u}}_l^2\right] \times \exp\left[-\sum_{j=1}^{3} \overline{a}'_j \overline{\mathbf{w}}_j^2\right] \times \prod_{i=1}^{3} \left(\frac{\overline{a}'_i}{\pi}\right)^{1/2} \exp\left[i\mathbf{k}'\mathrm{z}\right]$$

$$\times \exp\left[-\sum_{l=1}^{3} \overline{a}''_l \overline{u}_l^2\right] \times \exp\left[-\sum_{j=1}^{3} \overline{a}''_j \overline{\mathbf{w}}_j^2\right] \times \prod_{i=1}^{3} \left(\frac{\overline{a}''_i}{\pi}\right)^{1/2} \exp\left[i\mathbf{k}''\mathrm{z}\right], \tag{2.167}$$

with $\overline{a}'_i = \frac{1}{2}a\left[1 + \gamma^4(\overline{v}'_i)^2\right]$ and $\overline{a}''_i = \frac{1}{2}a\left[1 + \gamma^4(\overline{v}''_i)^2\right]$.

In comparison with previous calculations [cf. Stumpf (1986a, 1986b)], the evaluation of the lepton-boson coupling term [Eq. (4.23)] can be considerably changed by the introduction of velocity-dependent states.

CHAPTER 3

Dynamics of the Extended Standard Model

3.1 INTRODUCTORY COMMENTS

The extended Standard Model is a theoretical construction. It is based on the realization of weak mappings within the spinor field model, and it is designed to describe the effective dynamics of composite bosons and composite fermions and their mutual interactions under symmetry-breaking conditions.

In this connection, weak interactions play a central role. Weak interactions or electroweak interactions, respectively, are transmitted by electroweak vector bosons, and the latter belong to representations of non-abelian gauge theories which in their physical reactions are governed by rather complicated dynamics. In view of this, the central question is:

Is the concept of the extended Standard Model physically meaningful, and are the corresponding theoretical results correct which are gained by evaluation of weak mappings, correct—that is, do these calculations lead to results in agreement with phenomenological theory?

In this chapter, we concentrate upon the discussion of boson dynamics and postpone the treatment of the physics of composite bosons and composite fermions (and particularly leptons) to the next chapter. This is because composite fermion physics depends crucially on the results of boson dynamics.

Concerning composite boson dynamics, in section 2.3 of Chapter 2 in this part, the investigations and calculations of Grimm (1994a, 1994b) and Fuss (2004) were mentioned.

For conserved symmetries, composite bosons, and pointlike fermions, Grimm elaborated an extensive group theoretical formalism for the application of chain rule mappings. His results were obtained by strict antisymmetrization of the boson wave functions. Under CP symmetry breaking, this assumption is no longer guaranteed. Nevertheless, for conserved symmetries, these calculations have led to a complete agreement of the spinor field mappings with phenomenology and have opened new insights into further group theoretical consequences.

The work of Fuss was done on the same assumptions as that of Grimm, but it is mainly concerned with the verification of the inherent spinor field regularization, even for higher-order terms. Furthermore, Fuss devoted very detailed investigations into the problem of the consistency of mapping formalism. The latter problem is nontrivial because the chain rule postulate [Eq. (1.27)] of Chapter 1 cannot be completely evaluated analogous to the quantum mechanical case.

There is no other paper where, for conserved symmetries, the weak mapping and the regularization of the boson theory has been presented in greater detail and transparency. Considering these two theses as test objects for the comparison with phenomenological theories, one can say that the corresponding evidence was demonstrated and that one reaches an agreement with the phenomenological theories, provided that the higher-order terms of the chain rule mappings are tiny.

Both theses offer convincing arguments for the spinor field concept. However, they were too extensive to be published completely in journals.

From this, it follows that in the contribution at issue the dynamics of the extended Standard Model cannot be offered like the contents of a textbook—i.e., with well-ordered propositions, proofs, and comments.

The subject can only be reviewed in order to acquaint the interested reader with the matter.

3.2 THEORY WITH COMPOSITE ELECTROWEAK BOSONS

In a first step, we concentrate on the effects of composite bosons without modifications of fermion states by symmetry breaking. Even under this restrictive assumption, the conventional Standard Model cannot be elementary; rather, it must become an effective theory (i.e., an extended Standard Model).

To treat such problems, an effective functional energy operator $\tilde{\mathcal{H}}$ was formulated in functional space. Formally, this operator was reviewed in section 1.4 of this part. It reads as

$$\tilde{\mathcal{H}} = \mathcal{H}_f + \mathcal{H}_b^1 + \mathcal{H}_b^2 + \mathcal{H}_b^3 + \mathcal{H}_{bf}^1 + \mathcal{H}_{bf}^2, \tag{3.1}$$

where the various terms of Eq. (3.1) are defined by Eqs. (1.33) and (1.34) and the indices and arguments of the composite particle sources in Eq. (1.26) characterize the effective (macroscopic) degrees of freedom of these particles in Chapter 1.

From the mapping formulas [Eqs. (1.33) and (1.34)], it further follows that these indices and arguments must coincide with the quantities that characterize the test functions of bosons and fermions. For instance, the test functions of vector boson states [Eq. (2.146)] of Chapter 2 are made of vector potential parts and field strength parts with the possibility of decomposing the latter parts into test functions for both electric and for magnetic fields.

In Eq. (2.146), the superspin-isospin states ($S^a + T^a$) charaterize the macroscopic $SU(2) \otimes U(1)$ group representations for CP symmetry breaking, while ($\gamma^\mu C$) refers to the electric vector potentials and $i(\gamma^5\gamma^\mu C)$ to the magnetic axial vector potentials. The field part ($\Sigma^{\mu\nu} C$) can be decomposed into two parts that refer to electric and magnetic field strengths. Owing to the antisymmetry of $\Sigma^{\mu\nu} = 1/2[\gamma^\mu\gamma^\nu]$ and the resulting decomposition,

$$\Sigma^{\mu\nu} = \Sigma^{0h} + \Sigma^{hj} = i\alpha^h + \varepsilon^{hjk}\Sigma^k, \tag{3.2}$$

one can define the electric and magnetic fields. Hence, according to Eq. (2.146), the boson sources (functional operators) and their duals have to be introduced as follows:

$$\begin{aligned}\{\mathbf{b}_k\} &:= \left\{\mathbf{b}^A_{\mu,a}(\mathbf{k}), \mathbf{b}^G_{\mu,a}(\mathbf{k}), \mathbf{b}^E_{h,a}(\mathbf{k}), \mathbf{b}^B_{k,a}(\mathbf{k})\right\}\\ \{\mathbf{d}^b_k\} &:= \left\{\mathbf{d}^A_{\mu,a}(\mathbf{k}), \mathbf{d}^G_{\mu,a}(\mathbf{k}), \mathbf{d}^E_{h,a}(\mathbf{k}), \mathbf{d}^B_{k,a}(\mathbf{k})\right\},\end{aligned} \tag{3.3}$$

with quantum numbers corresponding to Eq. (2.146). In a similar manner, the fermion sources can be defined. Furthermore, in the following application, the Fourier transforms of the functional boson and fermion operators are used, referred to the argument $\mathbf{z}$ in order to be in agreement with the terms in the literature.

The calculation of a corresponding effective theory then proceeds in two steps:

1. The matrix elements [Eq. (1.34)] have to be calculated.
2. The effective theory has to be evaluated.

In accordance with the initial remarks, the composite boson model will be based on calculations in Stumpf (2005), where composite bosons and composite fermions have been assumed but the composite fermions remain spectators of the boson dynamics. In these calculations, boson test functions for broken CP symmetry have been used, whereas the composite spectator fermions are characterized by test functions for conserved CP symmetry. This choice of the test functions violates the condition of uniform transformation properties for test functions but allows one to draw the focus of the attention to the boson dynamics.

The effective fermion states can then be represented by Dirac spinors. By the latter proceding, a better understanding of the boson theory is possible, insofar as the effective fermionic spectator currents appear in the effective boson equations.

Later in this volume, in Chapter 4, the crucial coupling term that is responsible for the evidence of existing magnetic monopoles will be treated on a firm footing (i.e., with wave functions of bosons as well as of fermions, which both result from CP symmetry breaking). To be in conformity with the phenomenological field definitions, it is convenient to carry out a canonical transformation of the functional algebra for the G-vector potentials and E-fields, which is defined by

$$\mathbf{b}_{la}^{G}(\mathbf{z}) = i\mathbf{b}_{la}^{G}(\mathbf{z})' \ ; \mathbf{d}_{la}^{G}(\mathbf{z}) = -i\mathbf{d}_{la}^{G}(\mathbf{z})'$$
$$\mathbf{b}_{la}^{E}(\mathbf{z}) = -\mathbf{b}_{la}^{E}(\mathbf{z})'; \mathbf{d}_{la}^{E}(\mathbf{z}) = -\mathbf{d}_{la}^{E}(\mathbf{z})', \tag{3.4}$$

while the other algebra elements for the **A**-vector potentials and the **B**-fields remain unchanged.

After having performed the evaluation of Eq. (3.1), the primes of the new sources will be disregarded in the following discussion. With Eq. (3.4), one gets the explicit expressions for the various terms of Eq. (3.1), where the identification of the terms can be done by Eqs. (1.33) and (1.34) if one keeps to Eqs. (3.3) and (3.4) and to Eqs. (106), (45), (48), (55), and (104) in Stumpf (2005), which refer to the final formulas [but cf. also Stumpf (2006)]. In the composite boson theory, we start with the Dirac energy of the effective spectator fermions with quantum number κ

$$\mathcal{H}_f = \int d^3z\, \mathbf{f}(\mathbf{z}|B_1 b_1 \alpha_1)\left[-i(\gamma^0\gamma^k)\partial_k^z + m\gamma^0\right]_{\alpha_1\alpha_2} \mathbf{d}^f(\mathbf{z}|B_1 b_1 \alpha_2). \tag{3.5}$$

where the index κ is decomposed into (B,b) which characterize the superspin (B) and isospin (b) of the fermions.

While the kinetic energy is completely explained by kinematical reasons, in this formula, the rest mass of the pointlike fermion depends on its inner structure. This mass is of mechnical origin; i.e., it is the consequence of the self-interaction due to the basic spinor field.

If the fermion has an electric charge, a magnetic charge, or additional dipole-moments, then the associated self-energy has to be added to the mechanical mass. This problem cannot be satisfactorily treated for pointlike fermions; rather, it needs a finite spatial extension; i.e., a truly composite fermion where the regularization of the spinor field will give finite values of the self-mass in any case.

Corresponding field theoretic calculations of this phenomenon are not available at present but are important for further research into the properties of the magnetic monopole. In Chapter 4, later in this volume, we apply a phenomenological method for the treatment of these field-induced contributions to the self-mass of composite bosons, as well as of composite fermions.

The other contributions to the energy [Eq. (3.1)] are matrix elements with composite boson wave functions and spectator fermion states. They can be adopted from the previously cited formulas in Stumpf (2005), and it should be noted that these matrix elements have been calculated as the leading terms of the functional Hamilton operator in Borne, Lochak, and Stumpf [2001, Eq. (5.74)].

The remaining nonleading higher-order terms are characterized by the appearence of dressed dual states in Borne, Lochak, and Stumpf [2001,

Eq. (5.74)], which due to Eq. (1.36) can become infinitely small by a suitable choice of the regularization constants. Hence, the following formulas are the physically relevant terms of the composite boson dynamics:

$$\begin{aligned}\mathcal{H}_b^1 =& i\int d^3z\mathbf{b}_{la}^A(\mathbf{z})\left[c_1\varepsilon_{lkm}\partial_k^z\mathbf{d}_{ma}^G(\mathbf{z}) - c_2\mathbf{d}_{la}^E(\mathbf{z})\right]\\&-i\int d^3z\mathbf{b}_{la}^G(\mathbf{z})\left[c_1\varepsilon_{lkm}\partial_k^z\mathbf{d}_{ma}^A(\mathbf{z}) - c_3\mathbf{d}_{la}^B(\mathbf{z})\right]\\&+i\int d^3z\mathbf{b}_{la}^E(\mathbf{z})\left[\varepsilon_{lkm}\partial_k^z\mathbf{d}_{ma}^B(\mathbf{z}) + c_2\mathbf{d}_{la}^A(\mathbf{z})\right]\\&-i\int d^3z\mathbf{b}_{la}^B(\mathbf{z})\left[\varepsilon_{lkm}\partial_k^z\mathbf{d}_{ma}^E(\mathbf{z}) + c_3\mathbf{d}_{la}^G(\mathbf{z})\right]\end{aligned} \tag{3.6}$$

$$\begin{aligned}\mathcal{H}_b^2 =& -i\int d^3z\widehat{f}^A c_4\mathbf{b}_{la}^E(z)\mathbf{d}_{la}^A(z)\\&+i\int d^3z\widehat{f}^G c_4\mathbf{b}_{la}^B(z)\mathbf{d}_{la}^G(z)\end{aligned} \tag{3.7}$$

$$\begin{aligned}\mathcal{H}_b^3 =& \overline{\eta}_{abc}\varepsilon_{lkm}\Big\{64\widehat{f}^A\int d^3z\Big[k_1\mathbf{b}_{l,a}^A(\mathbf{z})\mathbf{d}_{k,b}^A(\mathbf{z})\mathbf{d}_{m,c}^G(\mathbf{z}) + k_2\mathbf{b}_{l,a}^E(\mathbf{z})\mathbf{d}_{k,b}^A(\mathbf{z})\mathbf{d}_{m,c}^B(\mathbf{z})\\&-k_2'\mathbf{b}_{l,a}^B(\mathbf{z})\mathbf{d}_{k,b}^A(\mathbf{z})\mathbf{d}_{m,c}^E(\mathbf{z}) - k_3\mathbf{b}_{l,a}^G(\mathbf{z})\mathbf{d}_{k,b}^A(\mathbf{z})\mathbf{d}_{m,c}^A(\mathbf{z})\Big]\\&+64\widehat{f}^G\int d^3z\Big[k_4\mathbf{b}_{l,a}^A(\mathbf{z})\mathbf{d}_{k,b}^G(\mathbf{z})\mathbf{d}_{m,c}^A(\mathbf{z}) + k_5\mathbf{b}_{l,a}^E(\mathbf{z})\mathbf{d}_{k,b}^G(\mathbf{z})\mathbf{d}_{m,c}^E(\mathbf{z})\\&+k_5'\mathbf{b}_{l,a}^B(\mathbf{z})\mathbf{d}_{k,b}^G(\mathbf{z})\mathbf{d}_{m,c}^B(\mathbf{z}) + k_6\mathbf{b}_{l,a}^G(\mathbf{z})\mathbf{d}_{k,b}^G(\mathbf{z})\mathbf{d}_{m,c}^G(\mathbf{z})\Big]\Big\}\end{aligned} \tag{3.8}$$

$$\begin{aligned}\mathcal{H}_{bf}^1 =& -K_1\int d^3z(\gamma^0\gamma^k)_{nm}(T^0\gamma^5)_{lj}\mathbf{f}_{nl}(\mathbf{z})\mathbf{d}_{k0}^A(\mathbf{z})\mathbf{d}_{mj}^f(\mathbf{z})\\&+iK_1\int d^3z(\gamma^0\gamma^k\gamma^5)_{nm}(S^0\gamma^5)_{lj}\mathbf{f}_{nl}(\mathbf{z})\mathbf{d}_{k0}^G(\mathbf{z})\mathbf{d}_{mj}^f(\mathbf{z})\\&+\frac{1}{3}K_1\sum_{b=1}^{3}\int d^3z(\gamma^0\gamma^k)_{nm}(T^b\gamma^5)_{lj}\mathbf{f}_{nl}(\mathbf{z})\mathbf{d}_{kb}^A(\mathbf{z})\mathbf{d}_{mj}^f(\mathbf{z})\\&-i\frac{1}{3}K_1\sum_{b=1}^{3}\int d^3z(\gamma^0\gamma^k\gamma^5)_{nm}(S^b\gamma^5)_{lj}\mathbf{f}_{nl}(\mathbf{z})\mathbf{d}_{kb}^G(\mathbf{z})\mathbf{d}_{mj}^f(\mathbf{z})\end{aligned} \tag{3.9}$$

$$\begin{aligned}\mathcal{H}_{bf}^2 =& iKt(0)^4\int d^3z\boldsymbol{\Theta}_{B_1b_1,B_2b_2}^a\Big[-2f^E(\gamma^kC)_{\mu_1\mu_2}\mathbf{b}^E(\mathbf{z}|a,k)\\&+f^Bi(\gamma^5\gamma^kC)_{\mu_1\mu_2}\mathbf{b}^B(\mathbf{z}|a,k)\Big]\mathbf{d}^f(\mathbf{z}|B_1,b_1,\mu_1)\mathbf{d}^f(\mathbf{z}|B_2,b_2,\mu_2)\end{aligned} \tag{3.10}$$

It should be emphasized that the input of Eqs. (3.5)–(3.10) is solely the spinor field model (Stumpf, 2005, section 2) and its sets of single-bosonic and single-fermionic states. These calculations were done under the premise that the most important contributions to the effect of CP symmetry breaking are the parafermi boson wave functions, whereas the very small contribution of the symmetry-breaking term in the propagator was ignored even though it is essential from the theoretical point of view.

In Borne, Lochak, and Stumpf (2001, section 7.5), the details of the evaluation of the boson matrix elements and the resulting effective canonical equations of motion can be found. A full account of the evaluation technique is delivered by Fuss (2004), which is based on conserved symmetry but can be easily extended to broken symmetry.

3.3 EFFECTIVE CANONICAL EQUATIONS OF MOTION

A physical interpretation of the associated effective functional energy equation [Eq. (1.28)] can be achieved by considering its classical limit, which can be attained by application of the ansatz [Eqs. (2.140) and (2.141); cf. also Borne, Lochak, and Stumpf (2001, section 7.5, p. 242].

In Eqs. (3.11)–(3.14), the quantities E_{la} and B_{la}, $l = 1, 2, 3$ and $a = 1, 2, 3, 0$ (= 4) represent the $SU(2) \otimes U(1)$ field strengths, while A_{la} and G_{la} are the electric and magnetic vector potentials in *temporal gauge*. This gauge can be consistently justified as a general constraint, even if the original $SU(2)$ invariance is broken; cf. Borne, Lochak, and Stumpf (2001, p. 235). Such vector potentials were introduced by de Broglie (1940, 1942) and by Cabibbo and Ferrari (1962) in electrodynamics and the following set of equations represents an electroweak generalization of this approach:

$$\begin{aligned} i\dot{A}_{la}(z) = {} & ic_1\varepsilon_{lkm}\partial_k^z G_{ma}(z) - ic_2 E_{la}(z) \\ & + \overline{\eta}_{abc}\varepsilon_{lkm}\left[\widehat{f}^{A} k_1 A_{kb}(\mathrm{z})G_{mc}(\mathrm{z}) + \widehat{f}^{G} k_4 G_{kb}(\mathrm{z})A_{mc}(\mathrm{z})\right] \end{aligned} \tag{3.11}$$

$$\begin{aligned} i\dot{G}_{la}(z) = {} & -ic_1\varepsilon_{lkm}\partial_k^z A_{ma}(z) + ic_3 B_{la}(z) \\ & + \overline{\eta}_{abc}\varepsilon_{lkm}\left[-\widehat{f}^{A} k_3 A_{kb}(\mathrm{z})A_{mc}(\mathrm{z}) + \widehat{f}^{G} k_6 G_{kb}(\mathrm{z})G_{mc}(\mathrm{z})\right] \end{aligned} \tag{3.12}$$

$$\begin{aligned} i\dot{E}_{la}(z) = {} & i\varepsilon_{lkm}\partial_k^z B_{ma}(z) + i\left(c_2 - \widehat{f}^{A} c_4\right)A_{la}(z) \\ & + \overline{\eta}_{abc}\varepsilon_{lkm}\left[\widehat{f}^{A} k_2 A_{kb}(\mathrm{z})B_{mc}(\mathrm{z}) + \widehat{f}^{G} k_5 G_{kb}(\mathrm{z})E_{mc}(\mathrm{z})\right] \\ & - iK'\Theta^a_{B_1,b_1,B_2,b_2}\widehat{f}^{E}\left(\gamma^l C\right)^{+}_{\alpha_1\alpha_2}\psi_{B_1,b_1,\alpha_1}(z)\psi_{B_2,b_2,\alpha_2}(\mathrm{z}) \end{aligned} \tag{3.13}$$

$$
\begin{aligned}
i\dot{B}_{la}(z) &= i\varepsilon_{lkm}\partial_k^z E_{ma}(z) - i(c_3 - \widehat{f}^{\,G} c_4)\, G_{la}(z) \\
&+ \overline{\eta}_{abc}\varepsilon_{lkm}\left[-\widehat{f}^{\,A} k'_2 A_{kb}(z) E_{mc}(z) + \widehat{f}^{\,G} k'_5 G_{kb}(z) B_{mc}(z) \right] \\
&- i(K'/2)\Theta^a_{B_1,b_1,B_2,b_2} \widehat{f}^{\,B} (i\gamma^5\gamma^l C)^+_{\alpha_1\alpha_2} \psi_{B_1,b_1,\alpha_1}(z)\psi_{B_2,b_2,\alpha_2}(z).
\end{aligned} \tag{3.14}
$$

The factor 64 in Eq. (3.8) has been included in the definition of the constants k_i. The latter are defined in Stumpf (2005, section 5).

For the fermion fields from Eqs. (3.5) and (3.9), the following equations of motion can be derived:

$$
\begin{aligned}
&i\dot{\psi}_{\alpha l}(z) = \left[-i(\gamma^0\gamma^k)_{\alpha\beta}\partial_k^z + \gamma^0_{\alpha\beta} m \right] \psi_{\beta l}(z) \\
&- K_1 \left[(\gamma^0\gamma^k)_{\alpha\beta} (T^0\gamma^5)_{ln} A_{k0}(z) - i(\gamma^0\gamma^k\gamma^5)_{\alpha\beta} (S^0\gamma^5)_{ln} G_{k0}(z) \right] \psi_{\beta n}(z) \\
&+ \frac{1}{3} K_1 \sum_{b=1}^{3} \left[(\gamma^0\gamma^k)_{\alpha\beta} (T^b\gamma^5)_{ln} A_{kb}(z) - i(\gamma^0\gamma^k\gamma^5)_{\alpha\beta} (S^b\gamma^5)_{ln} G_{kb}(z) \right] \psi_{\beta n}(z),
\end{aligned} \tag{3.15}
$$

where the indices l, n refer to the phenomenological numeration of the lepton states. This means that the field quantities $\psi_{\alpha,l}$ are superspinors of the phenomenological theory and ought not to be confused with the spinor field operators of the basic spinor field model in the background.

By comparison of Eqs. (3.13) and (3.14) with the phenomenological theory, it follows that the last term of Eq. (3.13) or (3.14), respectively, has to be identified with the electric or magnetic currents. In these terms, the superspinor-isospinor denotation of Eq. (1.4) was chosen in order to agree with the denotation in Stumpf (2005).

For instance, in Eq. (3.13), the last term has to be identified with the electric current j_l^a which leads to

$$
j_l^a = \left(\Theta^a_{B_1,b_1,B_2,b_2} \right)^D (\gamma^l C)^+_{\alpha_1\alpha_2} \psi_{B_1,b_1,\alpha_1} \psi_{B_2,b_2,\alpha_2}, \tag{3.16}
$$

with the superspinor-isospinor decomposition $\kappa = B,\, b$

$$
(\Theta^n)^D_{\kappa_1\kappa_2} = \frac{1}{2} (i\sigma^2 + \sigma^1)_{B_1B_2} \left[(\sigma^a)^T c^T \right]_{b_1b_2} \equiv \delta_{B_1 1}\delta_{B_2 2} \left[(\sigma^a)^T c^T \right]_{b_1b_2}. \tag{3.17}
$$

Eq. (3.16) is the result of the evaluation of the term W_4 in Eq. (1.34). For broken CP symmetry, the calculation leads to currents for pointlike fermions although composite fermions are involved in it.

To understand the formulas [Eqs. (3.16) and (3.17)] we refer to Stumpf (2005, eqs. 115-120). Furthermore from Stumpf [2012, Eqs. (18), (19)], one obtains the following interpretation of the phenomenological Dirac spinors referred to the subscript κ

$$\psi_{2,\alpha} \equiv e_\alpha^+; \quad \psi_{1,\alpha} \equiv \overline{\nu}_\alpha; \quad \psi_{4,\alpha} \equiv \nu_\alpha; \quad \psi_{3,\alpha} \equiv e_\alpha^-, \tag{3.18}$$

where the symbols on the right side of the equation mean the corresponding spinor wave function. According to their construction, these spinors are referred to as the D representation. Then, with Eqs. (3.16) and (3.17), Eq. (3.16) reads as

$$\begin{aligned} j_l^a &= \frac{1}{2}\delta_{2B_1}\delta_{1B_2}\left[(\sigma^a)^T c^T\right]_{b_1 b_2}\left(\gamma^l C\right)^+_{\alpha_1\alpha_2}\psi_{B_1 b_1\alpha_1}\psi_{B_2 b_2\alpha_2} \\ &= \frac{1}{2}\left[(\sigma^a)^T c^T\right]_{b_1 b_2}\left(\gamma^l C\right)^+_{\alpha_1\alpha_2}\psi_{2b_1\alpha_1}\psi_{1b_2\alpha_2}. \end{aligned} \tag{3.19}$$

The phenomenological fields in the S representation are defined by

$$\psi_{4,\alpha} \equiv \psi^S_{4,\alpha} \equiv \nu_\alpha; \quad \psi_{3,\alpha} \equiv \psi^S_{3,\alpha} \equiv e_\alpha^-. \tag{3.20}$$

The charge-conjugated counterparts can be generated by the transformation $\psi_{1,b,\alpha} = c^T_{b,b'}\psi^S_{1,b',\alpha}$. Therefore, Eq. (3.19) can be rewritten in the following form:

$$\begin{aligned} j_l^a &= \frac{1}{2}\left[(\sigma^a)^T c^T c^T\right]_{b_1 b_2}\left(\gamma^l C\right)^+_{\alpha_1\alpha_2}\psi^S_{2,b_1\alpha_1}\psi^S_{1,b_2\alpha_2} \\ &= -\frac{1}{2}[(\sigma^a)]^T_{b_1 b_2}\left(\gamma^l C\right)^+_{\alpha_1\alpha_2}\psi_{b_1\alpha_1}\psi^c_{b_2\alpha_2}. \end{aligned} \tag{3.21}$$

With $(\gamma^l C)$, its Hermitean conjugate is symmetric too. Thus, Eq. (3.19) reads equivalently as

$$j_l^a = \left(\psi^c_{b_2\alpha_2}\right)^T (\sigma^a)_{b_1 b_2}\left(\gamma^l C\right)^+_{\alpha_1\alpha_2}\psi_{b_1\alpha_1}. \tag{3.22}$$

In the last step, one uses $(\psi^c)^T = \overline{\psi} C^T$ and obtains from Eq. (3.21) the $U(1)$ and $SU(2)$- currents:

$$j_l^a \equiv -\frac{1}{2}\overline{\psi}_{b_1\alpha_1}\sigma^a_{b_1 b_2}\gamma^l_{\alpha_1\alpha_2}\psi_{b_2\alpha_2} \tag{3.23}$$

In the same way, one can proceed to get the magnetic currents J_l^a. The factors (1/2) will be absorbed in the coupling constants; i.e., normalization of the states is irrelevant.

Finally, we rearrange the field equations into their final form. Neglecting for simplicity the coupling between $SU(2)$-fields and $U(1)$-fields from Stumpf [2005, Eq. (53)], it follows that $\overline{\eta}_{abc} := i\varepsilon_{abc}$. Furthermore, we define $c_1 = 1$, $\widehat{f}^A = \widehat{f}^G$, $k_2 = k_2'$, $k_2 = k_5'$, $(c_2 - \widehat{f}^A c_4) =: \mu_A^2$ and $(c_3 - \widehat{f}^A c_4) =: \mu_G^2$, and express the current coupling constants g_e and g_m by the original constants in Eqs. (3.13) and (3.14).

Insertion of these definitions into Eqs. (3.11)–(3.14) and canceling out i yields the following set of field equations:

$$\begin{aligned}\dot{A}_{la}(\mathbf{z}) = &- \varepsilon_{lkm}\partial_k^z G_{ma}(\mathbf{z}) - c_2 E_{la}(\mathbf{z}) \\ &+ \varepsilon_{abc}\varepsilon_{lkm}\widehat{f}^A [k_1 A_{kb}(\mathbf{z}) G'_{mc}(\mathbf{z}) + k_4 G_{kb}(\mathbf{z}) A_{mc}(\mathbf{z})]\end{aligned} \tag{3.24}$$

$$\begin{aligned}\dot{G}_{la}(\mathbf{z}) = &- \varepsilon_{lkm}\partial_k^z A_{ma}(\mathbf{z}) + c_2 B_{la}(\mathbf{z}) \\ &+ \varepsilon_{abc}\varepsilon_{lkm}\widehat{f}^A [k_3 A_{kb}(\mathbf{z}) A_{mc}(\mathbf{z}) - k_6 G_{kb}(\mathbf{z}) G_{ma}(\mathbf{z})]\end{aligned} \tag{3.25}$$

$$\begin{aligned}\dot{E}_{la}(\mathbf{z}) = &\varepsilon_{lkm}\partial_k^z B_{ma}(\mathbf{z}) + g_e j_l^a + \mu_A^2 A_{la} \\ &\varepsilon_{abc}\varepsilon_{lkm}\widehat{f}^A [k_2 A_{kb}(\mathbf{z}) B'_{mc}(\mathbf{z}) + k_5 G_{kb}(\mathbf{z}) E'_{mc}(\mathbf{z})]\end{aligned} \tag{3.26}$$

$$\begin{aligned}\dot{E}_{la}(\mathbf{z}) = &- \varepsilon_{lkm}\partial_k^z E_{ma}(\mathbf{z}) + i g_m J_l^a - \mu_G^2 G_{la} \\ &- \varepsilon_{abc}\varepsilon_{lkm}\widehat{f}^A [k_2 A_{kb}(\mathbf{z}) E_{mc}(\mathbf{z}) - k_5 G_{kb}(\mathbf{z}) B'_{mc}(\mathbf{z})].\end{aligned} \tag{3.27}$$

For $a = 0$, all terms with ε_{abc} vanish; i.e., one gets the $U(1)$ field equations.

To complete the theory of vector fields, their constraints have to be formulated (electric and magnetic Gauss's law). In the canonical version of the theory, these constraints need not be postulated; rather, they can be derived from the vector field equations in combination with the spinor field equation; cf., for instance, Borne, Lochak, and Stumpf (2001, section 8.2). This derivation will not be spelled out here because it is not along the lines of this investigation.

In the next step, we rearrange the Dirac equation [Eq. (3.15)] into the conventional form. For the interpretation of Eq. (3.15), it is important to realize that the $(T\gamma^5)$ and $(S\gamma^5)$ matrices in the equation arise from matrix elements between two three-parton states that characterize the superspin-isospin part of the composite leptons; see Stumpf [2005, Eqs. (68) and (69)]. The latter matrix elements can be equivalently calculated in the isospinor-superspinor S-basis or D-basis of the elementary partons. It is convenient to do this in the D-basis. The calculation yields for $a = 1, 2, 3$

$$\left(S^a\gamma^5\right)^D_{ln} = \begin{pmatrix} \sigma^a & 0 \\ 0 & -\sigma^a \end{pmatrix}; \quad \left(T^a\gamma^5\right)^D_{ln} = \begin{pmatrix} \sigma^a & 0 \\ 0 & \sigma^a \end{pmatrix}, \tag{3.28}$$

and for a = 0,

$$\left(S^0\gamma^5\right)^D_{ln} = \begin{pmatrix} 1 & 0 \\ 0 & 1 \end{pmatrix}; \quad \left(T^0\gamma^5\right)^D_{ln} = \begin{pmatrix} 1 & 0 \\ 0 & -1 \end{pmatrix}, \tag{3.29}$$

where the indices l, n are referred to the state numbers of Eq. (3.15). Insertion of Eqs. (3.28) and (3.29) into Eq. (3.15) shows that this equation can be decomposed into two separate equations for ψ_1, ψ_2 and ψ_3, ψ_4. In particular, for $(\psi_3, \psi_4) = (\nu, e^-)$, one obtains after multiplication of Eq. (3.15) by γ^0 in spin-space the following equation:

$$\begin{aligned} &\left[-i\gamma^\mu\partial_\mu + m\right]\psi_l + \frac{1}{2}\left[g\sigma^a_{ln}\gamma^k A_{ka} + g'\sigma^0_{ln}\gamma^k A_{k0}\right]\psi_n \\ &+ i\frac{1}{2}\left[g\sigma^a_{ln}\left(\gamma^k\gamma^5\right)G_{ka} + g'\sigma^0_{ln}\left(\gamma^k\gamma^5\right)G_{k0}\right]\psi_n = 0. \end{aligned} \tag{3.30}$$

The corresponding equation for (ψ_1, ψ_2) is redundant and will not be explicitly given here for the sake of brevity.

3.4 A CONSISTENCY TEST OF THE BOSON THEORY

For the treatment of a combined boson-fermion theory, a self-evident postulate has to be that boson and fermion wave functions have the same symmetry properties. Leaving aside for the moment the construction of fermion functions, one can ask whether this postulate is sufficient to guarantee the consistency of the weak mapping method as a whole. To answer this question, one can postulate an effective Lagrangian that describes both electroweak electric and magnetic bosons under CP symmetry breaking conditions. Then, by definition of the Lagrange formalism, the associated field theory is consistent and can be compared with the results of the preceding sections to prove the consistency.

To facilitate the distinction between coordinate indices and superspin-isospin indices, we introduce the η-tensor by the following definition [cf. Stumpf 2005, Eq. (53)]:

$$\eta_{abc} = \varepsilon_{abc} = \varepsilon^{abc}.$$

Likewise, the electroweak field tensor in terms of the vector fields is required. We follow the definition used in the treatment of gauge theories by differential forms [Goeckeler and Schuecker, 1989, Eq. (4.6); Baez and Muniain, 1994, p. 70), which reads for antisymmetric $F^a_{\mu\nu}$

$$E^a_k = -F^a_{0k}; \quad B^a_k = \frac{1}{2}\varepsilon_{kij}F^a_{ij}, \tag{3.31}$$

where the metric is defined by $g_{kk} = diag\ (1, -1, -1, -1)$.

This definition of the fields is consistent with that used previously in this chapter in section 3.2. Furthermore, for the currents, the following definitions are introduced:

$$j^a_\mu := \overline{\psi}\sigma^a\gamma_\mu\psi = \left(j^a_\mu\right)^+; \quad J^a_\mu := \overline{\psi}\sigma^a\gamma^5\gamma_\mu\psi = \left(J^a_\mu\right)^+ \quad a = 0,1,2,3, \tag{3.32}$$

where the minus sign in Eq. (3.23) is absorbed in the coupling constant.

As a test model, we generalize the Lagrangian for nonabelian vector fields [cf. Goeckeler and Schuecker 1989, Eq. (4.83)] to the case of nonabelian electric and magnetic fields.

$$\begin{aligned} \mathcal{L} := &-\frac{1}{4}F^a_{\mu\nu}\eta^{\mu\varrho}F^a_{\varrho\kappa} + \frac{i}{2}\left[\overline{\psi}\gamma^\mu\partial_\mu\psi + (\partial_\mu\overline{\psi})\gamma^\mu\psi\right] - m\overline{\psi}\psi \\ &- g_\chi A^a_\mu j^\mu_a - ig_\pi G^a_\mu J^\mu_a + \frac{1}{2}\mu^2_A A^a_\mu\eta^{\mu\varrho}A^a_\varrho + \frac{1}{2}\mu^2_G G^a_\mu\eta^{\mu\varrho}G^a_\varrho. \end{aligned} \tag{3.33}$$

In Eq. (3.33), the field strength tensor is postulated to be

$$\begin{aligned} F^a_{\mu\nu} := &\ \partial_\mu A^a_\nu - \partial_\nu A^a_\mu - \varepsilon_{\mu\nu\varrho\sigma}\eta^{\varrho\varrho'}\eta^{\sigma\sigma'}\partial_{\varrho'}G^a_{\sigma'} \\ &+ \eta^{abc}\left(g_1 A^b_\mu A^c_\nu + g_2 G^b_\mu G^c_\nu + g_3\varepsilon_{\mu\nu\varrho\sigma}\eta^{\varrho\varrho'}\eta^{\sigma\sigma'}A^b_{\varrho'}G^c_{\sigma'}\right), \end{aligned} \tag{3.34}$$

which, owing to the required covariance under SU(2) transformations, contains mixtures of electric and magnetic potentials [cf. Stumpf (2006, section 4)], which is the essential effect of CP symmetry breaking in the extended Standard Model.

In order to guarantee a consistent comparison with the results of the preceding calculations, the Lagrangian density and its associated equations of motion are exclusively expressed in terms of covariant fields. By means of the Lagrangian formalism, these equations of motion can be derived from Eqs. (3.33) and (3.34). Furthermore, the Lagrangian [Eq. (3.33)] is invariant under global $SU(2) \otimes U(1)$ transformations (Stumpf, 2006) sect. 4 whereas owing to the masses in Eq. (3.33), no local gauge invariance is possible.

We start with the fermion equation. Its derivation is trivial. From Eq. (3.33), one obtains the Dirac equation

$$i\partial_\mu \gamma^\mu \psi - g_\chi A^a_\mu \gamma \sigma^a \gamma^\mu \psi - i g_\pi G^a_\mu \sigma^a \gamma^5 \gamma^\mu \psi + m\psi = 0. \tag{3.35}$$

As far as the vector fields are concerned, we assume that Eqs. (3.33) and (3.34) are evaluated in temporal gauge as described in section 3.3. This gauge must be compatible with the field dynamics, even if in Eq. (3.33) the gauge invariance is lost, because the conjugate momenta of A^a_0 and G^a_0, $a = 0, 1, 2, 3$ vanish identically; i.e., A^a_0 and G^a_0 are not genuine independent field variables.

We first study the consequences of the field tensor definition [Eq. (3.34)].

From Eq. (3.34), one obtains the equation for the E-fields, which in temporal gauge read as

$$-E^a_k := F^a_{0k} = \partial_0 A^a_k - \varepsilon_{0k\varrho\sigma}\eta^{\varrho\varrho'}\eta^{\sigma\sigma'}\partial_{\varrho'} G^a_{\sigma'} + g_3 \eta^{abc} \varepsilon_{0k\varrho\sigma}\eta^{\varrho\varrho'}\eta^{\sigma\sigma'} A^b_{\varrho'} G^c_{\sigma'}, \tag{3.36}$$

or, equivalently,

$$\partial_0 A^a_k = -E^a_k + \varepsilon_{kij}\partial_i G^a_j - \frac{1}{2} g_3 \eta^{abc} \varepsilon_{kij}\left(A^b_i G^c_j + G^b_i A^c_j\right). \tag{3.37}$$

Similarly, one obtains the equation for the B-fields from Eq. (3.34) in the form

$$\begin{aligned} B^a_k := \frac{1}{2}\varepsilon_{ijk} F^a_{ij} &= \frac{1}{2}\varepsilon_{ijk}\left(\partial_i A^a_j - \partial_j A^a_i\right) \\ &- \frac{1}{2}\varepsilon_{ijk}\varepsilon_{ij\varrho\sigma}\eta^{\varrho\varrho'}\eta^{\sigma\sigma'}\partial_{\varrho'} G^a_{\sigma'} \\ &+ \frac{1}{2}\eta^{abc}\varepsilon_{ijk}\left(g_1 A^b_i A^c_j - g_2 G^b_i G^c_j\right) \\ &+ \frac{1}{2} g_3 \eta^{abc} \varepsilon_{ijk}\varepsilon_{ij\varrho\sigma} A^b_\varrho G^c_\sigma \end{aligned} \tag{3.38}$$

for $i, j, k = 1, 2, 3$. Owing to this restriction, the last term in Eq. (3.38) vanishes in temporal gauge and after some rearrangements, Eq. (3.38) goes over into

$$\partial_0 G^a_k = B^a_k + \varepsilon_{ijk}\partial_i A^a_j - \frac{1}{2}\eta^{abc}\varepsilon_{ijk}\left(g_1 A^b_i A^c_j - g_2 G^b_i G^c_j\right). \tag{3.39}$$

Next, we consider the Lagrangian equations of motion for the A-fields:

$$\partial_\mu\left(\frac{\delta\mathcal{L}}{\delta(\partial_\mu A^d_{\nu'})}\right)-\frac{\delta\mathcal{L}}{\delta A^d_{\nu'}}\equiv$$

$$-\partial_{\mu'}\eta^{\mu\mu'}\eta^{\nu\nu'}F^d_{\mu\nu}+g_\chi\eta^{\lambda\nu'}j^d_\lambda-\mu^2_A\eta^{\lambda\nu'}A^d_\lambda$$

$$+\frac{1}{2}\eta^{abc}\left(g_1A^b_\mu\delta_{cd'}\delta_{\nu\nu'}+g_1A^c_\nu\delta_{bd'}\delta_{\mu\nu'}+g_3\varepsilon_{\mu\nu\varepsilon\sigma}\delta_{\varrho'\nu'}\delta_{bd'}\eta^{\varepsilon\varrho'}\eta^{\sigma\sigma'}G^c_{\sigma'}\right)\eta^{\mu\varrho}\eta^{\nu k}F^a_{\varrho\kappa}=0 \tag{3.40}$$

For v' = k, a rearrangement of Eq. (3.40) leads to

$$-\partial_0E^{d'}_k=-\varepsilon_{ljk}\partial_lB^{d'}_j-\eta^{bad'}\varepsilon_{ljk}\left(g_1A^b_lB^a_j+g_3G^b_lE^a_j\right)+g_\chi j^{d'}_k-\mu^2_AA^{d'}_k. \tag{3.41}$$

The same procedure can be performed for the G-fields. The Lagrangian equations of motion read in this case as

$$\partial_\lambda\left(\frac{\delta\mathcal{L}}{\delta(\partial_\lambda G^d_{\kappa'})}\right)-\frac{\delta\mathcal{L}}{\delta G^d_{\kappa'}}\equiv$$

$$\frac{1}{2}\varepsilon_{\mu\nu\delta\varepsilon}\eta^{\mu\varrho}\eta^{\nu\kappa}\eta^{\delta\lambda}\eta^{\varepsilon\kappa'}\partial_\lambda F^d_{\varrho\kappa}+ig_\pi\eta^{\kappa'\lambda}J^d_\lambda-\mu^2_G\eta^{\kappa'\lambda}G^d_\lambda$$

$$\frac{1}{2}\eta^{abc}\left(-g_2G^b_\mu\delta_{cd'}\delta_{\nu\kappa'}+g_2G^c_\nu\delta_{bd'}\delta_{\mu\kappa'}+g_3\varepsilon_{\mu\nu\varepsilon\sigma}\eta^{\varepsilon\varrho'}\eta^{\varrho\varrho'}A^b_{\varrho'}\delta_{\sigma'\kappa'}\delta_{cd'}\right)\eta^{\mu\varrho}\eta^{\nu k}F^a_{\varrho\kappa}=0. \tag{3.42}$$

For $\sigma'=h$, only one of the remaining indices μ, ν, ϱ, can adopt the value zero. Therefore, the sum over μ, ν, ϱ can be resolved into partial sums over ν, ϱ for $\mu=0$: ϱ for $\nu=0$ and μ, ν for $\varrho=0$. This leads to

$$\begin{aligned}&-\frac{1}{2}\varepsilon_{0ijk}\partial_jF^{d'}_{0i}-\frac{1}{2}\varepsilon_{i0jh}\partial_jF^{d'}_{i0}-\frac{1}{2}\varepsilon_{ij0h}\partial_0F^{d'}_{ij}-ig_\pi J^{d'}_h+\mu^2_GG^{d'}_h=\\&-\frac{1}{2}\left(-g_2G^b_lF^a_{lh}\eta^{abd'}-g_2G^c_lF^a_{hl}\eta^{ad'c}+g_3\varepsilon_{\mu\nu lh}A^b_l\eta^{\mu g\delta}\eta^{\nu\varepsilon'}F^a_{\delta\varepsilon'}\eta^{abd'}\right).\end{aligned} \tag{3.43}$$

If in Eq. (3.43) the field strength tensors are expressed by the vector fields, Eq. (3.43) can be reformulated in the following form:

$$\partial_0B^{d'}_h=-\varepsilon_{ijh}\partial_iE^{d'}_j-\varepsilon_{jih}\eta^{bad'}\left(g_2G^b_jB^a_i-g_3A^b_jE^a_i\right)-ig_\pi J^{d'}_h+\mu^2_GG^{d'}_h. \tag{3.44}$$

In addition, the Lagrange formalism allows the derivation of the electric and magnetic Gauss's law. For $\nu' = 0$, it follows the electric Gauss's law from Eq. (3.40), while for $\sigma' = 0$ from Eq. (3.42), the magnetic Gauss's law can be derived. It is also possible to derive pseudoconservation laws from Eqs. (3.40) and (3.42) for the currents. Because we are concentrating on the discussion of the canonical equations of motions, we refer to the comment about constraints at the end of section 3.3 and do not express these constraints explicitly at this point.

In borderline cases, this theory should pass into the description of conventional physics in order to be physically acceptable. To achieve this it is necessary to impose two additional conditions:

The Lagrangian density should lead to equations of motion that are identical in two scenarios:

i) With the equations of motion of a $SU(2) \otimes U(1)$ gauge theory if the magnetic vector potential G vanishes and the vector bosons are assumed to be massless

ii) With the equations of free massive electroweak bosons if their interactions are switched off

We consider case **i**) first. In this case, Eqs. (3.37), (3.39), (3.41), and (3.44) yield

$$\partial_0 A_k^a = -E_k^a; \quad B_k^a = \varepsilon_{ijk}\partial_i A_j^a + \frac{1}{2}\varepsilon_{ijk}\eta^{abc} g_1 A_i^b A_j^c \tag{3.45}$$

$$\partial_0 E_k^a = \varepsilon_{ijk}\partial_i B_j^a + \varepsilon_{ijk}\eta^{abc} g_1 A_i^b B_j^c - g_\chi j_k^a \tag{3.46}$$

$$\partial_0 B_k^a = -\varepsilon_{ijk}\partial_i E_j^a + \varepsilon_{ijk}\eta^{abc} g_3 A_i^b E_j^c. \tag{3.47}$$

For a = 1, 2, 3, these equations are identical to the equations of a $SU(2)$-gauge theory [cf. Borne, Lochak, and Stumpf 2001, Eqs. (12.58) and (12.59)], in temporal gauge, if the relation $g_3 = -g_1$ holds. The agreement with a $U(1)$-gauge theory for $a = 0$ is trivial.

With respect to case **ii**), we consider Eq. (3.40) in Lorentz gauge, switching off all interactions. This gives

$$-\partial_\mu F_a^{\mu\nu} - \mu_A^2 A_a^\nu \equiv \left[\partial_\mu \partial^\mu + \mu_A^2\right] A_a^\nu = 0. \tag{3.48}$$

This is the equation of a massive vector field; cf. Gross [1999, Eq. (2.42)].

For the magnetic vector potential, one obtains in this case from Eq. (3.42)

$$\frac{1}{2}\varepsilon_{\mu\nu\varrho\sigma}\partial^{\varrho}F_{a}^{\mu\nu} - \mu_G^2 G_\sigma^a \equiv \left[\partial_\mu\partial^\mu - \mu_G^2\right] G_\sigma^a = 0, \tag{3.49}$$

where the right side of the equation is the result of some rearrangement. Obviously, one has to assume $\mu_G = i\overline{\mu}_G$ in order to get a physical magnetic vector boson. This means that right from the beginning in the Lagrangian (3.33) the corresponding mass term should have a negative sign in order to describe physical magnetic bosons. Furthermore, for simplicity, we assume that $g_2 = g_1$.

If these conditions are incorporated into Eqs. (3.37), (3.39), (3.41), and (3.44), one eventually gets the following set of equations:

$$\partial_0 A_k^a = -E_k^a + \varepsilon_{ijk}\partial_i G_j^a + \frac{1}{2} g_1 \eta^{abc}\varepsilon_{kij}\left(A_i^b G_j^c + G_i^b A_j^c\right) \tag{3.50}$$

$$\partial_0 G_k^a = B_k^a - \varepsilon_{kij}\partial_i A_j^a - \frac{1}{2} g_1 \eta^{abc}\varepsilon_{kij}\left(A_i^b A_j^c - G_i^b G_j^c\right) \tag{3.51}$$

$$\partial_0 E_k^a = \varepsilon_{kij}\partial_i B_j^a + g_1 \eta^{abc}\varepsilon_{kij}\left(A_i^b B_j^c - G_i^b E_j^c\right) - g_\chi j_k^a + \mu_A^2 A_k^a \tag{3.52}$$

$$\partial_0 B_k^a = -\varepsilon_{kij}\partial_i E_j^a - g_1 \varepsilon_{kij}\eta^{abc}\left(G_i^b B_j^c + A_i^b E_j^c\right) - i g_\pi J_k^a - \overline{\mu}_G^2 G_k^a. \tag{3.53}$$

To compare Eqs. (3.24)−(3.27) with Eqs. (3.50)−(3.53), the constants in the former equations have to be fixed. In Stumpf [2005, Eq. (54)], their values are expressed by the formation of various scalar products of the space parts of the boson wave functions. As these scalar products (with inclusion of their regularization) are defined in auxiliary space, they can adopt positive and negative values, in contrast to the normal expressions in physical state space.

While the algebraic structure of the boson wave functions (and, of course, of the fermion wave functions, too) is strictly set up, the space parts of these wave functions can be chosen only with a certain degree of arbitrariness, which reflects the lack of information about the influence of the field theoretic vacuum on the space structure of these states. Therefore, without using self-consistent calculation schemes for the boson wave functions, the corresponding scalar products [Stumpf, 2005, Eq. (54)] represent parameters of the theory that can be adapted in order to get plausible results. In the present case, we define

$$g_1 = \hat{f}^{A} k_1 = \hat{f}^{A} k_2 = \hat{f}^{A} k_3 = \hat{f}^{G} k_4 = -\hat{f}^{G} k_5 = \hat{f}^{G} k_6. \tag{3.54}$$

Theorem 1: If Eq. (3.54) is satisfied, and the masses and coupling constants are adapted, then the set [Eqs. (3.24)–(3.27)] is identical with Eqs. (3.50)–(3.53). The same holds for the corresponding fermion equations [Eqs. (3.30) and (3.35)].

In Fuss (2004), Chapter 8 is devoted to the consistent calculation and definition of the various coupling constants and the comparison of the resulting effective theory with the corresponding phenomenological theory.

Addendum: The effective field theory defined by the Lagrangian density [Eq. (3.33)] is limited to a finite range of energy. Above a certain energy threshold, it loses its meaning and has to be modified by form factors. In this way, one does not encounter the divergence difficulties of conventional field theories with Lagrangians of the type in Eq. (3.33).

3.5 FIELDS FOR CP AND ISOSPIN SYMMETRY BREAKING

The isospin invariance of the field equations belongs to the theoretical foundation of the spinor field model. But this symmetry is not realized in nature; thus, after its introduction, it has to be broken immediately in order to explain experiments. A similar situation holds for the Standard Model, where the isospin symmetry is broken by the introduction of the Higgs fields. But in contrast to the Standard Model, another method is adopted in the spinor field model. And this method can be followed even if one starts with an already broken symmetry (for instance, CP symmetry breaking). Independent of this proceeding, the relation of the spinor field to Higg's formalism will be discussed at the end of this section.

With regard to the combined CP and isospin symmetry breaking in the spinor field model, an important difference between the proceeding in these particular cases has to be taken into account.

While for CP symmetry breaking, the original set of state vectors for conserved symmetries has to be supplied by new additional states, for isospin symmetry breaking, the structure of the state space is not changed before and after.

In the first case, the inclusion of new states changes the results of weak mapping qualitatively, whereas in the second case, only quantitative changes of the mass spectrum occur. Next, we refer to the corresponding calculations and give some comments that are essential for understanding the effect of this difference.

We decompose the propagator of the algebraic state definition into two parts:

$$F = F^0 + F^1, \tag{3.55}$$

where F^0 is the CP symmetry–breaking propagator that was given in Stumpf [2004d, Eq. (27)], while F^1 refers to that part that must be added to F^0 in order to reflect the change of the ground state caused by the isospin symmetry breaking.

Under these presuppositions, the functional energy operator [Eq. (3.1)] can also be decomposed into

$$\tilde{H} = \tilde{H}(F^0) + \tilde{H}(F^1), \tag{3.56}$$

as $\tilde{H}$, given by Eqs. (1.32),(1.33), and (1.34), only linearly depends on F. But the second most important point is that this superposition is compatible with weak mapping because both terms of Eq. (3.56) are referred to the *same* state space; i.e., we investigate the effect of isospin symmetry breaking on a state space that was already modified by CP symmetry breaking. Consequently, we add the contribution of $\tilde{H}(F^1)$ to the effective canonical equations of motion that were obtained from $\tilde{\mathcal{H}}(F^0)$.

For CP symmetry breaking, the effective canonical equations of motion are given by Eqs. (3.24)–(3.27) or (3.50)–(3.53), respectively, and the currents are defined by Eq. (3.32) with the coupling constants $g_A^a = g_G^a = g$, $a = 1, 2, 3$, and $g_A^0 = g_G^0 = g'$. The various constants of the effective theory have been eliminated by appropriate assumptions, and the definitions

$$\left(c_2 - f^A c_4\right) =: \mu_A^2 \quad \left(c_3 - f^G c_4\right) =: \mu_G^2 \tag{3.57}$$

are applied.

By the addition of F^1 to F^0, the modified propagator influences the fermion mass matrix and the current coupling to the boson fields, as can be read from the definition of the corresponding operators in Eqs. (1.32)–(1.34). For the latter, modifications are of minor interest for the intended application, so we will disregard them. We will consider only the effect on the boson masses. For details of this calculation, we refer to Stumpf (2006, section 4). As a result, one gets

$$\mu_A^2 A_k^a \rightarrow \left(\mu_A^2 \delta_{ab} - a_A \sigma_{ab}^2\right) A_k^b, \quad \mu_G^2 G_k^a \rightarrow \left(\mu_G^2 \delta_{ab} + a_A \sigma_{ab}^2\right) G_k^b \quad a = 1, 2 \tag{3.58}$$

and

$$\mu_A^2 A_k^a \rightarrow \left(\mu_A^2 \delta_{ab} + a_A \sigma_{ab}^1\right) A_k^b, \quad \mu_G^2 G_k^a \rightarrow \left(\mu_G^2 \delta_{ab} - a_A \sigma_{ab}^1\right) G_k^b \quad a = 3, 0. \tag{3.59}$$

Due to the self-regularization of the spinor field, all coefficients in Eqs. (3.56) and (3.57) are finite without any further manipulation. Then, in accordance with Eqs. (3.58) and (3.59), the associated equations for the electroweak electric and magnetic fields [Eqs. (3.26) and(3.27)] can be subdivided into two sets:

$$\begin{aligned} \dot{E}_{la}(z) = \varepsilon_{lkm} \partial_k^z B_{ma}(z) + g_e j_l^a + \left(\mu_A^2 \delta_{ab} - a_A \sigma_{ab}^2\right) A_{la} \text{ and} \\ \varepsilon_{abc} \varepsilon_{lkm} \widehat{f}^A \left[k_2 A_{kb}(z) B'_{mc}(z) + k_5 G_{kb}(z) E'_{mc}(z)\right] \end{aligned} \tag{3.60}$$

$$\begin{aligned} \dot{B}_{la}(z) = -\varepsilon_{lkm} \partial_k^z E_{ma}(z) + i g_m J_l^a + \left(\mu_G^2 \delta_{ab} + a_A \sigma_{ab}^2\right) G_{la} \\ - \varepsilon_{abc} \varepsilon_{lkm} \widehat{f}^A \left[k_2 A_{kb}(z) E_{mc}(z) - k_5 G_{kb}(z) B'_{mc}(z)\right] \end{aligned} \tag{3.61}$$

for a = 1, 2, and

$$\begin{aligned} \dot{E}_{la}(z) = \varepsilon_{lkm} \partial_k^z B_{ma}(z) + g_e j_l^a + \left(\mu_A^2 \delta_{ab} + a_A \sigma_{ab}^1\right) A_{la} \text{ and} \\ \varepsilon_{abc} \varepsilon_{lkm} \widehat{f}^A \left[k_2 A_{kb}(z) B'_{mc}(z) + k_5 G_{kb}(z) E'_{mc}(z)\right] \end{aligned} \tag{3.62}$$

$$\begin{aligned} \dot{B}_{la}(z) = -\varepsilon_{lkm} \partial_k^z E_{ma}(z) + i g_m J_l^a + \left(\mu_G^2 \delta_{ab} + a_A \sigma_{ab}^1\right) G_{la} \\ - \varepsilon_{abc} \varepsilon_{lkm} \widehat{f}^A \left[k_2 A_{kb}(z) E_{mc}(z) - k_5 G_{kb}(z) B'_{mc}(z)\right] \end{aligned} \tag{3.63}$$

for a = 3, 0.

The nondiagonal form of the mass matrices in Eqs. (3.58) and (3.59) indicates that the associated vector fields in Eqs. (3.60)−(3.63) are not identifiable with observable fields because in connection with these equations for scattering processes, no asymptotic free fields can be defined.

Hence, the transition to observable fields can be achieved only by appropriate transformations that diagonalize these mass matrices. Such transformations are consistent with these field equations only if all the field quantities (including potentials and currents) are transformed in the same way.

As the charge of the complex vector bosons depends on their status as ingoing or outgoing particles, we use a new notation to avoid any association to their charge. From now on, we define W and W^+ bosons, where the plus sign means Hermitian conjugation [cf., for instance, Gross 1999, Eq. (15.102), or Mandl and Shaw 1993, Eq. (12.43)]. This gives the following scheme for the transformed (i.e., the physical quantities):

$$\begin{matrix} \tilde{\mathbf{E}}_1 & \tilde{\mathbf{E}}_2 & \tilde{\mathbf{E}}_3 & \tilde{\mathbf{E}}_0 \\ \tilde{\mathbf{B}}_1 & \tilde{\mathbf{B}}_2 & \tilde{\mathbf{B}}_3 & \tilde{\mathbf{B}}_0 \\ \tilde{\mathbf{A}}_1 := \mathbf{W} & \tilde{\mathbf{A}}_2 := \mathbf{W}^+ & \tilde{\mathbf{A}}_3 := \mathbf{Z} & \tilde{\mathbf{A}}_0 := \mathbf{A} \\ \tilde{\mathbf{G}}_1 := M & \tilde{\mathbf{G}}_2 := \mathbf{M}^+ & \tilde{\mathbf{G}}_3 := \mathbf{X} & \tilde{\mathbf{G}}_0 := \mathbf{G} \end{matrix}, \tag{3.64}$$

where for both the electric and the magnetic potentials, only *one* set of electric and magnetic fields exists.

These physical potentials and fields are assumed to diagonalize the mass matrices and are related to the original quantities by the transformations

$$E_k^a = y_{ab}\tilde{E}_k^a; \quad B_k^a = y_{ab}\tilde{B}_k^b; \quad A_k^a = y_{ab}\tilde{A}_k^b; \quad G_k^a = y_{ab}\tilde{G}_k^b; \quad a = 1,2$$

$$or \quad \tilde{A}_k^a = y_{ab}^+ A_k^b, \quad \tilde{G}_k^a = y_{ab}^+ G_k^b \quad \text{and} \tag{3.65}$$

$$E_k^a = z_{ab}\tilde{E}_k^b; \quad B_k^a = z_{ab}\tilde{B}_k^b; \quad A_k^a = z_{ab}\tilde{A}_k^b; \quad G_k^a = z_{ab}\tilde{G}_k^b; \quad a = 3,0, \tag{3.66}$$

where y_{ab} and z_{ab} are explicitly given by the expressions

$$y_{ab} = 2^{-1/2}\begin{pmatrix} 1 & 1 \\ i & -i \end{pmatrix} \quad z_{ab} = 2^{-1/2}\begin{pmatrix} 1 & 1 \\ -1 & 1 \end{pmatrix}. \tag{3.67}$$

The special form of z_{ab} in Eq. (3.67) results from $g = -g' = 1$ and can be interpreted as a Weinberg transformation of the phenomenological theory with a special Weinberg angle $\Theta = 45^0$. By a suitable choice of parameters, this idealized angle can be changed into a realistic value, which will not be further pursued here.

Apart from diagonalizing the mass matrices, the unitary transformation y_{ab} is identical to that transformation in phenomenological theory that leads to the charged vector boson states. The charge values of $\tilde{A}^1$ and $\tilde{A}^2$ are $q = -1$ and $q = 1$, respectively, and coincide with those in the literature.

The transformed currents are defined by

$$\bar{j}_k^a = y_{ab}^+ g_A^b j_k^b, \quad a = 1,2 \quad \bar{j}_k^a = z_{ab}^{-1} g_A^b j_k^b, \quad a = 3,0, \tag{3.68}$$

with an analogous transformation for J_k^a.

After application of the transformations [Eqs. (3.65) and (3.66)] to Eqs. (3.60)–(3.63), the transformed field equations contain the transformed versions of the mass tensors [Eqs. (3.58) and (3.59)], which are diagonal. The corresponding diagonal elements m_A^a and m_G^a and a = 1, 2, 3, 0, associated with the fields [Eq. (3.64)], are given by the formulas

$$\left(m_A^a\right)^2 = \begin{pmatrix} \mu_A^2 - a_A \\ \mu_A^2 + a_A \\ \mu_A^2 - a_A \\ \mu_A^2 + a_A \end{pmatrix} \quad \left(m_G^a\right)^2 = \begin{pmatrix} \mu_G^2 + a_A \\ \mu_G^2 - a_A \\ \mu_G^2 + a_A \\ \mu_G^2 - a_A \end{pmatrix}. \tag{3.69}$$

For comparison with Eq. (3.69), we give the diagonal elements of the mass tensor for electric bosons if only isospin symmetry is broken; see Borne, Lochak, and Stumpf [2001, Eqs. (8.74)–(8.82)]:

$$\left(m_A^a\right)^2 = \begin{pmatrix} \mu_A^2 \\ \mu_A^2 \\ \mu_A^2 + \mu K \\ \mu_A^2 - \mu K \end{pmatrix}. \tag{3.70}$$

In contrast to Eq. (3.69), in Eq. (3.70), the mass values of the charged vector bosons are invariant under charge conjugation. Because isospin symmetry is broken in both cases, the differences between the mass matrices [Eqs. (3.70) and (3.69)] must result from CP symmetry breaking.

But these differences are only indirectly generated by the small CP violating term in the propagator. Owing to this term, the permutation invariance of the boson equation is broken, which leads to parafermionic boson eigenstates being the real cause for this considerable change in the mass values.

Finally, we transform the fermion equation [Eq. (3.30)] to physical fields. In this connection, attention must be paid to the fact that isospin symmetry breaking acts also on the lepton masses, which we express by the introduction of a diagonal lepton mass tensor $\mathbf{m}$ in isospace. Then, with Eqs. (3.65)–(3.67), one gets

$$\left[-i\gamma^\mu \partial\mu + m\right]\psi + \frac{1}{2} g \widehat{\sigma}^a \left[\gamma^\mu \tilde{A}_k^a + i\left(\gamma^k \gamma^5\right) \tilde{G}_k^a\right]\psi = 0, \tag{3.71}$$

where

$$\widehat{\sigma}^a := y_{abg}^T{}^b \sigma^b, \quad a = 1, 2, \quad \widehat{\sigma}^a := z_{abg}^T{}^b \sigma^b, \quad a = 3, 0 \tag{3.72}$$

holds. These definitions differ from those of the transformed Pauli matrices in the field equations, which are laid down by

$$\tilde{\sigma}^a := y_{abg}^+{}^b \sigma^b, \quad a = 1, 2, \quad \tilde{\sigma}^a := z_{abg}^T{}^b \sigma^b, \quad a = 3, 0, \tag{3.73}$$

with the corresponding currents

$$\tilde{j}^{a}_{\mu} := \frac{1}{2}\overline{\psi}\tilde{\sigma}^{a}\gamma_{\mu}\psi \quad \tilde{j}^{a}_{\mu} := \frac{1}{2}\overline{\psi}\tilde{\sigma}^{a}(\gamma^{5}\gamma_{\mu})\psi. \tag{3.74}$$

If electric and magnetic bosons occur, an important question is whether for both species, separate conservation laws can be derived. This is not the case. Independent of this symmetry breaking, only the conservation of the vector current (i.e., of the electric charge) can be established, and because the fermions are massive, no conservation of the axial current is possible.

Proposition 1: For combined CP and isospin symmetry breaking, the effective canonical equations of motion for the fields and the lepton Lagrangian [Eq. (3.71)] are forminvariant under the U(1)-group of charge transformations.

Proof: The U(1)-phase transformations of the whole system are fixed by the phase transformations of the charged currents in Eq. (3.73), while the neutral currents remain invariant under such transformations. To verify this statement, we consider the Lagrangian density associated to the lepton equation [Eq. (3.70)]:

$$L = \overline{\psi}[-i\gamma^{\mu}\partial_{\mu} + \mathbf{m}]\sigma^{0}\psi + \frac{1}{2}g\widehat{j}^{a}_{k}\tilde{A}^{a}_{k} + \frac{1}{2}g\mathrm{i}\widehat{J}^{a}_{k}\tilde{G}^{a}_{k}, \tag{3.75}$$

with

$$\widehat{j}^{a}_{k} := \overline{\psi}\widehat{\sigma}^{a}\gamma^{k}\psi \quad \widehat{j}^{a}_{k} := \overline{\psi}\widehat{\sigma}^{a}(\gamma^{k}\gamma^{5})\psi. \tag{3.76}$$

The corresponding Lagrangian density for Eq. (3.30) contains the currents j^{a}_{k} and J^{a}_{k}. According to the definitions $j := \widehat{j}^{1}$ and $J := \widehat{J}^{2}$, the hermitian conjugation of these quantities leads to $j^{+} = \widehat{j}^{1}$ and $J^{+} = \widehat{J}^{1}$. Therefore, one gets $(\widehat{j}^{1}\tilde{\mathrm{A}}^{1} + \widehat{j}^{2}\tilde{\mathrm{A}}^{2}) = (\mathrm{j}^{+}\mathrm{W} + \mathrm{j}\mathrm{W}^{+})$ and an analogous relation for the magnetic coupling.

Let us assume that in Eq. (3.30), the current j is transformed by a U(1)-phase transformation $\mathrm{j} = \mathrm{U}\,\mathrm{j}$. Then $j^{+'} = U^{-1}j^{+}$ must hold. Furthermore, the currents act as external forces in the field equations [Eq. (3.60)−(3.63)], which enforces the field quantities to transform in the same way. In particular, the W-potentials of Eq. (3.64) and their associated fields are transformed with U, while their hermitean conjugates are transformed with U^{-1}. From this, it follows that the Lagrangian density [Eq. (3.74)] is invariant under U(1)-phase transformations. By similar arguments, the form invariance of Eqs. (3.60)−(3.63) under these transformations can be verified. ◇

Finally, it should be pointed out that the complex Higg's doublet can be replaced by a real singlet and a real triplet, which both can be represented in terms of spinor field composites [cf. Grimm (1994b, section 6.3.1)].

CHAPTER 4

Magnetic Monopoles and Discharges

4.1 SUPERSONIC SPARK DISCHARGES IN WATER

For understanding the appearance of magnetic monopoles, it is necessary to first study the effects of a discharge in water, a discussion adopted from Stumpf (2011a).

The experimental standard arrangement of Urutskoiev (2004) is enclosed in a cylindrical hollow steel case that contains the electrical equipment needed for the discharge. The inner wall of the hollow steel case is of polyethilen. Within this container, the concentrical electrode (cathode) is formed by the outer hollow cylinder, while the inner hollow cylinder acts as the inner electrode (anode). (In various experiments, the form of the anode is modified, which does not influence the results essentially.)

The interior of the steel case, including the hollow cylinders of the electrodes, is filled with highly purified water. The geometrical characteristics are as follows: The diameter of the external electrode is 12 mm, the diameter of the inner electrode 6 mm, the height of the cylinder 5 cm, the diameter of the inner wall 20 mm, and the volume filled with 14−18 cm^3 of water. In terms of charge, the electric energy of the capacitor is given by $W^e = 50$ k J, and it can be expressed by $W^e = 1/2(U_1 - U_2)Q$, with $(U_1 - U_2) = 5$kV (Urutskoiev, 2004).

For the number of electrons N_e, one gets with $Q = N_e e$ the value $N_e = 1,25 \times 10^{20}$. This number of electrons is required for charge compensation during the discharge. If the decline of the tension of the capacitor during the discharge is ignored, the mean kinetic energy of the electrons is 5 keV.

The investigation of discharges was started in the early twentieth century by J. S. Townsend when he published his now classical theory about the mechanism of spark discharge (in gases) on the basis of measurements made at high values of E/p, the ratio of field strength to pressure, and low values of pd, the product of pressure and gap-length (Meek, 1973, p. 42). As for water in its various phases, the physical behavior can be described by van der Waal's equation (i.e., a gas equation), it seems to be reasonable to discuss discharges in water by means of the formalism of gas dynamics. In particular, the way in which a discharge can be triggered can be classified by the Townsend parameters, mainly by the pd value.

"At pressures about atmospheric (roughly), in gaps of about 1 cm and longer, that is of pd > 10^3 Torr cm, a spark discharge can occur. The voltages required for the breakdown at such high values of pd are quite high. running from tens to hundreds of kV" (Raizer, 1997, p. 324).

For the arrangement of Urutskoiev the mean voltage acting on the electrons is 5 kV; and for p = 760 Torr and d = 0.6 cm, one gets pd = 456 Torr cm, which obviously is too small for setting off sparks in gases. But water is a remarkable and extraordinary fluid. This can be illustrated by the following quotation.

"It has been known since Faraday's early researches that the presence of water vapor in discharge gaps greatly facilitates the passages of sparks" Morgan (1978, p. 665).

Sparks are localized discharges that run through thin channels in the medium. Such discharges are ignited by single electrons that accidentally are emitted from various places of the cathode, and it can be assumed that they occur in spite of the small pd value of Urutskoiev's arrangement, which is not a suitable parameter for sparks in water.

The extraordinary character of these sparks is enhanced by the observation that sparks are released a million times quicker in water than in gases (Morgan, 1978) and that by the discharge in water, considerably higher temperatures are attained than in gases. "The spark discharge is a rapid transient process, not a steady one, and is aptly described by the colloquial phrase 'a spark jumps'" (Raizer, 1997, p. 324). The understanding of the action of sparks is crucial for the explanation of the monopole creation, that is why the ideas behind the spark physics will be discussed and conclusion will be drawn.

"The fundamentals of the (enlarged) theory of spark discharges were developed by Loeb, Meek, and Raether above 1940. The theory is based on the concept of the growth of a thin ionized channel (streamer) between the electrodes. The streamer follows the positively charged trail left by the primary intensive avalanche" (Raizer, 1997, p. 327). Theoretically, this process was formulated by Drabkina (1951) and Braginski. The statements of Braginskis (1973) were verified in detail by several experiments, such as by Koppitz (1967).

Unfortunately, the investigations by Drabkina and Braginski have not been applied to water or water vapor. Nevertheless, using the knowledge imparted by these papers, some general conclusions can be drawn with respect to the possibility of low energy nuclear reactions and their connection with the creation of magnetic monopoles that results from the decay of unstable nuclei produced by the discharge.

The first statement is: While in gas discharges temperatures up to 10.000 degree Kelvin can be attained, by discharges in water or water vapor up to 50.000 Kelvin can be obtained, [Kop 67]. But even for such temperatures the thermal energy of a proton does not suffice to penetrate the Coulomb barrier of a Titanium nucleus. Titanium has six unoccupied shell model states open for a closure of a complete shell of the magic number 28 by the capture of additional nucleons. However, no neutrons have been observed during the experiments, Urutskoiev (2004, p. 1154), whereas protons are present due to the dissociation of water molecules. Hence it must be concluded that protons are the only nucleons that are available for the capture processes of the titanium nuclei.To initiate a proton capture into one of these unoccupied shell model states would at least require 10 keV thermal energy of the proton or temperatures of 15×10^8 degree Kelvin. (Basdevant, Rich & Spiro, 2004, p. 332, 338) that are unrealistic for low energy discharges.

Also a beam concentration is not possible. According to Braginskis theory spark channels cannot be sufficiently concentrated although in his theory the magnetic field pressure is incorporated, i.e. a possible pinch effect is not effective enough to produce higher energy densities.

But in sparks, another mechanism is active that is connected with the rise of the conductivity in the spark channel. Owing to this effect, the development of the spark is accompanied by a supersonic flow of the discharge current (Raizer, 1997; Braginski, 1973; Craggs, 1978), which manifests itself by a loud thunder of the corresponding supersonic pressure wave in experiments.

The sparks themselves are a rather complicated interplay of the primary electron avalanche and reflections of the waves on the electrodes. In any case, an avalanche is a primary and inescapable element of any breakdown mechanism (Raizer, 1997, p. 328). Such a mechanism is all the more complicated as the gap distance between the electrodes increases. In particular, the theory of lightning operates with various steps of the discharge. However, according to Raether's experiments, for $d = 3$ cm, one can assume that the primary avalanche has already caused the electrical breakdown (Raether, 1973, p. 28).

The data about the supersonic velocity of such a discharge depend on the various physical parameters and vary between 1.27×10^7 cm sec^{-1} in air (Raether, 1937) up to 5×10^9 cm sec^{-1} for streamer formation, which is connected with avalanche propagation. For comparison, at 20 K, the velocity of sound in water amounts to 1.4×10^5 cm sec^{-1} (Hering, Martin, & Stohrer, 1989, p. 493).

The first evidence of shock waves produced by spark discharges was reported in 1947 (Raizer, 1997, p. 344) but this quotation should obviously be completed by the reference to Raether (1937), see above. The theory of gas dynamics expansion of the spark channel, taking into account the shock wave and the energy release caused by time-dependent discharge currents, was first developed by Drabkina (1951, p. 345).

Of utmost interest in this discussion is the model of an avalanche, which is pictured in Figure 4.1 (Raizer, 1997, p. 330).

From this model, it follows that in one and the same discharge, not only electrons but also ions participate. For simplicity, we assume that the whole discharge takes place in only one initial spark channel.

The head of the avalanche is formed by electrons, which means that an enormous number of electrons will hit the anode on a hot spot. However, it should be noted that the duration of the breakdown of the avalanche is not directly correlated with the individual velocities of electrons and ions in the head of the avalanche. Without a detailed investigation of the energy transfer through the channel into its surroundings, we consider in a rough manner the energy for vaporizing the whole water content. For vaporizing 1 g = 1 cm^3 water, an energy transfer of 2257 J is necessary, which for 14 cm^3 amounts to 31, 598 J $\approx$ 31.6 kJ.

The vaporizing takes place at the constant temperature of 100 K. At this temperature, no nuclear reaction can be initiated. So there remains 20 kJ for the impact of electrons on the anode material.

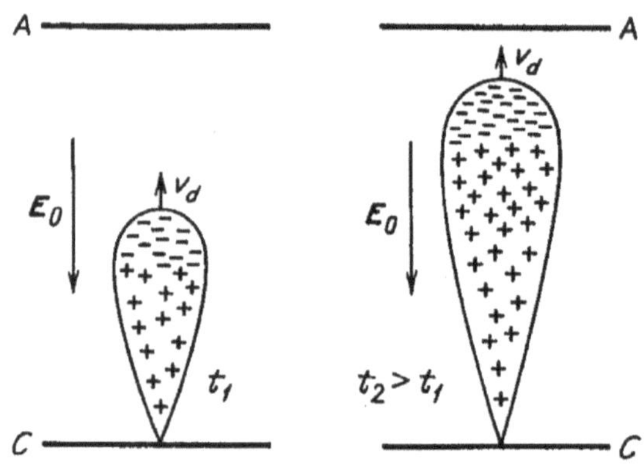

Figure 4.1 Shape and charge distribution of an electron avalanche at two consecutive moments in time. The arrows indicate the directions of external field E_0 and velocity v_d of the avalanche head (Raizer, 1997, p. 330).

Although the corresponding energy transfer from the electrons to the hot spot material and its surroundings will not suffice to induce local nuclear reactions, it will completely deprive the titanium atoms of their bound electrons; i.e., it leads to a melting process of a thin layer of Ti atoms in the anode.

The melting temperature of Ti is 1670°C. In this case. the corresponding energy that must be transferred to one atom of the layer is 3/2 kT, which corresponds to approximately 10^{-12} eV. If the hot spot is approximately a layer of 1 cm^2, it contains 10^{16} atoms. Then, for the melting of this layer, the electrons must transfer 10 keV of energy.

According to Figure 4.2, the discharge produces effects that are physically analogous to the production of soap bubbles consisting of a fluid soap film being blown up by air pressure. By electron bombardment of the surface of the titanium anode, which leads to surface melting, a thin film of titanium atoms is produced and the bubbles are blown up by the pressure of vaporized water.

The first stroke caused by electrons is followed by a second stroke, which hits the anode and which belongs to the trail of the avalanche consisting of ions. The latter ions are simply protons stemming from the dissociation of water vapor. Up to a certain degree, water is always dissociated in fluid and vapor. This dissociation leads to protons and OH^- ions. From the impact, the protons of the trail and the melted titanium film are in direct contact, and for this reason, a capture of protons by the titanium kernels cannot be excluded. The fact that protons can react solely with the titanium

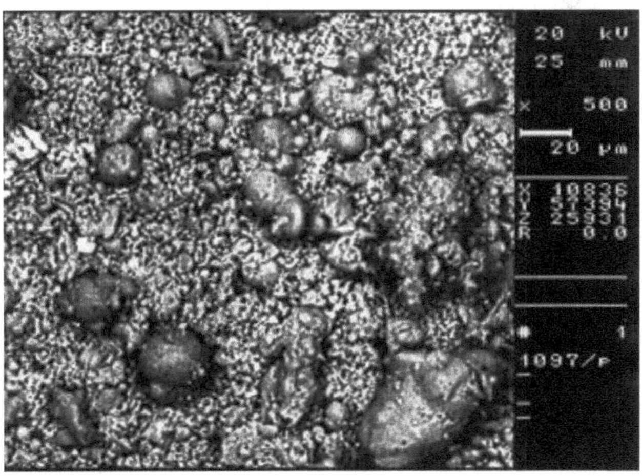

Figure 4.2 Typical micrograph of an arbitrary section of the sample of the titanium foil remainder after electric blasting. The characteristic scale is indicated in the lower-right corner (Urutskoiev, 2004).

kernels implies that for repeated captures, starting with the titanium kernels, only the states on the isotonic stepladder of Ti_{22}^{48} can be reached. The latter transfer reactions happen within 10^{-22} sec and fix in this way the time scale for this kind of reaction (direct reactions), which are due to strong forces. A definition and discussion of such direct reactions, which include transmutations, can be found in Mayer-Kuckuk (2002, section 7.7, p. 269).

But the effective occurrence of such captures depends on the ability of the protons to overcome the Coulomb barrier. Owing to the supersonic velocity of the avalanche, the protons of its trail have considerable energy. The observed supersonic velocities of avalanches in gases are $v = 5 \times 10^9$ cm sec^{-1} and may be even higher in water vapor. Such velocities lead to proton kinetic energies of $E_P > 1$ MeV, which are sufficient to pass the Coulomb barrier. The minimum energy for traversing the Coulomb barrier is E = 10 keV for protons (Basdevant, Rich, & Spiro, 2004, pp. 332, 164, 117, 168). The gradual absorbtion of two protons by the titanium kernel leads to Cr^{50}, which is nearly a stable element (apart from double β-decay).

After this generation of Cr^{50} from the trail of the avalanche, further protons hit the newly created element. The latter are also able to penetrate its Coulomb barrier and can initiate a further resonance capture. By such a resonance capture of a proton, the element Mn^{51} can be created with a decay time of 46.2 min for conserved symmetry. That is time enough to capture a further proton, which leads to Fe^{52} with a decay time of 45.9 sec or 8.27 h for conserved symmetry, too. All further captures produce increasingly unstable elements in the isotonic stepladder. Note that the weak decay times are modified by symmetry breaking and get smaller for the discharges under consideration.

So it is obvious that by the impact of the protons on the anode, unstable nuclei will be generated. Their weak decays will be accelerated by symmetry breaking and are accompanied by neutrino production. Such neutrinos might then be candidates to become magnetic monopoles.

4.2 PROCESSES CONNECTED WITH NEUTRINO EMISSION

Neutrinos can be created in weak decay processes where unstable nuclei undergo weak trasmutations assisted by exchange bosons. In contrast to the electromagnetic exchange boson, the weak vector bosons are not massless, but rather are heavy particles. Thus, the probability that a weak process can take place depends on the mass spectrum of these exchange bosons. This also holds if the mass spectrum of these bosons is modified by symmetry breaking. The appearance of neutrinos, therefore, depends on this mass spectrum [cf. Stumpf (2011a)].

For conserved symmetries, the transition probability for the exchange of a weakly charged vector boson is inversely proportional to the fourth power of its mass (Basdevant, Rich, & Spiro, 2004, Eqs. (4.96) and (4.157)], and for ordinary W-bosons with mass 80 GeV, this gives a very (or even infinitely) small value. Hence, in this case, the chance of an associated neutrino generation is vanishingly small.

In water, however, this situation is drastically changed if electroweak bosons are considered as composites. The corresponding mass spectrum is given by the mass formulas [Eq. (3.69); cf. also Stumpf (2008)], and the isospin symmetry-breaking parameter a_A, denoted as a in Stumpf (2008). In this formula for all mass values, only one symmetry-breaking parameter a_A occurs. In order to obtain physically meaningful results, the latter parameter has to be adjusted with respect to the most important vector boson (i.e., the photon). Then one obtains for $a_A = \mu_A^2$ a vanishing photon mass (i.e., $(m_A^0)^2 = 0$), and with this choice of a_A, the mass spectra [Eq. (3.69)] go over into the system:

$$\left(m_A^b\right)^2 = \begin{pmatrix} 2\mu_A^2 \\ 0 \\ 2\mu_A^2 \\ 0 \end{pmatrix} \quad \left(m_G^b\right)^2 = \begin{pmatrix} \mu_G^2 - \mu_A^2 \\ \mu_G^2 + \mu_A^2 \\ \mu_G^2 - \mu_A^2 \\ \mu_G^2 + \mu_A^2 \end{pmatrix} \quad b = \left(1, 2, 3, 0\right), \qquad (4.1)$$

with $b = (1,2,3,0)$, where 1 and 2 are the numbers of the charged vector bosons.

In the boson equations in Stumpf (2004b, 2008), the electromagnetic interactions are not directly covered; they appear later in effective theories only. Therefore, the pure spinor field interactions in these equations cannot be a good approximation for charged composites. Without extensive calculations, we are taking this into account by adding to the mass spectrum of the charged bosons in Eq. (4.1), which is exclusively generated by the spinorial interactions, an additional term of the elctromagnetic self-mass S_{SE}.

Because in Stumpf (2004b), the weak boson states were calculated without the participation of electromagnetic forces, there cannot be any distinction between electric and magnetic and charged and neutral vector bosons. Therefore, we assume $\mu_G = \mu_A$ for simplicity. This means that the mechanical mass of the charged G^1- and the A^2-bosons vanishes. Thus, in the case of symmetry breaking, the mass of these bosons can only have an electromagnetic origin.

In classical electrodynamics, a meaningful self-mass can be derived only if mechanical stresses (i.e., Poincare stressses) stabilize the charge distribution. In the spinor field model, these stabilizing forces are provided by the spinor field itself, which leads to the lowest-order bound states in Stumpf (2004b). But a detailed study of the self-mass problem has not been done. Therefore, we are forced to use a classical consideration for a preliminary calculation of this electromagnetic self-mass.

We calculate the electromagnetic mass of a charged vector boson by considering it as a classical, small, homogeneously charged sphere. Its electrostatic energy reads referred to the symmetry-conserving vacuum denoted by index V and fixed radius R_o (Fluegge, 1986):

$$S_{SE}^{V} = \frac{3}{5}\frac{e^2}{R_0} \tag{4.2}$$

where the radius R_0 is chosen in such a way that the electric self-energy for vanishing mechanical self-energy equals the empirical value of 80 MeV. Then, this boson charge is screened in water, which leads to

$$S_{SE}^{W} = \frac{1}{\varepsilon} S_{SE}^{V} \tag{4.3}$$

where the index W denotes the CP symmetry breaking water as ground state.

In section 4.1, it was shown that the protons in the trail of the avalanche can undergo direct reactions within 10^{-22} sec with titanium kernels during the supersonic impact on the anode. In particular, the elements

$$V_{23}^{49} \quad Cr_{24}^{50} \quad Mn_{25}^{51} \quad Fe_{26}^{52} \tag{4.4}$$

can be generated from Ti_{22}^{48} by closing nuclear shells or forming resonance states. Apart from Cr^{50}, which possibly shows the double β-decay without neutrino emission (which is of no interest in this context), the other elements are unstable by ordinary weak transitions, and this tendency is possibly increased by weak capture of avalanche electrons. The weak processes do not interfere with the direct proton reactions and go with another time scale than the latter ones. Although weak decays are much slower than direct reactions, they are of physical interest because they create the neutrinos that are the candidates for monopoles.

In this context, it is important to note that the question of whether one of the various weak transmutations can happen depends on the nuclear energy balance. Dominated by the masses of the nuclei, it is reasonable to assume that these energy balances undergo no major changes under CP symmetry breaking. Hence the results of nuclear science for conserved symmetries about the prevailing occurrence of certain weak processes hold in the case of CP symmetry breaking too. One gets (Basdevant, Rich, & Spiro, 2004, p. 474)

$$\begin{array}{ll} V_{23}^{49} \rightarrow Ti_{22}^{49} & \textit{electron capture} \\ Mn_{25}^{51} \rightarrow Cr_{24}^{51} & \beta^{+} \textit{decay} \\ Fe_{26}^{52} \rightarrow Mn_{25}^{52} & \beta^{+} \textit{decay} \end{array} \quad . \tag{4.5}$$

But while the nuclear energy balances remain unchanged under CP symmetry breaking, the corresponding transition probabilities are strongly influenced due to the modification of the weak boson masses for CP violation in water.

The reaction equation for electron capture reads as

$$e^{-} + p = \nu + n. \tag{4.6}$$

The avalanche electrons impinge on the titanium foil with a maximal velocity of about 5×10^{9}cm sec^{-1}. For these electrons, it holds that if $v \ll c$, these electrons dissipate their energy exclusively by inelastic collisions with the material of the anode and no bremsstrahlung is emitted [cf. Jackson 1999, Eq. (15.31)]. For the maximal velocity, this inequality is not very well satisfied, but for the average velocity of the avalanche electrons, it holds quite well.

Furthermore, without boson emission, the motion of the avalanche electrons is accompanied by their time-dependent electric and magnetic fields, which are subjected to the boundary condition for a transition between a dielectric (water) and an (ideal) conductor (titanium). So due to these conditions, the current flow in the anode, which is generated by the impinging electrons, is forced to run in a thin surface layer (known as the *skin effect*) (Jackson, 1999, Chapter 8).

The corresponding decay rate is given according to Basdevant, Rich, and Spiro (2004, p. 207, Eq. (4.96)]:

$$\lambda = \frac{c}{(\hbar c)^{4}} \left(\frac{1}{\pi}\right) (2,4G_{F})^{2} |\psi_{e}(0)|^{2} M^{2} Q_{ec}^{2} \tag{4.7}$$

for electron capture starting from any kind of bound states. For simplicity, we consider only superallowed transitions with $M^2 \sim 1$. The value of Q_{ec} is given by $(E + \Delta E)$ [Basdevant, Rich, & Shapiro (2004, p. 207)]. Furthermore, G_F must be related to the weak coupling constant g_1. This quantity is the weak counterpart of the electrical charge. The latter is defined by $e^2 = (\alpha_e \hbar c)$, while the weak coupling constant is given by $g_2^1 = (\alpha_w \hbar c)$, where α_e is the electrical fine structure constant and α_w is the weak fine structure constant, respectively. With $\alpha_w = g_w^2/(4\pi) = 1/29$ (Griffiths, 1996, Eqs. (10.38),(10.42)] one obtains

$$G_F = \frac{\sqrt{2}}{8}\left(\frac{g_w}{m_w c^2}\right)^2 (\hbar c)^3. \tag{4.8}$$

with m_w mass of the charged weak vector boson. Substitution of Eq. (4.8) into Eq. (4.7) yields

$$\lambda = (2,4)^2 2^{-3} |\psi_e(0)|^2 c \left(\frac{g_w}{m_w c^2}\right)^4 (hc)^2 Q_{ec}^2. \tag{4.9}$$

The value of the decay constant Eq. (4.9) for water can be related to the corresponding value for conserved symmetry (i.e., vacuum). Denoting

$$\lambda_W := \lambda(water), \quad \lambda_V := \lambda(vacuum)\,, \tag{4.10}$$

one gets

$$\lambda_W = \frac{\lambda_W}{\lambda_V}\lambda_V \tag{4.11}$$

With Eq. (4.9), the fraction in Eq. (4.11) can be explicitly evaluated if one takes into account the invariance properties of the quantities contained in Eq. (4.9) under CP symmetry breaking.

The invariance of the nuclear energy balance has already been discussed. For the coupling constant g_W, the following invariance argument can be given if one equivalently considers the invariance of g_1:

$$g_1^2 := \left(\frac{g_w^2}{4}\hbar c\right) = \left(\frac{1}{29}\hbar c\right) = \left(\frac{1}{29}197,3\right) MeV(fm) \tag{4.12}$$

Concerning this value of the weak coupling constant g_1, it should be noted that the coupling constant g_1 has the same dimension as the electrical charge e (i.e., g_1 must be proportional to e). Furthermore the U(1) symmetry, which allows the algebraic definition of charge, is not destroyed by electroweak symmetry breaking; i.e., both e and g_1 must have invariant values under electroweak symmetry breaking. So one can apply the value of Eq. (4.12) to the case of broken symmetry too.

In addition, if by definition for conserved and broken symmetry, the same electronic wave functions are applied in Eq. (4.9), and if one considers charge screening and defines $m_w \rightarrow m_V$ for vacuum and $m_w \rightarrow m_W$ for water, one obtains from Eqs. (4.9) and (4.11)

$$\lambda_W = \left(\frac{m_V c^2}{m_W c^2}\right)^4 \lambda_V \tag{4.13}$$

Now, in accordance with this exposition, it is

$$m_V c^2 = S_{SE} \quad m_W c^2 = \frac{1}{\varepsilon} S_{SE} \tag{4.14}$$

For V_{23}^{49}, one gets from Pfennig, Klewe-Nebenius, and Seelmann-Eggebert (1998) $\lambda_V = \tau^{-1} = (330\text{d})^{-1}$, and with $\varepsilon = 80$ for water [Eq. (4.13)] leads to

$$\lambda_W = \varepsilon^4 \lambda_V = \varepsilon^4 (60 \times 60 \times 24 \times 330)^{-1} sec^{-1} \rightarrow \tau_W = 10^{-1}\, 6.9\, sec \tag{4.15}$$

The very small value [Eq. (4.15)] refers to the time for the capture of one electron by one nucleus in the layer. But this value is not changed if, according to our assumptions, the electron spreads out over the whole layer. Furthermore, owing to the skin effect, the spreading of the electrons is confined to a thin surface layer of the anode, and so a remarkable dependence of the values [Eq. (4.15)] on the thickness of the anode foil is not to be expected.

Although no free neutrons are present in the discharge, for the neutrons and protons bound in the nuclear kernels, local neutron decay and local proton decay can occur. With respect to the accompanying neutrino production, the latter processes might be able to compete with neutrino emission by electron capture. Thus, to decide which process dominates the weak transitions, an estimate of neutron and proton decay has to be given. In contrast to electron capture, neutron decays and proton decays are internal processes within one nucleus; i.e., no extended wave functions of the neutron and proton are involved. For this case, standard formulas have been derived (Basdevant, Rich, & Shapiro, 2004, p. 202). According to Basdevant, Rich, and Shapiro [2004, p. 204, Eq. (4.88)].

$$\lambda\left(sec^{-1}\right) = \frac{2}{\pi^3 \hbar} \left(\frac{2,4 G_F}{(\hbar c)^3}\right)^2 |M|^2 \int_{m_e c^2}^{\Delta mc^2} \left(\Delta mc^2 - E\right)^2 \left(E^2 - m_e^2 c^4\right)^{1/2} E dE, \tag{4.16}$$

where Δmc^2 is the energy difference between initial and final nuclear masses.

For a rough estimate, we consider superallowed transitions $|M|^2 \sim 1$ and put $m_e c^2 \sim 0$. Then the decay constant gives an upper limit of possible decay processes, which suffices for a comparison with electron capture. With $a_\Delta \equiv \Delta mc^2$, Eq. (4.13) goes over into

$$\lambda\left(sec^{-1}\right) = \frac{8}{\pi^3 \hbar}\left(\frac{2,4G_F}{(\hbar c)^3}\right)^2 \int_0^{\Delta mc^2} (a_\Delta - E)^2 E^2 dE. \tag{4.17}$$

By substitution of Eq. (4.8) into Eq. (4.17) and integration over E, one obtains $\lambda = \lambda_w$ for w = V and w = W

$$\lambda_w\left(sec^{-1}\right) = \frac{2}{\pi^3 \hbar}(2,4)^2 \frac{2}{64}\left(\frac{g_w}{m_w c^2}\right)^4 (4\pi)^2 a_\Delta^5 \frac{1}{30}. \tag{4.18}$$

Eqs. (4.17) and (4.18) show that such decay processes depend on the nuclear energy balances, which already have been discussed. Earlier in this chapter, it has been explained that these energy balances are not modified by CP symmetry breaking compared with the symmetry-conserving case. Hence, the treatment of Eq. (4.18) can be done along the same lines as that of Eq. (4.9). Therefore, Eq. (4.11) can be applied to neutron decay too. In the same manner, the spontaneous weak proton decay within the nucleus can be treated. In both cases, the results of the chart of nuclides for conserved symmetries can be adopted. The step-by-step occupation of originally empty titanium (resonance) levels by protons leads to elements that are unstable against proton decay, and from the chart in Pfennig, Klewe-Nebenius, and Seelmann-Eggebert (1998), it follows that all elements of the isotonic stepladder above Cr_{24}^{50} show β^+ decay; i.e., in this isotonic stepladder, only proton decay occurs.

This process is characterized by the following weak transition:

$$p \rightarrow n + e^+ + \nu, \tag{4.19}$$

and one can apply Eq. (4.11). For conserved symmetry, the vacuum values are [cf. Pfennig, Klewe-Nebenius, and Seelmann-Eggebert (1998)]

$$Mn_{25}^{51} : \tau_V = 46.2\ min, \quad Fe_{26}^{52} : \tau_V = 45.9\ sec, \quad Co_{27}^{53} : \tau_V = 240\ msec,$$
$$Ni_{28}^{54} : \tau_V = 143\ msec \tag{4.20}$$

where τ_V is the vacuum decay time.

With Eq. (4.11), the following values for decay times in water result:

$$
\begin{aligned}
Mn_{25}^{51} &\rightarrow Cr_{24}^{51} & \tau_W &= 10^{-4} \times 1.066\,sec \\
Fe_{26}^{52} &\rightarrow Mn_{25}^{52} & \tau_W &= 10^{-6} \times 1.12\,sec \\
Co_{27}^{53} &\rightarrow Fe_{26}^{53} & \tau_W &= 10^{-9} \times 5.26\,sec \\
Ni_{28}^{54} &\rightarrow Co_{27}^{54} & \tau_W &= 10^{-9} \times 3.41\,sec
\end{aligned}
\tag{4.21}
$$

Before drawing any conclusion from Eq. (4.21), attention must be paid to the fact that the β^+ decay takes place via a two-step process. In the first step, a proton colliding with a nucleus is supposed to be captured by it to form a transmuted nucleus. After the proton has been absorbed by this nucleus, its energy is rapidly redistributed among the constituents of this nucleus and considerable time elapses. Then, in the second step, a decay mode develops, which finally leads to neutrino emission.

But it must be recognized that not all encounters between protons and nuclei end up with a transmuted nucleus formation. A considerable proportion of reactions takes place directly, without an intermediate stage, but at the same time, there is a sufficiently high probability for the formation of a transmuted nucleus, which leads to sharp resonance effects at low energies if the Coulomb barrier can be penetrated by the proton.

In a rough way, the probability for the formation of a transmuted nucleus can be expressed by

$$W = \frac{n_F \sigma_g}{F}, \tag{4.22}$$

where σ_g is the content of a surface area of a projection of the nucleus on a plane orthogonal to the incident proton beam; and n_F is the number of projections of all nuclei on F, which itself is the content of the area of a hot spot defined by the proton beam traversing this plane.

This simple consideration already shows the problem of defining numbers of generated neutrinos, as under discharges, the hot spots on the anode partly lose their crystallographic order; i.e., a definite number of target nucleus projections on the hot spot becomes impossible.

So the chart of nuclides only guarantees that the elements in the isotonic stepladder of Ti_{22}^{48} with considerable proton surplus can exist as resonance states with various finite lifetimes. However, a numerical calculation of the associated neutrino output cannot be based on this information, as the chart of nuclides gives no hint how many resonance states can be generated during the discharge. Thus, the question whether electron capture or compound decay dominates the neutrino production needs further studies in detail.

Independent of this problem, from Eq. (4.5), it clearly follows that the number of generated neutrinos is closely related to the number of transmuted elements. Indeed, the experiments of Urutskoiev, Liksonov, and Tsinov (2002) show that by the discharge, many transmuted elements are generated that are outside the stepladder of Ti_{22}^{48}. Hence, according to the results found in section 4.1, these elements must at least partly develop from the original resonance states by weak decay. Without going into detail about the occurrence of the transmuted elements in Urutskoiev (2004) to the elements in Eq. (4.5), one can say that such rearrangement of weak transitions is the source of neutrinos that are emitted under symmetry-breaking conditions.

A more detailed discussion of Urutskoiev's transmutation observations is imperative, but for the problem under consideration, it holds that from a theoretical point of view, a definite number of neutrinos being output is of minor interest for these processes. Principally, for proof of the existence of a magnetic monopole by symmetry breaking, it is only necessary to show that only a single neutrino is transformed into a magnetic monopole. This line of research will be followed in the next section.

4.3 HOW MAGNETIC MONOPOLES ARE LINKED TO DISCHARGES

From the theoretical point of view, the question whether magnetic monopoles do or do not exist can be answered by the calculation of the effective lepton-boson coupling term

$$H_{bf}^{1} = W_{2}^{qlp}\mathbf{f}_{q}\mathbf{d}_{l}^{b}\mathbf{d}_{p}^{f} = 3W_{I_1I_2I_3I_4}R_{II'I_1}^{q}C_{II'I_4}^{p}C_{I_2I_3}^{l}\mathbf{f}_{q}\mathbf{d}_{l}^{b}\mathbf{d}_{p}^{f}; \tag{4.23}$$

cf. Eq. (1.35). With Eq. (4.23), it is obvious that under symmetry breaking conditions, according to section 2.11 all wave functions in Eq. (4.23) must be adapted to this symmetry violation, and it has to be noted that the indices p, q, and l characterize the quantum numbers of the effective states that appear in the effective theory. In spite of the generality of Eq. (4.23), we first concentrate on the monopoles (i.e., the neutrino couplings to bosons), calculations that we partly adopt from Stumpf (2011b).

In the first step, the expression $W_{I_1I_2I_3I_4}C_{I_2I_3}^{k}\mathbf{d}_{k}^{b}$ has to be calculated. If $C_{I_2I_3}^{k}$ is projected on the vertex W, the third term drops out of the algebraic part of the vertex, and the same holds for the terms connected with $\mathbf{d}^{E}$ and $\mathbf{d}^{B}$ (see Appendix A for details). One obtains

$$W_{I_1 I_2 I_3 I_4} C^k_{I_2 I_3} \partial^b_k = \sum_{k} \sum_{I_2 I_3} \lambda_{i_1} B_{i_2 i_3 i_4} \int d^3 r_2 d^3 r_3 \delta(\mathbf{r}_1 - \mathbf{r}_2)\delta(\mathbf{r}_1 - \mathbf{r}_3)\delta(\mathbf{r}_1 - \mathbf{r}_4)\times$$

$$\sum_h \left[(\gamma^0 v^h)_{\beta_1 \beta_2} (v^h C)_{\beta_3 \beta_4} \delta_{\varrho_1 \varrho_2} \gamma^5_{\varrho_3 \varrho_4} - (\gamma^0 v^h)_{\beta_1 \beta_3} (v^h C)_{\beta_2 \beta_4} \delta_{\varrho_1 \varrho_3} \gamma^5_{\varrho_2 \varrho_4} \right] \times$$

$$\begin{aligned} &(T^b + S^b)_{\varrho_2 \varrho_3} \Big[f^A_{i_2 i_3}(\mathbf{r}_2 - \mathbf{r}_3|\mathbf{k})(\gamma^k C)_{\beta_2 \beta_3} \mathbf{d}^A_{k,b}(\mathbf{k}) \\ &+ f^G_{i_2 i_3}(\mathbf{r}_2 - \mathbf{r}_3|\mathrm{k})(\gamma^k \gamma^5 C)_{\beta_2 \beta_3} \mathbf{d}^A_{k,b}(\mathbf{k}) \Big] \exp\left[-i\mathbf{k}\frac{1}{2}(\mathbf{r}_2 + \mathbf{r}_3) \right] \end{aligned} \tag{4.24}$$

in temporal gauge of the vector potentials; cf. Borne, Lochak, and Stumpf (2001, p. 194). From Eq. (4.24), the following results:

$$\begin{aligned} W_{I_1 I_2 I_3 I_4} C^k_{I_2 I_3} \partial^b_k = &-4 \Big[(\gamma^0 \gamma^k)_{\beta_1 \beta_1} (T^b \gamma^5)_{\varrho_1 \varrho_4} \widehat{f}^A(\mathbf{k}) \mathbf{d}^A_{k,b}(\mathbf{k}) \\ &+ (\gamma^0 \gamma^k \gamma^5)_{\beta_1 \beta_4} (S^b \gamma^5)_{\varrho_1 \varrho_4} \widehat{f}^G(\mathbf{k}) \mathbf{d}^G_{k,b}(\mathbf{k}) \Big] \lambda_{i_1} B_{i_4} \delta(\mathbf{r}_1 - \mathbf{r}_4) \exp[-i\mathbf{k}\mathbf{r}_1], \end{aligned} \tag{4.25}$$

where $\widehat{f}^A$ and $\widehat{f}^G$ are the values of the boson functions [Eq. (2.146)] at the origin.

If one substitutes the wave functions $C_{I_1 I_2 I_3}$ together with the dual fermion functions $R_{I_1 I_2 I_3}$ [cf. Stumpf (2004c, section 6)], into Eq. (4.23), integrates over $\mathbf{r}_4$, and renames $\mathbf{r}_1$ to $\mathbf{r}''$, then one gets with Eq. (4.25):

$$\begin{aligned} \mathcal{H}^1_{bf} = &-12 \int \cdots \int d^3k\, d^3k'\, d^3k''\, d^3r\, d^3r'\, d^3r'' \\ &\times \Big[(\gamma^0 \gamma^k)_{\beta_1 \beta_4} (T^b \gamma^5)_{\varrho_1 \varrho_4} \widehat{f}^A(\mathbf{k}) \mathbf{d}^A_{k,b}(\mathbf{k}) \\ &+ (\gamma^0 \gamma^k \gamma^5)_{\beta_1 \beta_4} (S^b \gamma^5)_{\varrho_1 \varrho_4} \widehat{f}^G(\mathbf{k}) \mathbf{d}^G_{k,b}(\mathbf{k}) \Big] \\ &\times \sum_{i_1 i_4 i} \lambda_{i_1} R^{\varrho\varrho'\varrho_1}_{\beta\beta'\beta_1}(\mathbf{r}, \mathbf{r}', \mathbf{r}'|\mathbf{k}', l, n)_{ii'i_1} \exp\left[i\mathbf{k}'\frac{1}{3}(\mathbf{r} + \mathbf{r}' + \mathbf{r}'') \right] \\ &\times C^{\varrho_4 \varrho\varrho'}_{\beta_4 \beta\beta'}(\mathbf{r}'', \mathbf{r}, \mathbf{r}'|\mathbf{k}'', j, m)_{ii'i_4} \exp\left[-i\mathbf{k}''\frac{1}{3}(\mathbf{r} + \mathbf{r}' + \mathbf{r}'') \right] \\ &\times \exp[-i\mathbf{k}\mathbf{r}''] \mathbf{f}_{ln}(\mathbf{k}') \mathbf{d}^f_{jm}(\mathbf{k}'') \end{aligned} \tag{4.26}$$

Note that in Eq. (4.23), the summation convention has been used, which in Eqs. (4.24) and (4.26) is explicitly expressed by integrations.

Introduction of center of mass coordinates

$$z = \frac{1}{3}(\mathbf{r} + \mathbf{r}' + \mathbf{r}''); \quad u = \mathbf{r}' - \mathbf{r}; \quad w = \mathbf{r}'' - \mathbf{r}' \tag{4.27}$$

and

$$\mathbf{r} = \mathbf{z} - \frac{2}{3}\mathbf{u} - \frac{1}{3}\mathbf{w}; \quad \mathbf{r}' = \mathbf{z} + \frac{1}{3}\mathbf{u} - \frac{1}{3}\mathbf{w}; \quad \mathbf{r}'' = \mathbf{z} + \frac{1}{3}\mathbf{u} + \frac{2}{3}\mathbf{w} \tag{4.28}$$

in Eq. (4.26) yields the expression

$$\begin{aligned}
\mathcal{H}^1_{bf} = & -12 \int \cdots \int d^3k\, d^3k'\, d^3k''\, d^3z\, d^3u\, d^3w \\
& \times \Big[(\gamma^0\gamma^k)_{\beta_1\beta_4} (T^b\gamma^5)_{\varrho_1\varrho_4} \widehat{f}^A(\mathbf{k}) \mathbf{d}^A_{k,b}(\mathbf{k}) \\
& + (\gamma^0\gamma^k\gamma^5)_{\beta_1\beta_4} (S^b\gamma^5)_{\varrho_1\varrho_4} \widehat{f}^G(\mathbf{k}) \mathbf{d}^G_{k,b}(\mathbf{k}) \Big] \\
& \times \sum_{i_1 i_4 i} \lambda_{i_1} R^{\varrho\varrho'\varrho_1}_{\beta\beta'\beta_1}(\mathbf{u}, \mathbf{w}|\mathbf{k}', l, n)_{ii'i_1} C^{\varrho\varrho'\varrho_4}_{\beta\beta'\beta_4}(\mathbf{u}, \mathbf{w}|\mathbf{k}'', j, m)_{ii'i_4} \mathbf{f}_{ln}(\mathbf{k}') \mathbf{d}^f_{jm}(\mathbf{k}'') \\
& \times \exp(i\mathbf{k}'\mathbf{z}) \exp(-i\mathbf{k}''\mathbf{z}) \exp\left[-i\mathbf{k}\left(\mathbf{z} + \frac{1}{3}\mathbf{u} + \frac{2}{3}\mathbf{w} \right) \right]
\end{aligned} \tag{4.29}$$

The further evaluation depends upon the parafermi wave functions [Eqs. (2.147) and (2.149)]. In particular, with Eq. (2.147) and with

$$Y^s \in \{(T^a\gamma^5), (S^a\gamma^5), a = 0, 1, 2, 3\} \quad X^t \in \{(\gamma^0\gamma^k), (\gamma^0\gamma^k\gamma^5), k = 0, 1, 2, 3\}, \tag{4.30}$$

the parts containing the wave functions in Eq. (4.29) can be written as follows:

$$\begin{aligned}
& R^{\varrho\varrho'\varrho_1}_{\beta\beta'\beta_1}(\mathbf{u}, \mathbf{w}|\mathbf{k}', l, n)_{ii'i_1} Y^s_{\varrho_1\varrho_4} X^t_{\beta_1\beta_4} C^{\varrho\varrho'\varrho_4}_{\beta\beta'\beta_4}(\mathbf{u}, \mathbf{w}|\mathbf{k}'', j, m) \equiv \sum_{i_1 i_2 i_3 i_4} \times \\
& \left\{ (\Theta^l)_{\varrho\varrho'\varrho_1} \left[\Omega^n_{\beta\beta'\beta_1} \psi^*(\mathbf{u}, \mathbf{w}) \right] \right\}_{i_1 i_2 i_3} Y^s_{\varrho_1\varrho_4} X^t_{\beta_1\beta_4} \left\{ (\Theta^j)_{\varrho\varrho'\varrho_4} \left[\Omega^m_{\beta\beta'\beta_4} \psi(\mathbf{u}, \mathbf{w}) \right] \right\}_{i_1 i_2 i_4},
\end{aligned} \tag{4.31}$$

where the summation over the auxiliary indices is explicitly indicated.

In this context, attention has to be paid to the meaning of the indices n, l, j, and m, which represent the quantum numbers of the states involved and which should not be confused with the ordinary algebraic indices of spin, superspin, and isospin, which are denoted by Greek letters.

In section 2.8, of chapter 2 of this part it was emphasized that within the algebraic Schroedinger representation on a spacelike hyperplane, neither antisymmetric nor symmetric superspin-isospin and spin-orbital combinations are admitted for broken symmetry.

This leads to the introduction of single-product representations in superspin-isospin and in spin-space. In accordance with Eq. (2.148) and the comment regarding the choice of the matrix C in section 2.8, we thus introduce the following states:

$$\Omega^n_{\alpha_1\alpha_2\alpha_3} := \xi^n_{\alpha_1} C_{\alpha_2 a_3}, \quad \Theta^j_{\kappa_1\kappa_2\kappa_3} := \zeta^l_{k_1}\Gamma_{\kappa_2\kappa_3}, \tag{4.32}$$

where in addition to the unit spinors ξ^n_α in spin space, unit spinors ζ^l_k in superspin-isospin space are introduced.

With

$$\zeta^l_{\rho_1} \otimes \zeta^l_{\rho_4} := \Theta^l_{\rho\rho'\rho_1}\Theta^j_{\rho\rho'\rho_4} \quad \xi^n_{\beta_1} \otimes \xi^m_{\beta_4} := \Omega^n_{\beta\beta'\beta_1}\Omega^m_{\beta\beta'\beta_4}, \tag{4.33}$$

one gets

$$\xi^n_{\beta_1} X^t_{\beta_1\beta_4}\xi^m_{\beta_4} = X^t_{nm} \quad \zeta^l_{\rho_1} Y^s_{\rho_1\rho_4}\zeta^j_{\rho_4} = Y^s_{lj}, \tag{4.34}$$

and if the orbital parts in Eq. (4.31) are collected in only one term χ, then Eq. (4.29) can be written as

$$\begin{aligned}
&\mathcal{H}^1_{bf} = -12\int\cdots\int d^3k\, d^3k'\, d^3k''\, d^3z\, d^3u\, d^3w\times \\
&\left[(\gamma^0\gamma^k)_{nm}(T^b\gamma^5)_{lj}\,\widehat{f}^A(\mathbf{k})\mathbf{d}^A_{k,b}(\mathbf{k}) + (\gamma^0\gamma^k\gamma^5)_{nm}(S^b\gamma^5)_{lj}\,\widehat{f}^G(\mathbf{k})\mathbf{d}^G_{k,b}(\mathbf{k})\right] \\
&\times\ \chi(\mathbf{u},\mathbf{w}|\mathbf{k}',\mathbf{k}'',n,m,l,j)\mathbf{f}_{ln}(\mathbf{k}')\mathbf{d}^f_{jm}(\mathbf{k}'') \\
&\exp(i\mathbf{k}'\mathbf{z})\exp(-i\mathbf{k}''\mathbf{z})\exp\left[-i\mathbf{k}\left(\mathbf{z}+\frac{1}{3}\mathbf{u}+\frac{2}{3}\mathbf{w}\right)\right]
\end{aligned} \tag{4.35}$$

Fourier transformation of the functional operators $\mathbf{d}^A(\mathbf{k})$, $\mathbf{d}^G(\mathbf{k})$, $\mathbf{f}(\mathbf{k}')$, $\mathbf{d}^f(\mathbf{k}'')$ then gives for Eq. (4.35)

$$\mathcal{H}_{bf}^{1} = -12 \int \cdots \int d^3k\, d^3k'\, d^3k''\, d^3z\, d^3u\, d^3w\, d^3x\, d^3p\, d^3y \times$$

$$\Big[(\gamma^0\gamma^k)_{nm} (T^b\gamma^5)_{lj} \widehat{f}^A(\mathbf{k}) \tilde{\mathbf{d}}^A_{k,b}(\mathbf{X}) + (\gamma^0\gamma^k\gamma^5)_{nm} (S^b\gamma^5)_{lj} \widehat{f}^G(\mathbf{k}) \tilde{\mathbf{d}}^G_{k,b}(\mathbf{x}) \Big]$$

$$\chi(\mathbf{u},\mathbf{w}|\mathbf{k}',\mathbf{k}'',n,m,l,j) \tilde{\mathbf{f}}_{ln}(\mathbf{y}) \tilde{\mathbf{d}}^f_{jm}(\mathbf{p})$$

$$\exp[i\mathbf{k}'(\mathbf{z}+\mathbf{y})] \exp[-i\mathbf{k}''(\mathbf{z}+\mathbf{p})] \exp\left[-i\mathbf{k}\left(\mathbf{z} + \frac{1}{3}\mathbf{u} + \frac{2}{3}\mathbf{w} + \mathbf{x} \right) \right], \tag{4.36}$$

where the transformed functional operators are denoted by tilde.

A reduction of this expression can be achieved if assumptions about the form factors of the bosons are made.

By careful calculation, it was demonstrated in Fuss (2004) that for conserved symmetries, the dependence of $\widehat{f}^A(\mathbf{k})$ on $\mathbf{k}$ drops out. If one transfers this to the symmetry-breaking boson functions and extends it to the magnetic boson value at the origin, a further simplification of Eq. (4.36) can be achieved.

But before proceeding to further calculation, the formalism has to be adapted to the concept of excited leptons. In order to indicate this functional sources $\mathbf{f}^*(\mathbf{x})$ and their duals $\mathbf{d}^*(\mathbf{x})$ are introduced for the excited lepton states. Analogous to ordinary leptons these leptons must possess a functional Dirac Hamiltonian, which reads

$$\mathcal{H}_f^* = \int d^3z\, \mathbf{f}^*_\alpha(z) \left[-i(\gamma^0\gamma^k)\partial_k^z + m^*\gamma^0 \right]_{\alpha\beta} \mathbf{d}^{f*}_\beta(\mathbf{z}) \tag{4.37}$$

and is a necessary energy contribution to the adequate formulation of the corresponding effective theory.

Furthermore, we need the coupling of the leptons to the boson fields. The general formula for this lepton-boson coupling is given by Eq. (4.23). The symbolic state numbers and corresponding quantum numbers q, p,and k run over all relevant particle states of the system, including the hypothetical excited neutrino states. But attention must be paid to the fact that the excited neutrinos are the consequence of symmetry breaking and represent the lowest excited energy states in the neutrino sector of the first generation.

A characteristic property of Eq. (4.23) is that in this equation in principle any q can interact with any p and any k. Among these states are leptons with orbital excitations that represent particles of the higher generations and which are *not* to be considered as the effects of symmetry breaking. On the whole, the former will be called *excited leptons,* in contrast to leptonic states, which due to symmetry breaking, are also called *excited* in a restricted sense. In reality, this enormous assemplage of interaction terms can be reduced by appropriate evaluation [for instance, to the interaction terms in Eq. (4.36)].

To separate the orbitally excited lepton states of the higher generation from the leptonic states for broken symmetry, we postulate that higher-generation lepton states are orthogonal to the lepton states of the first generation. By this obvious postulate, no mixture between first generation states that are excited and excited states of higher generations can occur, and if one simply replaces the functional sources for leptons of the first generation by the star functionals, one obtains from Eq. (4.36) the following expression:

$$\begin{aligned}
\mathcal{H}_{bf}^{1*} &= -12c^b \int \ldots \int d^3k d^3k' d^3k'' d^3z d^3u d^3w d^3x d^3y d^3p \times \\
&\Big[\xi_\alpha^n (\gamma^0\gamma^k)_{\alpha\beta} \xi_\beta^m (T^b\gamma^5)_{lj} \widehat{f}^A(0) \tilde{\mathbf{d}}_{kb}^A(\mathbf{x}) \\
&+ \xi_\alpha^n (\gamma^0\gamma^k\gamma^5)_{\alpha\beta} \xi_\beta^m (S^b\gamma^5)_{lj} \widehat{f}^G(0) \tilde{\mathbf{d}}_{kb}^G(\mathbf{x})\Big] \\
&\Upsilon(\mathrm{u}, \mathrm{w}|\mathrm{k}', \mathrm{k}'', n, m, l, j) \tilde{\mathbf{f}}_{ln}^*(\mathrm{y}) \tilde{\mathbf{d}}_{jm}^{f*}((\mathbf{p}) \\
&\exp[i\mathbf{k}'(\mathbf{z}+\mathbf{y})] \exp[-i\mathbf{k}''(\mathbf{z}+\mathbf{p})] \exp\left[-i\mathbf{k}\left(\mathbf{z}+\frac{1}{3}\mathbf{u}+\frac{2}{3}\mathbf{w}+\mathbf{x}\right)\right]
\end{aligned} \tag{4.38}$$

With the transformation to the new variable $\mathbf{s} = \mathbf{z} + 1/3\mathbf{u} + 2/3\mathbf{w}$ and the abovementioned replacements, Eq. (4.38) reads for the neutrino-boson coupling

$$\begin{aligned}
\mathcal{H}_{bf}^{1*} &= -12c^b \int \ldots \int d^3k d^3k' d^3k'' d^3s d^3u d^3w d^3x d^3y d^3p \times \\
&\Big[(\gamma^0\gamma^k)_{\alpha\beta} \zeta(v^*)_\rho (T^b\gamma^5)_{\rho\rho'} \zeta(v^*)_{\rho'} \widehat{f}^A(0) \tilde{\mathbf{d}}_{kb}^A(\mathbf{x}) \\
&+ (\gamma^0\gamma^k\gamma^5)_{\alpha\beta} \zeta(v^*)_\rho (S^b\gamma^5)_{\rho\rho'} \zeta(v^*)_{\rho'} \widehat{f}^G(0) \tilde{\mathbf{d}}_{kb}^G(\mathbf{x})\Big] \Upsilon(\mathbf{u}, \mathbf{w}) \tilde{\mathbf{f}}_{2\alpha}^*(\mathbf{y}) \tilde{\mathbf{d}}_{2\beta}^{f*}(\mathbf{p}) \\
&\exp\left[ik'\left(\mathbf{s}-\frac{1}{3}\mathbf{u}-\frac{2}{3}\mathbf{w}+\mathbf{y}\right)\right] \exp\left[-i\mathbf{k}''\left(\mathbf{s}-\frac{1}{3}\mathbf{u}-\frac{2}{3}\mathbf{w}+\mathbf{p}\right)\right] \exp[-i\mathbf{k}(\mathbf{s}+\mathbf{x})]
\end{aligned} \tag{4.39}$$

The further evaluation depends on information about the neutrino wave functions. The three-body, GBBW-bound state equations of the composite neutrino were discussed in section 2.11, and the reasons for developing this should be repeated once more: the matric element [Eq. (4.23)] describes the electroweak coupling between leptons and bosons, including the photons. In these processes, nucleons and leptons are involved, but the nucleons are considered as spectators that are assumed to have been prepared in advance to be able to start electroweak reactions. Therefore, the leptons are the active players in the resulting processes, which are general β decays (i.e., electron capture and β^+ and β^- decay). In these decays, leptons with high velocities appear, so that their treatment generally requires states in relativistic formulation. Without repeating the extensive calculations, we adopt the results of section 2.11 and (2.12) with inclusion of the definition of appropriate test functions.

After transformation of Eq. (4.38) to the new variable $\mathbf{s} = \mathbf{z} + \frac{1}{3}\mathbf{u} + \frac{2}{3}\mathbf{w}$, the revised version starts from the central formula [Eq. (4.39)] referring to the moving frame where Eq. (2.167) was substituted, which gives

$$
\begin{aligned}
\mathcal{H}_{bf}^{1*} = &-12c^b \int \ldots \int d^3k d^3k' d^3k'' d^3s d^3\overline{u} d^3\overline{w} d^3x d^3y d^3p \\
&\times \Big[(\gamma^0\gamma^k)_{\alpha\beta} \zeta(v^*)_\rho (T^b\gamma^5)_{\rho\rho'} \zeta(v^*)_{\rho'} \widehat{f}^A(0) \tilde{\mathbf{d}}_{kb}^A(\mathbf{x}) \\
&+ (\gamma^0\gamma^k\gamma^5)_{\alpha\beta} \zeta(v^*)_\rho (S^b\gamma^5)_{\rho\rho'} \zeta(v^*)_{\rho'} \widehat{f}^G(0) \tilde{\mathbf{d}}_{kb}^G(\mathbf{x}) \Big] \Upsilon(\overline{\mathbf{u}}, \overline{\mathbf{w}}) \tilde{\mathbf{f}}_{2\alpha}^*(\mathbf{y}) \tilde{\mathbf{d}}_{2\beta}^{f*}(\mathbf{p}) \\
&\exp\left[i\mathbf{k}'\left(\mathbf{s} - \frac{1}{3}\overline{\mathbf{u}} - \frac{2}{3}\overline{\mathbf{w}} + \mathbf{y}\right)\right] \exp\left[-i\mathbf{k}''\left(\mathbf{s} - \frac{1}{3}\overline{\mathbf{u}} - \frac{2}{3}\overline{\mathbf{w}} + \mathbf{p}\right)\right] \exp\left[-i\mathbf{k}(\mathbf{s} + \mathbf{x})\right]
\end{aligned}
\tag{4.40}
$$

In this matrix element, it is possible to separate a part that can exactly be integrated and which, with $\widehat{a}_l := \frac{1}{2}(\overline{a}'_l + \overline{a}''_l)$, is defined by the product of the following integrals:

$$
\begin{aligned}
&\int d\overline{u}_l \exp\left[-i(k'_l - k''_l)\frac{1}{3}\overline{u}_l - \widehat{a}_l \overline{u}_l^2 \right] \times \int d\overline{w}_l \exp\left[-i(k'_l - k''_l)\frac{2}{3}\overline{w}_l - \widehat{a}_l \overline{w}_l^2 \right] \left(\frac{\overline{a}'_l \overline{a}''_l}{\pi^2} \right)^{1/2} \\
&= \exp\left[-\left(\frac{5}{9\widehat{a}_l}\right)(k'_l - k''_l)^2 \left(\frac{\overline{a}'_l \overline{a}''_l}{\widehat{a}_l^2} \right)^{1/2} \right] \qquad l = 1, 2, 3
\end{aligned}
\tag{4.41}
$$

Afterward, the resulting formula is integrated over **k** (boson three-momentum) and s, and one obtains from Eq. (4.40) the matrix element

$$\mathcal{H}_{bf}^{1*} = -12c^b \int \dots \int d^3k' d^3k'' d^3x d^3y d^3p \times$$

$$\Big[\left(\gamma^0\gamma^k\right)_{\alpha\beta} \zeta(v^*)_\rho \left(T^b\gamma^5\right)_{\rho\rho'} \zeta(v^*)_{\rho'} \widehat{f}^A(0) \tilde{\mathbf{d}}_{kb}^A(\mathbf{x})$$

$$+ \left(\gamma^0\gamma^k\gamma^5\right)_{\alpha\beta} \zeta(v^*)_\rho \left(S^b\gamma^5\right)_{\rho\rho'} \zeta(v^*)_{\rho'} \widehat{f}^G(0) \tilde{\mathbf{d}}_{kb}^G(\mathbf{x}) \Big] \qquad (4.42)$$

$$\tilde{\mathbf{f}}_{2\alpha}^*(\mathrm{y}) \tilde{\mathbf{d}}_{2\beta}^{f*}(\mathbf{p}) \exp[i\mathbf{k}'(-\mathbf{x}+\mathbf{y})] \exp[-i\mathbf{k}''(-\mathbf{x}+\mathbf{p})] \times$$

$$\prod_{l=1}^{3} \exp\left[-\frac{5}{9\widehat{a}_l}(k_l' - k_l'')^2\right] \left(\frac{\overline{a}_l'\overline{a}_l''}{\widehat{a}_l^2}\right)^{1/2}.$$

In Eq. (4.42), integrations remain over the wave vectors of the lepton states and their duals k″ and k′, while the internal structure constants of the lepton states are given by the quantities a, which depend on the lepton velocities $\mathbf{v}'$ and $\mathbf{v}''$.

By means of the transformation $\mathbf{k}' = \mathbf{v} + \mathbf{h}$ and $\mathbf{k}'' = \mathbf{v}$, Eq. (4.42) passes into a form that allows exact integrations, provided that one neglects the dependence of the $\overline{a}$ coefficients on $\mathbf{v}$. One obtains

$$\mathcal{H}_{bf}^{1*} = -12c^b \int \dots \int d^3v d^3h d^3x d^3y d^3p \times$$

$$\Big[\left(\gamma^0\gamma^k\right)_{\alpha\beta} \zeta(v^*)_\rho \left(T^b\gamma^5\right)_{\rho\rho'} \zeta(v^*)_{\rho'} \widehat{f}^A(0) \tilde{\mathbf{d}}_{kb}^A(\mathbf{x}) \Big]$$

$$+ \left(\gamma^0\gamma^k\gamma^5\right)_{\alpha\beta} \zeta(v^*)_\rho \left(S^b\gamma^5\right)_{\rho\rho'} \zeta(v^*)_{\rho'} \widehat{f}^G(0) \tilde{\mathbf{d}}_{kb}^G(\mathbf{x}) \Big] \qquad (4.43)$$

$$\tilde{\mathbf{f}}_{2\alpha}^*(\mathbf{y}) \tilde{\mathbf{d}}_{2\beta}^{f*}(\mathbf{p}) \exp[i\mathbf{v}(\mathbf{y}-\mathbf{p})] \exp[-i\mathbf{h}(\mathbf{x}+\mathbf{y})] \times$$

$$\prod_{l=1}^{3} \exp\left[-\frac{5}{9\widehat{a}_l}(h_l)^2\right] \left(\frac{\overline{a}_l'\overline{a}_l''}{\widehat{a}_l^2}\right)^{1/2}.$$

Integration over **v** yields

$$\mathcal{H}_{bf}^{1*} = -12c^b \int \ldots \int d^3hd^3xd^3yd^3p\delta(\mathbf{y}-\mathbf{p})(2\pi)^3 \times$$

$$\Big[(\gamma^0\gamma^k)_{\alpha\beta}\zeta(v^*)_\rho(T^b\gamma^5)_{\rho\rho'}\zeta(v^*)_{\rho'}\widehat{f}^A(0)\tilde{\mathbf{d}}_{kb}^A(\mathbf{x})\Big]$$

$$+ (\gamma^0\gamma^k\gamma^5)_{\alpha\beta}\zeta(v^*)_\rho(S^b\gamma^5)_{\rho\rho'}\zeta(v^*)_{\rho'}\widehat{f}^G(0)\tilde{\mathbf{d}}_{kb}^G(\mathbf{x})\Big] \quad (4.44)$$

$$\tilde{\mathbf{f}}_{2\alpha}^*(\mathrm{y})\tilde{\mathbf{d}}_{2\beta}^{f*}(\mathbf{p})\exp[i\mathbf{h}(\mathbf{x}+\mathbf{y})] \times$$

$$\prod_{l=1}^{3} \exp\left[-\frac{5}{9\widehat{a}_l}(h_l)^2\right]\left(\frac{\overline{a}_l'\overline{a}_l''}{\widehat{a}_l^2}\right)^{1/2}.$$

Integration over **p** and **h** leads to

$$\mathcal{H}_{bf}^{1*} = -12c^b \int \ldots \int d^3xd^3y(2\pi)^3 \times$$

$$\Big[(\gamma^0\gamma^k)_{\alpha\beta}\zeta(v^*)_\rho(T^b\gamma^5)_{\rho\rho'}\zeta(v^*)_{\rho'}\widehat{f}^A(0)\tilde{\mathbf{d}}_{kb}^A(\mathbf{x})\Big]$$

$$+ (\gamma^0\gamma^k\gamma^5)_{\alpha\beta}\zeta(v^*)_\rho(S^b\gamma^5)_{\rho\rho'}\zeta(v^*)_{\rho'}\widehat{f}^G(0)\tilde{\mathbf{d}}_{kb}^G(\mathbf{x})\Big] \quad (4.45)$$

$$\tilde{\mathbf{f}}_{2\alpha}^*(\mathbf{y})\tilde{\mathbf{d}}_{2\beta}^{f*}(\mathbf{y}) \times$$

$$\prod_{l=1}^{3}\left(\frac{\pi 9\widehat{a}_l}{5}\right)^{1/2} \exp\left[-\frac{9\widehat{a}_l}{5}(x_1+y_1)^2\right],$$

where due to the equality of the fermion state and its dual, the factor $\left(\frac{\overline{a}_l'\overline{a}_l''}{\widehat{a}_l^2}\right)^{1/2} = 1$ and is therefore omitted in Eq. (4.45). Later, these approximations will be justified in a self-consistent way.

But there is a problem: in Eq. (4.24), the boson functions are physical wave functions [i.e., regularized ones in accordance with section 1.2 of chapter 1 of this part and Stumpf (2000) and Eq. (4.59)]. And therefore, one would guess that this also should hold for the fermion wave functions. But owing to Eq. (4.31), this is not true. The solution of this problem will be discussed in the next two sections.

4.4 CHANGES AND INVARIANTS OF THE COUPLING TERM

The coupling term [Eq. (4.23)] is part of an effective functional energy equation which according to Eq. (1.29) can be formulated in the algebraic Schroedinger representation.

In the case at issue, we confine ourselves to the fermionic part of this energy equation which with Eqs. (4.23) and (4.37) reads as

$$E\big|\mathcal{P}(\mathbf{f},\mathbf{b},a)\big\rangle = \left(\mathcal{H}_f^* + \mathcal{H}_{bf}^{1*}\right)\big|\mathcal{P}(\mathbf{f},\mathbf{b},a)\big\rangle. \quad (4.46)$$

Analogous to the derivation of effective classical equations from a functional energy equation in Borne, Lochak, and Stumpf [2001, Eqs. (8.71)–(8.73)], an effective Dirac equation can be derived from Eq. (4.46), if use is made of Eq. (4.45).

In this equation, the coupling to the charged vector bosons is not relevant to the monopol existence problem and will be omitted. Then one gets for the effective spinor amplitude of the excited neutrino $\psi^*(y)$ the reduced effective equation

$$\begin{aligned}
&i\partial_t\psi_{2\alpha}^*(\mathbf{y}) = \\
&\left[-i\left(\gamma^0\gamma^k\right)_{\alpha\beta}\partial_k + \gamma^0_{\alpha\beta}m^*\right]\psi_{2\beta}^*(\mathbf{y}) \\
&+ (2\pi)^3\int d^3x\Big\{+g_A^*\left(\gamma^0\gamma^k\right)_{\alpha\beta}\zeta_\rho(v^*)\left(T^0\gamma^5\right)^D_{\rho\rho'}\zeta_{\rho'}(v^*)A_k^0(\mathbf{x}) \\
&- g_A'^*\left(\gamma^0\gamma^k\right)_{\alpha\beta}\zeta_\rho(v^*)\left(T^3\gamma^5\right)^D_{\rho\rho'}\zeta_{\rho'}(v^*)A_k^3(\mathbf{x})\Big\}\times \\
&\prod_{l=1}^{3}\left(\frac{\pi 9\widehat{a}_l}{5}\right)^{1/2}\exp\left[-(x_1-y_1)^2\left(\frac{9\widehat{a}_l}{5}\right)\right]\psi_{2\beta}^*(\mathbf{y}) \\
&+ (2\pi)^3\int d^3x\Big\{+ig_G^*\left(\gamma^0\gamma^k\gamma^5\right)_{\alpha\beta}\zeta_\rho(v^*)\left(S^0\gamma^5\right)^D_{\rho\rho'}\zeta_{\rho'}(v^*)G_k^0(\mathbf{x}) \\
&- ig_G'^*\left(\gamma^0\gamma^k\gamma^5\right)_{\alpha\beta}\zeta_\rho(v^*)\left(S^3\gamma^5\right)^D_{\rho\rho'}\zeta_{\rho'}(v^*)G_k^3(\mathbf{x})\Big\}\times \\
&\prod_{l=1}^{3}\left(\frac{\pi 9\widehat{a}_l}{5}\right)^{1/2}\exp\left[-(x_1-y_1)^2\left(\frac{9\widehat{a}_l}{5}\right)\right]\psi_{2\beta}^*(\mathbf{y}).
\end{aligned} \quad (4.47)$$

This equation admits a simple interpretation. It holds that

$$\left(\frac{9\widehat{a}_l}{5\pi}\right)^{-1/2}\int_{-\infty}^{+\infty} dx_1 \exp\left[-(x_l+y_l)^2\left(\frac{9\widehat{a}_l}{5}\right)\right] = 1, \tag{4.48}$$

and thus one can define a three-dimensional probability density function in x-space:

$$\varrho(\mathbf{x},\mathbf{y}) := \prod_{l=1}^{3}\left(\frac{9\widehat{a}_l}{5\pi}\right)^{-3/2}\exp\left[-[x_1-y_1]^2\left(\frac{9\widehat{a}_l}{5}\right)\right], \tag{4.49}$$

with the expectation value for $A_k^0(\mathbf{x})$ defined by

$$\overline{A}_k^0(\mathbf{y}) = \int d^3x A_k^0(\mathbf{x})\varrho(\mathbf{x},\mathbf{y}) \tag{4.50}$$

and corresponding expectation values for the other field potentials.

The main point is that the normalization [Eq. (4.48)] holds for any value of $\overline{a}_l$; i.e., the preceding calculations do not depend on special values of this quantity and the other quantities, which consistently justifies the approximation applied to the derivation of Eqs. (4.43)–(4.45).

If these expectation values are inserted in Eq. (4.47), with simultaneously redefined coupling constants $\overline{g}$ this equation goes over into

$$\begin{aligned}
&i\partial_t\psi_{2\alpha}^*(\mathbf{y}) = \\
&\left[-i\left(\gamma^0\gamma^k\right)_{\alpha\beta}\partial_k + \gamma_{\alpha\beta}^0 m^*\right]\psi_{2\beta}^*(\mathbf{y}) \\
&+\left(2^{1/2}\pi\right)^6\Big\{+\overline{g}_A\left(\gamma^0\gamma^k\right)_{\alpha\beta}\zeta_p(v^*)\left(T^0\gamma^5\right)_{\rho\rho'}^D\zeta_{\rho'}(v^*)\overline{A}_k^0(\mathbf{y}) \\
&-\overline{g}_A^{\prime *}\left(\gamma^0\gamma^k\right)_{\alpha\beta}\zeta_p(v^*)\left(T^3\gamma^5\right)_{\rho\rho'}^D\zeta_{\rho'}(v^*)\overline{A}_k^3(\mathbf{y})\Big\}\psi_{2\beta}^*(\mathbf{y}) \\
&+\left(2^{1/2}\pi\right)^6\Big\{+i\overline{g}_G^*\left(\gamma^0\gamma^k\gamma^5\right)_{\alpha\beta}\zeta_\rho(v^*)\left(S^0\gamma^5\right)_{\rho\rho'}^D\zeta_{p'}(v^*)\overline{G}_k^0(\mathbf{y}) \\
&-i\overline{g}_G^{\prime *}\left(\gamma^0\gamma^k\gamma^5\right)_{\alpha\beta}\zeta_\rho(v^*)\left(S^3\gamma^5\right)_{\rho\rho'}^D\zeta_{\rho'}(v^*)\overline{G}_k^3(\mathbf{y})\Big\}\psi_{2\beta}^*(\mathbf{y})
\end{aligned} \tag{4.51}$$

Compared with lepton-boson couplings in conventional theory instead of local vector potentials in Eq. (4.51), expectation values of vector potentials over the density distribution of the composite particle appear, which manifest changes that undergo the couplings in the transition from a point-particle theory to a composite particle theory.

On the other hand, all coupling constants in Eq. (4.51) are invariants of this transition, as all effects of this transition from the point-particle theory to a composite particle theory are incorporated into the expectation values of the vector potentials.

4.5 REGULARIZATION AND PROBABILITY CONSERVATION

A general discussion of these topics is complicated. Therefore, in this section we will confine ourselves to a thorough examination of a simple example that shows some essential features of this method. The example is provided by the boson states, which are exact solutions of the corresponding (hard-core) GBBW equations for conserved and broken symmetry.

For the time-ordered, antisymmetric amplitudes [Eq. (2.1)], the following set of covariant GBBW-boson equations can be derived:

$$\left[D^{\mu}_{Z_1X_1}\partial_u(x_1)-m_{Z_1X_1}\right]\varphi_{X_1Z_2}(x_1x_2)=3U_{Z_1X_2X_3X_4}F_{X_4Z_2}(x_1-x_2)\varphi_{X_2X_3}(x_1,x_1), \tag{4.52}$$

where, owing to antisymmetry simultaneously with Eq. (4.52), the equation

$$\left[D^{\mu}_{Z_2X_2}\partial_u(x_2)-m_{Z_2X_2}\right]\varphi_{Z_1X_2}(x_1x_2)=3U_{Z_2X_2X_3X_4}F_{X_4Z_1}(x_2-x_1)\varphi_{X_2X_3}(x_2,x_2) \tag{4.53}$$

must hold. Here, it must be emphasized that the operators $\partial_\mu(x)$ are ordinary partial derivatives and no functional duals of the functional sources.

Furthermore, in order to derive a probability interpretation for the bosonic wave functions, the energy representation is needed (Hamilton formalism, algebraic Schroedinger representation). cf. eqs. (1.17) and (1.18) for the general definition and the paragraph "relativistic covariance" in explanatory notes in the introduction. The essential point is that by the algebraic Schroedinger representation the description of a quantum field theory is complete. For the case at issue we refer to proposition 9, section 2.12 of this part and to proposition 5 in Stumpf (2002).

Starting with the covariant equations [Eqs. (4.52) and (4.53)], the transition into this representation was explained in detail in Stumpf and Borne (1994, propositions 5.2 and 5.3, p. 110). Thus, we can briefly argue that Eqs. (4.52) and (4.53) can be rewritten as

$$\begin{aligned}&\left[\delta_{Z_1X_1}\partial_0(t_1)+\left(D^0D^k\right)_{Z_1X_1}\partial_k(x_1)-\left(D^0m\right)_{Z_1X_1}\right]\varphi_{X_1Z_2}(x_1,x_2)=\\&\quad 3D^0_{Z_2X_1}U_{Z_1X_2X_3X_4}F_{X_4Z_2}(x_1-x_2)\varphi_{X_2X_3}(x_1,x)\end{aligned} \tag{4.54}$$

and

$$\left[\delta_{Z_2X_1}\partial_0(t_2) + \left(D^0D^k\right)_{Z_2X_1}\partial_k(x_2) - \left(D^0m\right)_{Z_2X_1}\right]\varphi_{Z_1X_1}(x_1,x_2) = \\ -3D^0_{Z_2X_1}U_{Z_1X_2X_3X_4}F_{X_4Z_1}(x_2 - x_1)\varphi_{X_2X_3}(x_1,x_1). \tag{4.55}$$

Subsequent addition of the resulting equations yields

$$\begin{aligned} &[\partial_0(t_1) + \partial_0(t_2)\,\varphi_{Z_1Z_2}(x_1,x_2) = \\ &\left[+D^0D^k\right)_{Z_1X_1}\partial_k(x_1) - \left(D^0m\right)_{Z_1X_1}\right]\varphi_{X_1Z_2}(x_1,x_2) \\ &+3D^0_{Z_1X_1}U_{Z_1X_2X_2X_4}F_{X_4Z_2}(x_1 - x_2)\varphi_{X_2X_3}(x_1,x_1) \\ &\left[+\left(D^0D^k\right)_{Z_2X_1}\partial_k(x_2) - \left(D^0m\right)_{Z_2X_1}\right]\varphi_{Z_1X_1}(x_1x_2) \\ &3D^0_{Z_2X_1}U_{Z_1X_2X_3X_4}F_{X_4Z_1}(x_2 - x_1)\varphi_{X_2X_3}(x_2,x_2). \end{aligned} \tag{4.56}$$

We now introduce the center of mass time $t_{cm} = t$ and the relative time t_r:

$$t_{cm} = \frac{1}{2}(t_1 + t_2) \equiv t, \quad t_r = t_1 - t_2, \tag{4.57}$$

and without going into subtle detail, we carry out the limit to equal times; i.e., the center of mass time in Eq. (4.56). This gives

$$\begin{aligned} &\partial_t\varphi_{Z_1Z_2}(\mathbf{r}_1,\mathbf{r}_2,t) = \\ &\left[+D^0D^k\right)_{Z_1X_1}\partial_k(\mathbf{r}_1) - \left(D^0m\right)_{Z_1X_1}\right]\varphi_{X_1Z_2}(\mathbf{r}_1,\mathbf{r}_2,t) \\ &+3D^0_{Z_1X_1}U_{Z_1X_2X_3X_4}F_{X_4Z_2}(\mathbf{r}_1 - \mathbf{r}_2,t)\varphi_{X_2X_3}(\mathbf{r}_1,\mathbf{r}_2,t) \\ &\left[+\left(D^0D^k\right)_{Z_2X_1}\partial_k(\mathbf{r}_2) - \left(D^0m\right)_{Z_2X_1}\right]\varphi_{Z_1X_1}(\mathbf{r}_1,\mathbf{r}_2,t)\times \\ &3D^0_{Z_2X_1}U_{Z_1X_2X_3X_4}F_{X_4Z_1}(\mathbf{r}_2 - \mathbf{r}_1,t)\varphi_{X_2X_3}(\mathbf{r}_2,\mathbf{r}_2,t) \end{aligned} \tag{4.58}$$

For the subsequent calculations, we decompose the index $Z := (\alpha, \kappa, i)$ in Eqs. (4.52) and (4.53) into $Z = (z, i)$. With the definition

$$\widehat{\varphi}_{z_1z_2}(\mathbf{r}_1,\mathbf{r}_2,t) := \sum_{i_1i_2}\varphi_{Z_1Z_2}(\mathbf{r}_1,\mathbf{r}_2), \tag{4.59}$$

and after summation over i_1, i_2 in Eq. (4.58), one obtains the eigenvalue equation

$$
\begin{aligned}
&\partial_t \widehat{\varphi}_{z_1 z_2}(\mathbf{r}_1, \mathbf{r}_2, t) = \\
&D^0_{z_1 x_1} D^k_{x_1 x_2} \partial_k(\mathbf{r}_1) \widehat{\varphi}_{x_2 z_2}(\mathbf{r}_1, \mathbf{r}_2, t) + D^0_{z_2 x_1} D^k_{x_1 x_2} \partial_k(\mathbf{r}_2) \widehat{\varphi}_{z_1 x_2}(\mathbf{r}_1, \mathbf{r}_2, t) \\
&- \sum_{i_1 i_2} \left[D^0_{Z_1 X_1} m_{X_1 X_2} \varphi_{X_2 Z_2}(\mathbf{r}_1, \mathbf{r}_2, t) + D^0_{Z_1 X_1} m_{X_1 X_2} \varphi_{X_2 Z_2}(\mathbf{r}_1, \mathbf{r}_2, t) \right] \\
&- 3 \sum_{i_1 i_2} \left[D^0_{Z_1 X_1} U_{X_1 X_2 X_3 X_4} F_{X_4 Z_2}(\mathbf{r}_1 - \mathbf{r}_2) \varphi_{X_2 X_3}(\mathbf{r}_1, \mathbf{r}_1, t) \right] \\
&- D^0_{Z_2 X_1} U_{X_1 X_2 X_3 X_4} F_{X_4 Z_1}(\mathbf{r}_2 - \mathbf{r}_1) \varphi_{X_2 X_3}(\mathbf{r}_2, \mathbf{r}_2) t \big]
\end{aligned}
\tag{4.60}
$$

Because $\sum_{i_1 i_2} D^0_{z_1 x_1} \delta_{i_1 i_2} \lambda_{i_2} = 0$, the two terms in the last bracket in Eq. (4.60) vanish and the equation goes to

$$
\begin{aligned}
&\partial_t \widehat{\varphi}_{z_1 z_2}(\mathbf{r}_1, \mathbf{r}_2, t) = \\
&D^0_{z_1 x_1} D^k_{x_1 x_2} \partial_k(\mathbf{r}_1) \widehat{\varphi}_{x_2 z_2}(\mathbf{r}_1, \mathbf{r}_2, t) + D^0_{z_2 x_1} D^k_{x_1 x_2} \partial_k(\mathbf{r}_2) \widehat{\varphi}_{z_1 x_2}(\mathbf{r}_1, \mathbf{r}_2, t) \\
&- \sum_{i_1 i_2} \left[D^0_{Z_1 X_1} m_{X_1 X_2} \varphi_{X_2 Z_2}(\mathbf{r}_1, \mathbf{r}_2, t) + D^0_{Z_1 X_1} m_{X_1 X_2} \varphi_{X_2 Z_2}(\mathbf{r}_1, \mathbf{r}_2, t) \right].
\end{aligned}
\tag{4.61}
$$

From this equation, one can derive a current conservation law. For brevity, we suppress all indices and coordinates except for the auxiliary field indices. With $\alpha^k(1) := \alpha^k \otimes 1$ and $\alpha^k(2) := 1 \otimes \alpha^k$ in spin space, one obtains the following from Eq. (4.61):

$$
\begin{aligned}
\partial_t \widehat{\varphi}_{z_1 z_2}(\mathbf{r}_1, \mathbf{r}_2, t) &= -\sum_{l=1}^{2} \alpha^k(l) \partial_k(\mathbf{r}_l) \widehat{\varphi}(\mathbf{r}_1, \mathbf{r}_2, t) \\
&- i \sum_{i_1 i_2} \left[\beta(1) m_{i1} \varphi_{i_2 i_2}(\mathbf{r}_1, \mathbf{r}_2, t) + \beta(2) m_{i_2} \varphi_{i_2 i_2}(\mathbf{r}_1, \mathbf{r}_2, t) \right],
\end{aligned}
\tag{4.62}
$$

while the Hermitian conjugate of this equation reads as

$$
\begin{aligned}
\partial_t \widehat{\varphi}_{z_1 z_2}(\mathbf{r}_1, \mathbf{r}_2, t)^+ &= -\sum_{l=1}^{2} \partial_k(\mathbf{r}_l) \widehat{\varphi}(\mathbf{r}_1, \mathbf{r}_2, t)^+ \alpha^k(l) \\
&+ i \sum_{i_1 i_2} \left[m_{i1} \varphi_{i_2 i_2}(\mathbf{r}_1, \mathbf{r}_2, t)^+ \beta(1) + m_{i_2} \varphi_{i_2 i_2}(\mathbf{r}_1, \mathbf{r}_2, t)^+ \beta(2) \right].
\end{aligned}
\tag{4.63}
$$

Now multiply Eq. (4.62) on the left side by $\widehat{\varphi}^+$ and Eq. (4.63) on the right side by $\widehat{\varphi}$. Note that in analogy to the formation of the Dirac probability density the expression $\widehat{\psi}^+ \widehat{\psi} = (\widehat{\psi} | \widehat{\psi})$ means a trace with respect to the algebraic indices, cf. Stumpf (2002) eq. (68). Addition of both expressions gives

$$\begin{aligned}&\partial_t(\widehat{\varphi}^+\widehat{\varphi}) + \partial_k^1(\widehat{\varphi}^+\alpha^k(1)\widehat{\varphi}) + \partial_k^2(\widehat{\varphi}^+\alpha^k(2)\widehat{\varphi})\\ &+ i\widehat{\varphi}^+\beta(1)\sum_{i_1 i_2}\varphi_{i_1 i_2}m_{i_1} - i\sum_{i_1 i_2}\varphi^+_{i_1 i_2}m_{i_1}\beta(1)\widehat{\varphi}\\ &+ i\widehat{\varphi}^+\beta(2)\sum_{i_1 i_2}\varphi_{i_1 i_2}m_{i_2} - i\sum_{i_1 i_2}\varphi^+_{i_1 i_2}m_{i_2}\beta(2)\widehat{\varphi}.\end{aligned} \tag{4.64}$$

Then with $m_i = m + \delta m_i$, Eq. (4.65) reads as

$$\begin{aligned}&\partial_t(\widehat{\varphi}^+\widehat{\varphi}) + \partial_k^1(\widehat{\varphi}^+\alpha^k(1)\widehat{\varphi}) + \partial_k^2(\widehat{\varphi}^+\alpha^k(2)\widehat{\varphi})\\ &\sum_{l=1}^{2}\left[\widehat{\varphi}^+\beta(l)\sum_{i_1 i_2}\varphi_{i_1 i_2}\delta m_{i_1} - \sum_{i_1 i_2}\varphi^+_{i_1 i_2}\delta m_{i_1}\beta(l)\widehat{\varphi}\right] = 0,\end{aligned} \tag{4.65}$$

which in the limit $\delta m_i \to 0$ leads to current conservation and, after integration over R^3, to probability conservation. The general case of probability conservation in the spinor field model was treated in Stumpf (2000).

For the study of the solutions of Eq. (4.52) and (4.53), it is convenient to replace these equations with the corresponding integral equation with the corresponding Green's function:

$$G_{z_1 z_2}(x_1 - x_2) := \delta_{i_1 i_2}\delta_{\kappa_1 \kappa_2}\left[\left(i\gamma^\mu\partial_\mu(x_1) + m_{i_1}\right)\right]_{\alpha_1\alpha_2}\Delta(x_1 - x_2, m_{i_1}) \tag{4.66}$$

Then, with the use of the summation convention, the solution of the integral equation

$$\begin{aligned}&\varphi_{\substack{i_1\ i_2\\ \kappa_1\ \kappa_2\\ \alpha_1\ \alpha_2}}(x_1, x_2)\\ &= 3\int G_{\substack{i_1\ j_1\\ \kappa_1\ \nu_1\\ \alpha_1\ \beta_1}}(x_1 - y)\lambda_{j_1} V_{\substack{\nu_1\ \nu_2\ \nu_3\ \nu_4\\ \beta_1\ \beta_2\ \beta_3\ \beta_4}}\sum_{j2} F_{\substack{j_2\ i_2\\ \nu_2\ \kappa_2\\ \beta_2\ \alpha_2}}(y - x_2)\times\\ &\sum_{i_3 i_4}\varphi_{\substack{i_3\ i_4\\ \nu_3\ \nu_4\\ \beta_3\ \beta_4}}(x_1, x_1)\end{aligned} \tag{4.67}$$

is a solution of Eqs. (4.52) and (4.53), provided that antisymmetry is secured.

In Eq. (4.67), two regularized propagator functions appear. With Eq. (4.66), we treat the simplest one. In four-dimensional space, it holds that

$$G_{\substack{i_1\ j_1\\ \kappa_1\ \nu_1\\ \alpha_1\ \beta_1}}(x_1-y|m_{j1})\lambda_{j1} = \begin{pmatrix} \lambda_1 & 0 & 0 \\ 0 & \lambda_2 & 0 \\ 0 & 0 & \lambda_3 \end{pmatrix} \cdot \begin{pmatrix} G(x_1-y|m_1) \\ G(x_1-y|m_2) \\ G(x_1-y|m_3) \end{pmatrix}_{\alpha_1\beta_1} \delta_{\kappa_1\nu_1}, \tag{4.68}$$

which can be evaluated to give

$$G^{reg}_{\substack{\kappa_1\ \nu_1\\ \alpha_1\ \beta_1}}(x_1-y) = \delta_{\kappa_1\nu_1}(2\pi)^{-4}\int d^4p\,\exp[-ip(x_1-y)]\left(\prod_{i=1}^{3}\gamma^{\mu}p_{\mu}-m_i^{-1}\right)_{\alpha_1\beta_1}. \tag{4.69}$$

In the same way, with the same result, the fermion propagator [Eq. (2.18)] can be treated; i.e., in both cases, the regularized functions no longer depend on the auxiliary field indices.

If one substitutes these regularized functions into Eq. (4.67), one obtains the result that φ itself does not depend on the auxuliary field indices. Expressing that by the definition

$$\widehat{\varphi}_{\substack{\kappa_1\ \kappa_2\\ \alpha_1\ \alpha_2}} = \sum_{i_1 i_2}\varphi_{\substack{i_1\ i_2\\ \kappa_1\ \kappa_2\\ \alpha_1\ \alpha_2}}, \tag{4.70}$$

one gets from Eq. (4.67) by summation over i_1, i_2 and observing the inherent summations of the vertex

$$\begin{aligned}&\varphi_{\substack{\kappa_1\ \kappa_2\\ \alpha_1\ \alpha_2}}(x_1,x_2)\\ &= 3\int G^{reg}_{\substack{\kappa_1\ \kappa_2\\ \alpha_1\ \beta_1}}(x_1-y)\lambda_{j_1} V_{\substack{\nu_1\ \nu_2\ \nu_3\ \nu_4\\ \beta_1\ \beta_2\ \beta_3\ \beta_4}} \sum_{j_2} F^{reg}_{\substack{\nu_2\ \kappa_2\\ \beta_2\ \alpha_2}}(y-x_2)\times \\ &\widehat{\varphi}_{\substack{\nu_3\ \nu_4\\ \beta_3\ \beta_4}}(x_1,x_1);\end{aligned} \tag{4.71}$$

i.e., all these summations produce self-regularization of Green's function and propagator and lead automatically to regularized wave functions of composite particles without any additional integration prescriptions.

This property has a direct influence on probability conservation. According to the general theory, $(\widehat{\psi}|\widehat{\psi})$ is a probability density. This means that

$$\sum_{\substack{i_1 i_2 \\ i'_1 i'_2}} \varphi_{\substack{i_1 i_2 \\ \kappa_1 \kappa_2 \\ \alpha_1 \alpha_2}}(\mathbf{r}_1, \mathbf{r}_2)\varphi^*_{\substack{i'_1 i'_2 \\ \kappa_1 \kappa_2 \\ \alpha_1 \alpha_2}}(\mathbf{r}_1, \mathbf{r}_2) = \widehat{\varrho}(\mathbf{r}_1, \mathbf{r}_2). \tag{4.72}$$

But due to the independence of the regularized functions on the auxiliary indices, one gets the fact that

$$\sum_{i_1, i_2} \varphi_{\substack{i_1 i_2 \\ \kappa_1 \kappa_2 \\ \alpha_1 \alpha_2}}(\mathbf{r}_1, \mathbf{r}_2)\varphi^*_{\substack{i_1 i_2 \\ \kappa_1 \kappa_2 \\ \alpha_1 \alpha_2}}(\mathrm{r}_1, \mathrm{r}_2) = \widehat{\varrho}(\mathbf{r}_1, \mathbf{r}_2)3^{-2} \tag{4.73}$$

is also a conserved quantity. If one extrapolates this result from $n = 2$ to $n = 3$, one can conclude that in Eq. (4.31), a probability density occurs, although the summation over the auxiliary indices does not seem to fit into the general formalism at first glance.

4.6 MAPPING WITH INCLUSION OF CHARGED LEPTON STATES

While in the preceding sections, the focus of the theory was on the investigation of magnetically active neutrinos, it must be acknowleged that it would not be a very good idea to further pursue this program without including the states of charged leptons into the formalism. This decision is required by the formalism itself, as the basic matrix element for the coupling of leptons to bosons [Eq. (4.23)] holds for all types of leptons without discrimination or priority treatment of selected lepton states. Only for practical reasons do we concentrate our investigation on the behavior of the states of the first lepton generation under symmetry breaking, which are assumed to be ground states in spite of this symmetry breaking. The only exception of this terminology is provided by the neutrinos, which under symmetry breaking aquire a new kind of property called the *magnetic charge*. Therefore, the property is called *excited*.

The hypothesis on which our approach is based proposes that the ground states of the first lepton generation differ only in their superspin-isospin part, while their spin parts and orbital parts are all the same. In addition, it should be noted that the parafermi wave functions of section 2.8 are factorized into products of superspin-isospin configurations and the spin-orbit configurations, which inevitably involves the application of G-conjugated spinors.

Furthermore, due to the lack of exact information about the lepton states, the latter states also will be based on test functions like the neutrino wave functions, which were introduced in section 2.11. This means that based on section 2.11, we can adopt all the calculations for our extended investigation program.

A convenient starting point for this adoption is the Dirac equation [Eq. (4.51)], where the neutrino indices can be replaced by the general lepton index l.

Three remarks should be made here about the meaning of this equation:

1. In this equation, the coupling terms of the leptons to the charged vector bosons have been omitted because this coupling is not relevant for our problem, as will be shown in the next section.
2. Originally, this equation was meant for the proof of excited neutrino states as monopoles. But one can easily see that this equation can likewise be applied to any other lepton state because, apart from the group theoretical modifications, the test functions for the orbital ground state wave functions are equal. By definition [Eq. (2.163)], these functions are assumed to be independent of the state quantum numbers. In this context, one should recall that in the conventional theory, leptons are considered as structureless point particles, so that Eq. (2.163) seems to be an acceptable assumption so long as no additional information about the solutions of the corresponding GBBW equation is available. In preceding papers, only some global deformation parameters have been introduced, as discussed later in this chapter
3. Analogous to the derivation of classical effective equations from a functional equation [cf. Borne, Lochak, and Stumpf 2001, Eqs. (8.71)–(8.73)], in Eq. (4.51), an effective Dirac equation for neutrinos was obtained. The generalization of Eq. (4.51) to lepton states of the first generation then reads as

$$
\begin{aligned}
i\partial_t \psi_{l\alpha}(x) &= \left[-i(\gamma^0\gamma^k)_{\alpha\beta}\partial_k + \gamma^0_{\alpha\beta} m_1^*\right]\psi_{l\beta}(x) \\
&\Big\{ + g_A^*(\gamma^0\gamma^k)_{\alpha\beta}\zeta_p(l)^D (T^0\gamma^5)^D_{\rho\rho'}\zeta_{\rho'}(l)^D A_k^0(x) \\
&- g_A'^*(\gamma^0\gamma^k)_{\alpha\beta}\zeta_p(l)^D (T^3\gamma^5)^D_{\rho\rho'}\zeta_{\rho'}(l)^D A_k^3(x)\Big\}\psi_{lb}(x) \\
&\Big\{ + ig_G^*(\gamma^0\gamma^k\gamma^5)_{\alpha\beta}\zeta_p(l)^D (S^0\gamma^5)^D_{\rho\rho'}\zeta_{\rho'}(l)^D G_k^0(x) \\
&- ig_g'^*(\gamma^0\gamma^k\gamma^5)_{\alpha\beta}\zeta_p(l)^D (S^3\gamma^5)^D_{\rho\rho'}\zeta_{\rho'}(l)^D G_k^3(x)\Big\}\psi_{lb}(x),
\end{aligned}
\tag{4.74}
$$

with the effective coupling constants

$$
g_Z^* := c^0\hat{f}^Z(0); \quad g_Z'^* := c^3\hat{f}^Z(0) \quad Z = A, G, \tag{4.75}
$$

where all other common factors are omitted because they are irrelevant for the following calculations. The constants c^b are elements of a set of deformation parameters c^1, c^2, c^3, $c^0 = 1$, which were introduced in preceding papers and in Eq. (4.38); however, they were not further pursued. Owing to the lack of exact information, these parameters characterize the differences of the various boson wave functions at the origin in a phenomenological way.

From Eq. (4.75), it follows that $g_A'^* = c^3 g_A^*$, and without further calculation, one can fit c^3 to the value that is used to define the universal electric Weinberg transformation. Let Θ_W^e be the universal (electric) Weinberg angle. Then, using Eq. (6.29) from Huang (1992), c^3 can be defined by

$$
\tan\,\Theta_W^e = \frac{g_A^*}{g_A'^*} = \frac{1}{c^3}, \tag{4.76}
$$

so long as one observes that in Eq. (6.23) of Huang (1992), the roles of g and g' are reversed as opposed to those in Eq. (4.74).

The (electric) Weinberg transformation can then be introduced by the definitions

$$
sin\Theta_W^e = \frac{g_A^*}{N_A} = \vartheta; \quad cos\Theta_W^e \frac{g_A'^*}{N_A} = \vartheta', \tag{4.77}
$$

with $N_A = (g_A^{*\mathbf{2}} + g_A'^{*\mathbf{2}})^{1/2}$, which leads to

$$
\begin{aligned}
\mathbf{A}^3 &= \vartheta'\mathbf{Z} + \vartheta\mathbf{A} \\
\mathbf{A}^0 &= -\vartheta\mathbf{Z} + \vartheta'\mathbf{A}
\end{aligned}
\tag{4.78}
$$

and corresponds to Huang [1992, Eq. (6.24)]. In Eq. (4.78), use has been made of the linear relation [Eq. (4.50)] so that any relation between the expectation values holds for the fields too.

With respect to the coupling of the leptons to the magnetic bosons, a similar transformation has to be applied, and one can conclude that the magnetic Weinberg angle Θ_W^m must be equal to the electric angle Θ_W^e. This is because for $f^{\mathrm{G}}(0) = \lambda f^{\mathrm{A}}(0)$, one obtains

$$g_G^* = \lambda g_A^* \quad g_G'^* = \lambda g_A'^*, \tag{4.79}$$

and consequently, it must hold that

$$\frac{1}{c^3} = \frac{g_G^*}{g_G'^*} = tan\Theta_W^m, \tag{4.80}$$

where, due to Eq. (4.80), the definitions in Eqs. (4.76) and (4.77) also hold if the electric coupling constants are replaced by the magnetic ones.

The appearance of magnetic potentials is new to the theory, and although the electric and magnetic Weinberg angles are equal, there is no compelling reason to apply the same Weinberg transformation to the magnetic potentials as to the electric ones. On the contrary, as a duality exists among electric and magnetic quantities, it is a heuristic working hypothesis to apply the dual Weinberg transformation to the magnetic quantities, which gives

$$\begin{aligned} \mathbf{G}^3 &= \vartheta'\mathbf{X} - \vartheta\mathbf{G} \\ \mathbf{G}^0 &= \vartheta\mathbf{X} + \vartheta'\mathbf{G}, \end{aligned} \tag{4.81}$$

and owing to $\Theta_W^e = \Theta_W^m$, one gets

$$\vartheta = \vartheta_A = \frac{g_A^*}{N_A} = \frac{g_G^*}{N_G} = \vartheta_G \quad \vartheta' = \vartheta'_A = \frac{g_A'^*}{N_A} = \frac{g_G'^*}{N_G} = \vartheta'_G. \tag{4.82}$$

These equalities define the **A-** and the **G-**representation, and one can formulate the transformation to the new variables in Eq. (4.74) by these relations.If one substitutes Eqs. (4.78) and (4.81) into Eq. (4.74), only the A and the G terms are of interest because they define the coupling of the electric and the magnetic bosons to the leptons. One gets from Eq. (4.74) that term of the electroweak lepton-boson coupling that expresses the coupling of the leptons to the genuine electromagnetic field; i.e., its potentials:

$$H_{bf}^{em}\psi_t = \frac{g_A^* g_A'^*}{N_A}\left[\zeta_\rho^D(l)\left(T^0\gamma^5\right)_{\rho Q'}^D \zeta_{\rho'}^D(l) - \zeta_\rho^D(l)\left(T^3\gamma^5\right)_{\rho\rho'}^D \zeta_{\rho'}^D(l)\right]\left(\gamma^0\gamma\right)_{\alpha\beta}\mathbf{A}(x)\psi_{l\beta}(x)$$

$$+\, i\frac{g_G^* g_G'^*}{N_G}\left[\zeta_\rho^D(l)\left(S^0\gamma^5\right)_{\rho\rho'}^D \zeta_{\rho'}^D(l) + \zeta_\rho^D(l)\left(S^3\gamma^5\right)_{\rho\rho'}^D \zeta_{\rho'}^D(l)\right]\left(\gamma^0\gamma\gamma^5\right)_{\alpha\beta}\mathbf{G}(x)\psi_{l\beta}(x) \tag{4.83}$$

It should be emphasized that the superspin-isospin parts in the brackets of Eq. (4.83) are real and that the signs and factors in front of the brackets are in agreement with the result of Lochak's quantum mechanical monopole theory.

4.7 DISCHARGE EFFECTS ON LEPTONIC DOUBLETS

The superspin-isospin matrices of Eq. (4.83) were given in Stumpf [2006, Eqs. (33) and (34)] in D-representation (i.e., G-conjugated representation). For convenience, they are explicitly reproduced here:

$$\left(S^0\gamma^5\right)^D = \begin{pmatrix} 1 & 0 \\ 0 & 1 \end{pmatrix} \quad \left(T^0\gamma^5\right)^D = \begin{pmatrix} 1 & 0 \\ 0 & -1 \end{pmatrix} \tag{4.84}$$

and

$$\left(S^3\gamma^5\right)^D = \begin{pmatrix} \sigma^3 & 0 \\ 0 & -\sigma^3 \end{pmatrix} \quad \left(T^3\gamma^5\right)^D = \begin{pmatrix} \sigma^3 & 0 \\ 0 & \sigma^3 \end{pmatrix}. \tag{4.85}$$

If these matrices are inserted into Eq. (4.83), one gets

$$\zeta_\rho^D(l)\begin{pmatrix} 0 & 0 & 0 & 0 \\ 0 & 2 & 0 & 0 \\ 0 & 0 & -2 & 0 \\ 0 & 0 & 0 & 0 \end{pmatrix}_{\rho\rho'} \zeta_{\rho'}^D(l)\frac{g_A^* g_A'^*}{N_A}\left(\gamma^0\gamma\right)_{\alpha\beta}\mathbf{A}(x)\psi_{l\beta}(x) \tag{4.86}$$

and

$$\zeta_\rho^D(l)\begin{pmatrix} 2 & 0 & 0 & 0 \\ 0 & 0 & 0 & 0 \\ 0 & 0 & 0 & 0 \\ 0 & 0 & 0 & 2 \end{pmatrix}_{\rho\rho'} \zeta_{\rho'}^D(l)\frac{g_G^* g_G'^*}{N_G}\left(\gamma^0\gamma\gamma^5\right)_{\alpha 3}\mathbf{G}(x)\psi_{l\beta}(x) \tag{4.87}$$

The superspin-isospin states $\zeta(l)$ are defined in Eq. (4.33). For broken symmetry, the superspin-isospin three-particle states Θ^l are reduced to a single product of these one-particle states (cf. section 2.8). But in addition to Eq. (4.33), the one-particle states are further restricted in the present case. For diagonal matrices in Eqs. (4.86) and (4.87), Eq. (4.33) changes into

$$\zeta_p(l) \otimes \zeta_{p'}(l) := \Theta^l_{kk'\rho}\Theta^l_{kk'\rho'}; \tag{4.88}$$

i.e., there remain only four superspin-isospin unit vectors for the state classification, as Θ^l itself is a direct product of such unit vectors.

However, before any further conclusion can be drawn, the following question must be answered: Does CP symmetry breaking change the conventional electroweak quantum numbers of the leptons? These numbers are defined by isospin t and isospin component t_3, fermion number f, hypercharge Y, and electric charge q. These quantities are linked to the numbering of representations of a $SU(2) \otimes U(1)$ group in superspin-isospin space by the eigenvalues of the group generators [cf. Borne, Lochak, & Stumpf 2001, Chapter 6)].

An answer to the above question is given by Proposition 6.5 (Borne, Lochak, & Stumpf, 2001, p. 165).

The normal ordered state functionals (i.e., the states referred to the modified ground state) satisfy the same group theoretical eigenvalue equations as the states for conserved symmetry, provided the propagator is invariant under the $SU(2) \otimes U(1)$ *group.*

For a proof of the invariance of F, the relevant terms are the operators $\gamma^5_{\kappa\kappa'}$ and $(\gamma^5\gamma^0)_{\kappa\kappa'}$ in superspin-isospin space where the latter is responsible for CP symmetry breaking. Both these terms are singlets of the superspin isospin group representations and leave F invariant under such transformations. Thus, for CP symmetry breaking, the conventional quantum numbers of leptons can be adopted to classify the vectors $\zeta_p(l)$.

On account of this result, it is no longer necessary to use the table in Borne, Lochak, and Stumpf (2001), (6.108). Rather, from Eq. (4.86), one can guess that the identities $\delta_{\rho,2} \equiv e^+$, $\delta_{\rho,3} \equiv e^-$, must hold and that from Eq. (4.87), the identities $\delta_{\rho,1} \equiv \bar{\nu}$, $\delta_{\rho,4} \equiv \nu$ can be concluded.

So in D-representation, one gets

$$\begin{pmatrix} \psi_{1\beta} \\ \psi_{2\beta} \end{pmatrix}^D \equiv \begin{pmatrix} \psi_\beta(\bar{\nu}) \\ \psi_\beta(e^+) \end{pmatrix}^D \quad \begin{pmatrix} \psi_{3\beta} \\ \psi_{4\beta} \end{pmatrix}^D \equiv \begin{pmatrix} \psi_\beta(e^-) \\ \psi_\beta(\nu) \end{pmatrix}^D. \tag{4.89}$$

The D-representation is not very well suited for physical classification of the state vectors because in the conventional Standard Model, the S-representation is preferred. By application of the transformation [Eq. (2.125)], one can return to the S-representation, which leads to the relations

$$\begin{pmatrix}\psi_{1\beta}\\ \psi_{2\beta}\end{pmatrix}^D \to \begin{pmatrix}\psi_\beta(\overline{\nu})\\ \psi_\beta(e^+)\end{pmatrix}^S \quad \begin{pmatrix}\psi_{3\beta}\\ \psi_{4\beta}\end{pmatrix}^D \to \begin{pmatrix}\psi_\beta(\nu)\\ \psi_\beta(e^-)\end{pmatrix}^S, \tag{4.90}$$

i.e., one gets the doublets of the first lepton generation in the conventional Standard Model form. From Eqs. (4.86), (4.87), and (4.90), it follows that the states $\psi^D_{1,\beta} \equiv \overline{\nu}$ $\psi^D_{4,\beta} \equiv \nu$ have equal magnetic charges and zero electric charge and the states $\psi^D_{2,\beta} \equiv e^+$ $\psi^D_{3,\beta} \equiv e^-$ have opposite electric charges and zero magnetic charge.

The strange behavior of the magnetic charge g under charge conjugation is in accordance with quantum mechanical investigations of Lochak (1985): "The most important feature appears in the formula (26): the charge conjugation does not change the sign of the magnetic charge g"; cf. also Lochak [2007, Eq. (7.2)].

In the literature, it has been reported that in the Standard Model with (nearly) vanishing neutrino masses, the lepton states are not influenced by CP symmetry breaking [Huang (1992, p. 116), Haller (1990, p. 47)]. Thus, by the application of this statement in Stumpf (2005), the calculations of the effects of CP symmetry breaking in an extended Standard Model have led to the transformation of charged lepton states into dyon states. But this result cannot be maintained in view of the new facts presented here and of the fact that the conventional Standard Model and the derivation of the extended Standard Model are based on different presuppositions and hence are not equivalent.

Furthermore, there is the problem of the generations. In the spinor field model, the set of states of the first lepton generation is characterized only by the set of different superspin-isospin states, while the orbit parts of the complete wave functions are assumed to be equal and are given approximately by the formula [Eq. (2.167)]. In contrast to this construction, the wave functions for the higher lepton generations are characterized by the same superspin-isospin configurations as those of the first generation, but the orbit parts of the wave functions must differ from the wave function of the first generation by orbital excitations.

As has been shown in previous papers, the same holds for all quark generations (with fermionic substructure), but with the difference that the orbital parts of the wave functions for the first generation must already have nontrivial excitation states that allow the distinction between leptons and quarks and that are the origin of the color degrees of quarks (Stumpf, 2004d and 1986a).

In the Dirac theory of a magnetic monopole, the famous Dirac relation between the magnetic pole strength and the electric pole strength can be deduced [cf. Felsager 1998, Eq. (9.34)]. Does this relation apply to the pole strengths of the lepton-boson couplings in Eqs. (4.86) and (4.87) too? The answer is clearly that the Dirac relation is not applicable because this relation depends on the presupposition of stable (conserved) electric as well as magnetic charges [cf. Felsager 1998, section 9.4)], which is not satisfied in our model. On the other hand, it should be possible to derive modified relations between electric and magnetic charges, which, however, in contrast to the universal Dirac relation, depend on the decay time of the magnetic monopoles and are not universal, but rather are connected with the experimental arrangement as was explained in Stumpf (2011a, section 3).

Another point should be emphasized again: The CP symmetry breaking depends on the presence of a suitable medium, such as water. Sparks in water tanks are bounded space-time phenomena; hence, such a discharge generates only a local CP symmetry breaking. If excited neutrinos leave the region of broken symmetry and enter the domain of ordinary conserved symmetry, the reason for being excited is dropped. Thus, the excited neutrinos will tend to rearrange themselves into ordinary neutrinos in order to adapt to the new, symmetry-conserving medium, such as air. As a result, the magnetic monopoles become unstable particles in dependence of boundary conditions, a fact that is in accordance with comments about experiments by Ivoilov (2006); Lochak (2005, 2008).

Finally, it has to be underlined that the purpose and the intention of this chapter is to review the theoretical foundations of the extended Standard Model, which includes the existence of magnetic monopoles. Of course, experimental consequences and technical applications related to the existence of such monopoles are highly interesting, but they exceed the scope of this discussion. Thus, for more information on this topic, see Lochak (2005, 2008, 2011).

4.8 THE ROLE OF ORDINARY NEUTRINOS

So far nothing has been said about the existence and appearance of ordinary neutrinos. In this respect one should bear in mind the experimental arrangement: CP-symmetry breaking occurs only in a cavity $\mathbf{V_W}$ filled with water, while outside this cavity air has to be assumed where no CP-symmetry breaking takes place. Hence the theoretical discussion is restricted to the description of physical effects in this cavity. The conclusion in this article is: any neutrino that is generated by β-emission in this cavity must be a magnetically excited neutrino.

However, there are lots of ordinary neutrinos which traverse **Vw** coming from the cosmic background or terrestrial reactors. Can they be influenced by the local symmetry breaking? These ordinary neutrinos have a high speed near c so that a rearrangement to symmetry breaking cannot occur because the time for traversing **Vw** is too small for a rearrangement and the energy transfer of the β-decay to the neutrino is not available i.e. ordinary neutrinos can traverse **Vw** without any modification by symmetry breaking. The conclusion must be:

Local symmetry breaking in V_w can influence onty local β decay in V_w and the local creation of the β-decay products. This fact does not prevent the occurrence of ordinary neutrinos. As was already stated by Lochak (2007) for massive excited neutrinos the magnetic charge is not conserved. The nonconservation of magnetic charge includes the loss of magnetic self-energy for the particles concerned. In a recently published paper by Stumpf (2015) it was shown that this magnetic self-energy loss manifests itself by the emission of electromagnetic radiation. After such a decay the excited neutrinos turned into ordinary neutrinos which have neither electric charge nor magnetic charge, but the same quantum numbers as the excited neutrinos with the difference that ordinary neutrinos have no coupling to the magnetic vector potential.

The electromagnetic radiation emitted during the decay of the magnetic monopoles is confined **Vw**. This radiation is a cavity radiation which must be considered for practical applications. For details cf. Stumpf (2015).

APPENDIX A

Note that the dynamical law of the spinor field [Eqs. (1.1), (1.2), and (1.9)] is invariant under symmetry breaking. As part of the basic dynamical law, therefore, the vertex is also invariant under symmetry breaking and as a consequence of the regularized anticommutators, one obtains [Borne, Lochak, & Stumpf, 2001, Eq. (2.71)]

$$\widehat{U}_{Z_1\, Z_2\, Z_3\, Z_4} := \widehat{U}_{Z_1[Z_2 Z_3 Z_4]} := .g\, \lambda_{i_1}\, B_{i_2 i_3 i_4}\, V_{Z_1[Z_2 Z_3 Z_4]}; \qquad (A.1)$$

i.e., the vertex operator is antisymmetrized in the indices Z_2, Z_3, and Z_4, which is the starting point of our calculation.

In the energy representation, the complete vertex term is then defined by Borne, Lochak, and Stumpf, [2001, Eq. (2.83)]:

$$W_{I_1 I_2 I_3 I_4} = g\lambda_{i_1} B_{i_2 i_3 i_4} D^0_{Z_1 Z'} V_{Z'[Z_2 Z_3 Z_4} \delta(\mathbf{r}_1 - \mathbf{r}_2)\delta(\mathbf{r}_1 - \mathbf{r}_3)\delta\left(\mathbf{r}_1 - \mathbf{r}_4\right), \quad \text{(A.2)}$$

where according to Borne, Lochak, and Stumpf [2001, Eq. (2.70)] and $\delta_{AA'}\delta_{\Lambda 1}\delta_{\Lambda' 2} \equiv \gamma^5(1-\gamma^0)$, the algebraic part reads as

$$V_{Z_1[Z_2 Z_3 Z_4]} =$$

$$\sum_{h=1}^{2}\Big\{ v^h_{\beta_1\beta_2}(v^h C)_{\beta_3\beta_4}\delta_{K_1K_2}\left[\gamma^5(1-\gamma^0)\right]_{K_3K_4} - v^h_{\beta_1\beta_2}(v^h C)_{\beta_4\beta_3}\delta_{K_1K_2}\left[\gamma^5(1-\gamma^0)\right]_{K_4K_3}$$

$$- v^h_{\beta_1\beta_3}(v^h C)_{\beta_2\beta_4}\delta_{K_1K_3}\left[\gamma^5(1-\gamma^0)\right]_{K_2K_4} + v^h_{\beta_1\beta_3}(v^h C)_{\beta_4\beta_2}\delta_{K_1K_3}\left[\gamma^5(1-\gamma^0)\right]_{K_4K_2}$$

$$- v^h_{\beta_1\beta_4}(v^h C)_{\beta_3\beta_2}\delta_{K_1K_4}\left[\gamma^5(1-\gamma^0)\right]_{K_3K_2} + v^h_{\beta_1\beta_4}(v^h C)_{\beta_2\beta_3}\delta_{K_1K_4}\left[\gamma^5(1-\gamma^0)\right]_{K_2K_3}\Big\} \quad \text{(A.3)}$$

With $v^1 = \delta_{\alpha_1\alpha_2} =$ and $v^2 = i\gamma^5_{\alpha_1\alpha_2}$, this vertex part can be evaluated to give

$$V_{Z_1[Z_2 Z_3 Z_4]} =$$

$$\delta_{\beta_1\beta_2} C_{\beta_3\beta_4}\delta_{K_1K_2}\gamma^5_{K_3K_4} - \delta_{\beta_1\beta_3} C_{\beta_2\beta_4}\delta_{K_1K_3}\gamma^5_{K_2K_4} - \delta_{\beta_1\beta_4} C_{\beta_3\beta_2}\delta_{K_1K_4}\gamma^5_{K_3K_2}$$

$$- \gamma^5_{\beta_1\beta_2}(\gamma^5 C)_{\beta_3\beta_4}\delta_{K_1K_2}\gamma^5_{K_3K_4} - \gamma^5_{\beta_1\beta_3}(\gamma^5 C)_{\beta_2\beta_4}\delta_{K_1K_3}\gamma^5_{K_2K_4} - \gamma^5_{\beta_1\beta_4}(\gamma^5 C)_{\beta_3\beta_2}\delta_{K_1K_4}\gamma^5_{K_3K_2} \quad \text{(A4)}$$

due to the symmetry of $\gamma^5$and the antisymmetry of $(\gamma^0\gamma^5)$.

For the subsequent conclusions, the symmetry properties of the Dirac algebra play an important role. The set of symmetric matrices is given by

$$(\gamma^\mu C) \quad \mu = 0,1,2,3 \quad (\Sigma^{\mu\nu} C) \quad \mu,\nu = 0,1,2,3 \;\; \mu < \nu, \qquad \text{(A.5)}$$

while the set of antisymmetric matrices is given by

$$\left(\gamma^5\gamma^\mu C\right) \quad \mu = 0,1,2,3 \quad \left(\gamma^5 C\right), \; (C) \qquad \text{(A.6)}$$

The expression that has to be calculated reads as

$$W_{I_1I_2I_3I_4}C^k_{I_2I_3}\mathrm{d}^b_k = \sum_{\mathrm{k}}\sum_{I_2I_3}\lambda_{i_1}B_{i_2i_3i_4}\int d^3r_2d^3r_3\delta(\mathbf{r}_1-\mathbf{r}_2)\delta(\mathbf{r}_1-\mathbf{r}_3)\delta(\mathbf{r}_1-\mathbf{r}_4)\times$$
$$\sum_h\Big[(\gamma^0\nu^h)_{\beta_1\beta_2}(\nu^hC)_{\beta_3\beta_4}\delta_{\varrho_1\varrho_2}\gamma^5_{\varrho_3\varrho_4}-(\gamma^0\nu^h)_{\beta_1\beta_3}(\nu^hC)_{\beta_2\beta_4}\delta_{\varrho_1\varrho_3}\gamma^5_{\varrho_2\varrho_4}$$
$$+(\gamma^0\nu^h)_{\beta_1\beta_4}(\nu^hC)_{\beta_2\beta_3}\delta_{\varrho_1\varrho_4}\gamma^5_{\varrho_2\varrho_3}\Big]\times$$
$$(T^b+S^b)_{\varrho_2\varrho_3}\Big[f^A_{i_2i_3}(\mathbf{r}_2-\mathbf{r}_3|\mathbf{k})(\gamma^kC)_{\beta_2\beta_3}\mathbf{d}^A_{k,b}(\mathbf{k})+f^G_{i_2i_3}(\mathbf{r}_2-\mathbf{r}_3|\mathrm{k})(\gamma^k\gamma^5C)_{\beta_2\beta_3}\mathbf{d}^G_{k,b}(\mathrm{k})$$
$$+f^{\mathrm{F}}(\mathbf{r}_2-\mathbf{r}_3|\mathbf{k})(\Sigma^{\mu\nu}C)_{\beta_2\beta_3}\mathbf{d}^F_{\mu\nu,b}(\mathbf{k})\Big]\times$$
$$\exp\left[-i\mathbf{k}\frac{1}{2}(\mathbf{r}_2+\mathbf{r}_3)\right] \tag{A.7}$$

The term with $\mathbf{d}^F_{\mu\nu,b}(\mathbf{k})$ completely drops out of this expression because

$$(\nu^hC)_{\beta_2\beta_3}(\Sigma^{\mu\nu}C)_{\beta_2\beta_3}=0,$$
$$\sum_h\Big[(\gamma^0\nu^h)_{\beta_1\beta_2}(\nu^hC)_{\beta_3\beta_4}\delta_{\varrho_1\varrho_2}\gamma^5_{\varrho_3\varrho_4}-(\gamma^0\nu^h)_{\beta_1\beta_3}(\nu^hC)_{\beta_2\beta_4}\delta_{\varrho_1\varrho_3}\gamma^5_{\varrho_2\varrho_4}\Big]\times(\Sigma^{\mu\nu}C)_{\beta_2\beta_3}=0. \tag{A.8}$$

Furthermore, it is

$$(\nu^hC)_{\beta_2\beta_3}(\gamma^kC)_{\beta_2\beta_3}=0,\quad(\nu^hC)_{\beta_2\beta_3}(\gamma^k\gamma^5C)_{\beta_2\beta_3}=0, \tag{A.9}$$

as the latter products ($h=1,2$) contain an odd number of gamma matrices. Hence Eq. (A.7) is reduced to the expression

$$W_{I_1I_2I_3I_4}C^k_{I_2I_{13}}\partial^b_k = \sum_{\mathrm{k}}\sum_{I_2I_3}\lambda_{i_1}B_{i_2i_3i_4}\int d^3r_2d^3r_3\delta(\mathbf{r}_1-\mathbf{r}_2)\delta(\mathbf{r}_1-\mathbf{r}_3)\delta(\mathbf{r}_1-\mathbf{r}_4)\times$$
$$\sum_h\Big[(\gamma^0\nu^h)_{\beta_1\beta_2}(\nu^hC)_{\beta_3\beta_4}\delta_{\varrho_1\varrho_2}\gamma^5_{\varrho_3\varrho_4}-(\gamma^0\nu^h)_{\beta_1\beta_3}(\nu^hC)_{\beta_2\beta_4}\delta_{\varrho_1\varrho_3}\gamma^5_{\varrho_2\varrho_4}\times$$
$$(T^b+S^b)_{\varrho_2\varrho_3}\Big[f^A_{i_2i_3}(\mathbf{r}_2-\mathbf{r}_3|\mathbf{k})(\gamma^kC)_{\beta_2\beta_3}\mathbf{d}^A_{k,b}(\mathbf{k})+f^G_{i_2i_3}(\mathbf{r}_2-\mathbf{r}_3|\mathbf{k})(\gamma^k\gamma^5C)_{\beta_2\beta_3}\mathbf{d}^G_{k,b}(\mathbf{k})\Big]$$
$$\exp\left[-i\mathbf{k}\frac{1}{2}(\mathbf{r}_2+\mathbf{r}_3)\right] \tag{A.10}$$

This formula is identical with Eq. (4.24). Integration over $\mathbf{r}_2$ and $\mathbf{r}_3$ yields

$$W_{I_1I_2I_3I_4}C^k_{I_2I_{13}}\partial^b_k = \sum_{\mathrm{k}}\sum_{I_2I_3}\lambda_{i_1}B_{i_4}\delta(\mathbf{r}_1-\mathbf{r}_4)\times$$
$$\sum_h\left[(\gamma^0\nu^h)_{\beta_1\beta_2}(\nu^hC)_{\beta_3\beta_4}\delta_{\varrho_1\varrho_2}\gamma^5_{\varrho_3\varrho_4}-(\gamma^0\nu^h)_{\beta_1\beta_3}(\nu^hC)_{\beta_2\beta_4}\delta_{\varrho_1\varrho_3}\gamma^5_{\varrho_2\varrho_4}\times\right.$$
$$\left.(T^b+S^b)_{\varrho_2\varrho_3}\left[\widehat{f}^A(0|\mathbf{k})(\gamma^kC)_{\beta_2\beta_3}\mathbf{d}^A_{k,b}(\mathbf{k})+\widehat{f}^G(0|\mathbf{k})(\gamma^k\gamma^5C)_{\beta_2\beta_3}\mathbf{d}^G_{k,b}(\mathbf{k})\right]\right.$$
$$\exp[-i\mathbf{k}\mathbf{r}_1], \tag{A.11}$$

where $\widehat{f}^A(0|\mathbf{k})$ and $\widehat{f}^G(0|\mathbf{k})$ are the regularized values of the boson wave functions at the origin.

Evaluation of the d^A-term in temporal gauge simplifies the subsequent calculation. In Eq. (A.11), there are two terms:

$$\begin{aligned}&[\gamma^0\nu^h\gamma^kC\nu^hC]_{\beta_1\beta_4}[(T^b+s^b)\gamma^5]_{\varrho_1\varrho_4}\\&-\left[\gamma^0\nu^h(\gamma^kC)^T\nu^hC\right]_{\beta_1\beta_4}\left[\left((T^b)^T+S^b\right)\gamma^5\right]_{\varrho_1\varrho_4}\\&=[\gamma^0\nu^h\gamma^kC\nu^hC]_{\beta_1\beta_4}[(T^b+s^b)\gamma^5]_{\varrho_1\varrho_4}\\&+[\gamma^0\nu^h\gamma^kC\nu^hC]_{\beta_1\beta_4}[(T^b-s^b)\gamma^5]_{\varrho_1\varrho_4},\end{aligned} \tag{A.12}$$

which yields

$$\begin{aligned}h&=1\quad -2(\gamma^0\gamma^k)_{\beta_1\beta_4}(T^b\gamma^5)_{\varrho_1\varrho_4}\\h&=2\quad -2(\gamma^0\gamma^5\gamma^kC\gamma^5C)_{\beta_1\beta_4}(T^b\gamma^5)_{\varrho_1\varrho_4}=-2(\gamma^0\gamma^k)_{\beta_1\beta_2}(T^b\gamma^5)_{\varrho_1\varrho_4}.\end{aligned} \tag{A.13}$$

In a similar way, one gets for the d^G-term in temporal gauge

$$\begin{aligned}&[\gamma^0\nu^h\gamma^kC\nu^hC]_{\beta_1\beta_4}[(T^b+s^b)\gamma^5]_{\varrho_1\varrho_4}\\&-\left[\gamma^0\nu^h(\gamma^kC)^T\nu^hC\right]_{\beta_1\beta_4}\left[\left((T^b)^T+S^b\right)\gamma^5\right]_{\varrho_1\varrho_4}\end{aligned} \tag{A.14}$$

and

$$\begin{aligned}h&=1\quad -2(\gamma^0\gamma^k\gamma^5)_{\beta_1\beta_4}(S^b\gamma^5)_{\varrho_1\varrho_4}\\h&=2\quad -2(\gamma^0\gamma^5\gamma^k\gamma^5C\gamma^5C)_{\beta_1\beta_4}(S^b\gamma^5)_{\varrho_1\varrho_4}=-2(\gamma^0\gamma^k\gamma^5)_{\beta_1\beta_2}(S^b\gamma^5)_{\varrho_1\varrho_4}.\end{aligned} \tag{A.15}$$

Substitution of Eqs. (A.13) and (A.15) into Eq. (A.11) eventually leads to Eq. (4.25).

REFERENCES

Baez, J., & Muniain, J. P. (1994). *Gauge Fields, Knots, and Gravity*. Singapore: World Scientific Inc.

Basdevant, J. L., Rich, J., & Spiro, M. (2004). *Fundamentals in Nuclear Physics*. New York: Springer.

Benner, M., & Rieckers, A. (2005). Spectral properties of weakly inhomogeneous BCS-models in different representations. *Zeitschrift für Naturforschung, 60a*, 343–365.

Bogolubov, N. N. (1991). *Selected Works. II Quantum and Classical Statistical Mechanics*. New York: Gordon and Breach Science Publishers.

Bopp, F. (1940). Eine lineare Theorie des Elektrons. *Annalen der Physik, 5*(38), 345–384.

Borne, T., Lochak, G., & Stumpf, H. (2001). *Nonperturbative Quantum Field Theory and the Structure of Matter*. Dordrecht: Kluver Inc.

Braginski, S. J. (1973). Theory of the development of a spark channel. In J. A. Rees (Ed.), *Electrical Breakdown in Gases* (pp. 188–200). London: Macmillan.

Bratelli, O., & Robinson, D. W. (1979). *Operator Algebras and Quantum Statistical Mechanics I*. New York: Springer Verlag.

de Broglie, L. (1940, 1942). *Une nouvelle theorie de la lumiere*. Hermann, Paris.

Buchmueller, W. (1985). Composite quarks and leptons. *Acta Phys. Austr. Suppl, 27*, 517–595.

Cabibbo, N., & Ferrari, E. (1962). Quantum electrodynamics with dirac monopoles. *Nuovo Cimento, 23*, 1147–1154.

Cheng, H. (1990). Why canonical quantization is the only known way to quantize correctly. *Lecture Notes in Physics, 361*, 3–14. Springer, Heidelberg.

Cheng, T.-P., & Li, L.-F. (1986). *Gauge Theory of Elementary Particle Physics*. Oxford: Clarendon Press.

Craggs, J. D. (1978). Spark channels. In J. M. Meek, & J. D. Craggs (Eds.), *Electrical Breakdown in Gases* (pp. 753–838). Chichester: Wiley & Sons.

Deser, S., & Teitelboim, C. (1976). Duality transformations of Abelian and non-Abelian gauge fields. *Physics Review D, 13*, 1592–1597.

Drabkina, S. I. (1951). Theory of the development of a spark discharge channel. *J. Exptl. Theoret. Phys. (USSR), 21*(4), 473–493.

Ellis, J. (1983). Testing GUTs: Where do monopoles fit, Nato Advanced Science, R.A. Carrigan, & W.P. Trower, eds, 102, 24.

Ellis, J. (2007). Beyond the standard model with the LHC. *Nature, 448*, 297–301.

Emch, G. (1972). *Algebraic Methods in Statistical Mechanics and Quantum Field Theory*. New York: Wiley Interscience.

Felsager, B. (1998). *Geometry, Particles, and Fields*. New York: Springer.

Filippov, D. V., & Urutskoiev, L. J. (2004). On the possibility of nuclear transformation in low-temperature plasma from the viewpoint of conservation laws. *Annales de la Fondation Louis de Broglie, 29*, 1187–1205.

Filippov, D. V., Rukhadze, A. A., & Urutskoiev, L. J. (2004). Effects of atomic electrons on nuclear stability and radio-active decay. *Annales de la Fondation Louis de Broglie, 29*, 1207–1217.

Flambaum, V. V., & Murray, D. W. (2000). Electrically charged magnetic monopoles as a source of time invariance violation and CP-violation. In X. H. Guo, M. Sevior, & A. W. Thomas (Eds.), *Proceedings of the Workshop on CP-Violation* (p. 87). Singapore: World Scientific.

Fluegge, S. (1986). *Rechenmethoden der Elektrodynamik*. Heidelberg: Springer.

Fonda, L., & Ghirardi, G. C. (1970). *Symmetry Principles in Quantum Physics*. New York: Dekker Inc.

Fuss, T. (2004). Quantenchromodynamik als effektive Theorie von Quarks und zusammenge-setzten Gluonen, Ph.D thesis, University Tübingen [in German].
Georgi, H., & Glashow, S. (1972). Unified weak and electromagnetic interactions without neutral currents. *Physics Review Letters, 28*, 1494–1497.
Goeckeler, M., & Schuecker, T. (1989). *Differential Geometry, Gauge Theories, and Gravity*. Cambridge: Cambridge University Press.
Goldstone, J., & Jackiw, R. (1978). Unconstrained temporal gauge for Yang-Mills theory. *Physics Letters, 74B*, 81–84.
Green, H. S. (1953). A generalized method of field quantization. *Physics Review, 90*, 270–273.
Greiner, W. (1979). *Quantenmechanik II, Symmetrien*. Deutsch Verlag, Thun.
Griffiths, D. (1996). *Einfuehrung in die Elementarteilchenphysik*. Berlin: Akademie Verlag.
Grimm, G. (1994a). Effective boson-fermion dynamics for subfermion models. *Zeitschrift für Naturforschung, 49a*, 649–662.
Grimm, G. (1994b). Elektroschwache Standard Theorie als Effektive Dynamik von Bindungszustaenden im Rahmen einer Nichtlinearen Spinortheorie. Ph.D thesis, University Tübingen.
Gross, F. (1999). *Relativistic Quantum Mechanics and Field Theory*. New York: Wiley Inc.
Grosser, D. (1983). On the decomposition of spinor fields which satisfy a nonlinear higher-order equation. *Zeitschrift für Naturforschung, 38a*, 1293–1295.
Haller, K. (1990). Canonical quantization of gauge theories in axial gauges. *Lecture Notes in Physics, 361*, 33–40. Springer, Heidelberg.
Hamermesh, M. (1962). *Group Theory*. London: Addison Wesley.
Heisenberg, W. (1966). *Introduction to the Unified Field Theory of Elementary Particles*. London: Interscience Publishers.
Heitler, W. (1949). *The Quantum Theory of Radiation*. London: Oxford University Press.
Hering, E., Martin, R., & Stohrer, M. (1989). *Physik für Ingenieure* (3rd ed.). Duesseldorf: VDI Verlag.
Huang, K. (1992). *Quarks, Leptons, and Gauge Fields* (2nd ed.). Singapore: World Scientific.
Ivoilov, N. G. (2006). Low-energy generation of the "strange" radiation. *Annales de la Fondation Louis de Broglie, 31*, 115–123.
Jackiw, R. (1980). Introduction to the Yang-Mills quantum theory. *Review of Modern Physics, 52*, 661–673.
Jackson, J. D. (1999). *Classical Electrodynamics* (3rd ed.). New York: Wiley.
Koppitz, J. (1967). Die radiale und axiale Entwicklung des Leuchtens im Funkenkanal, untersucht mit einer Wischkamera. *Zeitschrift für Naturforschung, 22a*, 1089–1097.
Kramer, P., & Stumpf, H. (unpublished). *Group Theoretical State Representation of the Partonic Substructure of Leptons and Quarks*.
Kramer, P., John, G., & Schenzle, D. (1981). *Group Theory and the Interaction of Composite Nuclear Systems*. Wiesbaden: Vieweg.
Lehner, G. (1990). *Elektromagnetische Feldtheorie*. Heidelberg: Springer.
Leibbrandt, G. (1987). Introduction to noncovariant gauges. *Review of Modern Physics, 59*, 1067–1119.
Lindner, M., & Weinheimer, C. (2011). Den Geisterteilchen auf der Spur. *Physics Journal, 10*(7), 31–37.
Lochak, G. (1983). Sur un monopole de masse nulle decrit par l'equation de Dirac, et sur une equation generale non lineaire qui contient des monopoles de spin 1/2 (1iere partie). *Annales de la Fondation Louis de Broglie, 8*, 345.
Lochak, G. (1984). Sur un monopole de masse nulle decrit par l'equation de Dirac, et sur une equation generale non lineaire qui contient des monopoles de spin 1/2 (2ieme partie). *Annales de la Fondation Louis de Broglie, 9*, 5–30.

Lochak, G. (1985). Wave equation for a magnetic monopole. *International Journal of Theoretical Physics, 24*, 1019–1050.
Lochak, G. (1995). Sur la presence d'un second photon dans la theorie de la lumiere de Louis de Broglie. *Annales de la Fondation Louis de Broglie, 20*, 111.
Lochak, G. (2003). L'equation de Dirac sur le cone de lumiere: Electrons de Majorana et monopoles magnetique. *Annales de la Fondation Louis de Broglie, 28*, 403–440.
Lochak, G. (2004). Un lepton magnetique capable d'intervenir dans les interactions faibles. *Annales de la Fondation Louis de Broglie, 29*, 1165–1175.
Lochak, G. (2005). Une nouvelle theorie du monopole magnetique avec un apercu sur les effets physiques, chimiques, biologiques et nucleaires. *Revue Electricite et Electronic, 9*, 23.
Lochak, G. (2007). The equation of a light leptonic magnetic monopole and its experimental aspects. *Zeitschrift für Naturforschung, 62a*, 231–246.
Lochak, G. (2008). Une nouvelle theorie du monopole magnetique avec un apercu sur les effets physiques, chimiques, biologiques et nucleaires. *Annales de la Fondation Louis de Broglie, 33*, 1–12.
Lochak, G. (2011). A propos de la catastrophe du Japon. Quelques reqmarques sur l'industrie nucleaire. *Annales de la Fondation Louis de Broglie, 36*, 187–193.
Lochak, G., & Urutskoiev, L. J. (2006). Low-energy nuclear reactions and the leptonic monopole. In J. P. Biberian (Ed.), *Condensed Matter Nuclear Science* (pp. 421–437). Singapore: World Scientific.
Lurie, D. (1968). *Particles and Fields*. New York: Interscience Publishers.
Lyons, L. (1983). An introduction to the possible substructure of quarks and leptons. *Progress in Particle and Nuclear Physics, 10*, 227–304.
Mandl, F., & Shaw, G. (1993). *Quantenfeldtheorie*. Wiesbaden: Aula Verlag.
Mayer-Kuckuk, T. (2002). *Kernphysik* (7th ed.). Stuttgart: Teubner.
Meek, J. M. (1973). A theory of spark discharge. In J. A. Rees (Ed.), *Electrical Breakdown in Gases* (pp. 42–51). London: Macmillan.
Milonni, P. W. (1994). *The Quantum Vacuum*. New York: Academic Press.
Mohapatra, R. N. (2003). *Unification and Supersymmetry* (3rd ed.). New York: Springer.
Morgan, G. (1978). Irradiation and time lags. In J. M. Meek, & J. D. Craggs (Eds.), *Electrical Breakdown in Gases* (pp. 655–688). Chichester: Wiley & Sons.
Nachtmann, O. (1986). *Elementarteilchenphysik, Phaenomene, und Konzepte*. Wiesbaden: Vieweg.
Ohnuki, Y., & Kamefuchi, S. (1982). *Quantum Field Theory and Parastatistics*. Heidelberg: Springer.
Pfennig, G., Klewe-Nebenius, H., & Seelmann-Eggebert, W. (1998). *Chart of the Nuclides* (6th ed.). Karlsruhe: Forschungszentrum Karlsruhe GmbH.
Pfister, W. (1990). Yang-Mills Dynamik als effective Theorie von vectoriellen Spinor-Isospinor Bindungszustaenden in einem Preonfeldmodell. Ph.D thesis, University of Tübingen [in German].
Pfister, W. (1993). Exact solutions of generalized three-particle Bargmann-Wigner equations in the strong coupling limit. *Nuovo Cimento, 107A*, 1523–1541.
Pfister, W. (1995a). Hidden quantum numbers of the group S(3) as a classification scheme of the generations of leptons and quarks in a subfermion model. *Nuovo Cimento, A, 108*, 1365–1389.
Pfister, W. (1995b). Mixed-symmetry solutions of generalized three-particle Bargmann-Wigner equations in the strong-coupling limit. *Nuovo Cimento A, 108*, 1427–1444.
Podolskyd, B. (1942). A generalized electrodynamics. Part I—Non-quantum. *Physics Review, 62*, 68–71.
Raether, H. (1937). Untersuchung der Elektronenlawine mit der Nebelkammer. *Zeitschrift für Physik, 107*, 91–110.

Raether, H. (1973). The development of kanal discharges. In J. A. Rees (Ed.), *Electrical Breakdown in Gases* (pp. 28–41). London: Macmillan.
Raizer, J. P. (1997). *Gas Discharge Physics*. New York: Springer.
Roman, P. (1960). *Theory of Elementary Particles*. Amsterdam: North Holland.
Schweber, S. S., Bethe, H. A., & de Hoffmann, F. (1956). *Mesons and Fields* (Vol. 1). New York: Row, Petersen. and Company.
Shnir, Y. (2005). *Magnetic Monopoles*. New York: Springer Verlag.
D'Souza, I. A., & Kalman, C. S. (1992). *Preons, Models of Leptons, Quarks, and Gauge Bosons as Composite Objects*. Singapore: World Scientific.
Stancu, F. (1996). *Group Theory in Subnuclear Physics*. Oxford: Oxford University Press.
Stumpf, H. (1982). On some physical properties of a unified lepton-hadron field model with composite bosons. *Zeitschrift für Naturforschung, 37a*, 1295–1299.
Stumpf, H. (1985a). Effective interactions of relativistic composite particles in unified nonlinear spinor field models. *Zeitschrift für Naturforschung, 40a*, 14–28.
Stumpf, H. (1985b). Effective interactions of relativistic composite particles in unified nonlinear spinor field models. *Zeitschrift für Naturforschung, 40a*, 183–190.
Stumpf, H. (1986a). Electroweak bosons, leptons, and Han-Nambu quarks in a unified spinor-isospinor preon field model. *Zeitschrift für Naturforschung, 41a*, 1399–1411.
Stumpf, H. (1986b). Yang-Mills dynamics as effective nonlinear field theory of composite vector bosons in unified spinor field models. *Zeitschrift für Naturforschung, 41a*, 683–703.
Stumpf, H. (2000). Covariant regularization of nonlinear spinorfield quantum theories and probability interpretation. *Zeitschrift für Naturforschung, 55a*, 415–432.
Stumpf, H. (2002). Eigenstates of generalized de Broglie–Bargmann-Wigner equations for photons with partonic substructure. *Zeitschrift für Naturforschung, 57a*, 723–736.
Stumpf, H. (2003a). PCT-invariance of generalized de Broglie–Bargmann-Wigner equations. *Zeitschrift für Naturforschung, 58a*, 481–490.
Stumpf, H. (2003b). Quantum numbers of eigenstates of generalized de Broglie–Bargmann-Wigner equations for fermions with partonic substructure. *Zeitschrift für Naturforschung, 58a*, 1–12.
Stumpf, H. (2003c). Symmetry properties of photon eigenstates of generalized de Broglie-Bargmann-Wigner equations. *Annales de la Fondation Louis de Broglie, 28*, 65–76.
Stumpf, H. (2004a). Are abelian and nonabelian gauge field quantum theories elementary? *Annales de la Fondation Louis de Broglie, 29*, 895–910.
Stumpf, H. (2004b). Simultaneous formation of electric and magnetic photon states by electroweak symmetry breaking. *Zeitschrift für Naturforschung, 59a*, 185–195.
Stumpf, H. (2004c). Symmetry properties and construction of relativistic composite fermion states. *Annales de la Fondation Louis de Broglie, 29*, 513–535.
Stumpf, H. (2004d). Why quarks are different from leptons. An explanation by a fermionic substructure of leptons and quarks. *Zeitschrift für Naturforschung, 59a*, 750–764.
Stumpf, H. (2005). A hypothetical mechanism of generating magnetically charged fermions by CP-symmetry breaking. *Zeitschrift für Naturforschung, 60a*, 696–718.
Stumpf, H. (2006). Change of electroweak nuclear reaction rates by CP- and isospin symmetry breaking. A model calculation. *Zeitschrift für Naturforschung, 61a*, 439–456.
Stumpf, H. (2008). Effect of vacuum properties on electroweak processes. A theoretical interpretation of experiments. *Zeitschrift für Naturforschung, 63a*, 301–317.
Stumpf, H. (2011a). Symmetry breaking by electric discharges in water and formation of light magnetic monopoles in an extended Standard Model. Part I. *Zeitschrift für Naturforschung, 66a*, 205–214.
Stumpf, H. (2011b). Symmetry breaking by electric discharges in water and formation of light magnetic monopoles in an extended Standard Model. Part II. *Zeitschrift für Naturforschung, 66a*, 329–338.

Stumpf, H. (2012). Symmetry breaking by electric discharges in water and formation of light magnetic monopoles in an extended Standard Model. Part III. *Zeitschrift für Naturforschung, 67a*, 173–179.

Stumpf, H. (2015). Radiative decay of Light Leptonic Magnetic Monopoles in an Extended Standard Model. *Zeitschrift für Naturforschung, 70*(3)a, 141–152.

Stumpf, H., & Borne, T. (1994). *Composite Particle Dynamics in Quantum Field Theory*. Wiesbaden: Vieweg.

Stumpf, H., & Pfister, W. (1996). Resolution and constraints and gauge equivalence in algerbraic Schroedinger representation in quantum electrodynamics. *Zeitschrift für Naturforschung, 51a*, 1054–1066.

Stumpf, H., & Pfister, W. (1997). Algebraic Schroedinger representation of quantum chro modynamics and resolution of constraints. *Zeitschrift für Naturforschung, 52a*, 220–240.

Tsai, E. C. (1990). Gauge invariance and BRS quantum field theory. *Lecture Notes in Physics, 361*, 41–51. Springer, Heidelberg.

Urutskoiev, L. J. (2004). Review of experimental results on low-energy transformation of nucleus. *Annales de la Fondation Louis de Broglie, 29*, 1149–1164.

Urutskoiev, L. J., Liksonov, V. I., & Tsinoev, V. G. (2002). Observation of transformation of chemical elements during electric discharge. *Annales de la Fondation Louis de Broglie, 27*, 701–726.

Volovik, G. E. (2003). *The Universe in a Helium Droplet*. Oxford: Clarendon Press.

Weinberg, S., & Witten, E. (1980). Limits on massless particles. *Physics Letters, 96B*, 59–62.

Contents of Volumes 151–188

Volume 151[1]

Claas Bontus and Thomas Köhler, Reconstruction algorithms for computed tomography
Laurent Busin, Nicolas Vandenbroucke and Ludovic Macaire, Color spaces and image segmentation
Glenn R. Easley and Flavia Colonna, Generalized discrete Radon transforms and applications to image processing
Tomáš Radlička, Lie agebraic methods in charged particle optics
Valerie Randle, Recent developments in electron backscatter diffraction

Volume 152

Nina S.T. Hirata, Stack filters: from definition to design algorithms
Sameen Ahmed Khan, The Foldy–Wouthuysen transformation technique in optics
Saverio Morfu, Patrick Marquié, Brice Nofiélé and Dominique Ginhac, Nonlinear systems for image processing
Tohru Nitta, Complex-valued neural network and complex-valued backpropagation learning algorithm
Jérôme Bobin, Jean-Luc Starck, Y. Moudden and M.J. Fadili, Blind source separation: the sparsity revoloution
Ray L. Withers, "Disorder": structured diffuse scattering and local crystal chemistry

Volume 153

Aberration-corrected Electron Microscopy
Harald Rose, History of direct aberration correction
Maximilian Haider, Heiko Müller and Stephan Uhlemann, Present and future hexapole aberration correctors for high-resolution electron microscopy
Ondrej L. Krivanek, Niklas Dellby, Robert J. Kyse, Matthew F. Murfitt, Christopher S. Own and Zoltan S. Szilagyi, Advances in aberration-corrected scanning transmission electron microscopy and electron energy-loss spectroscopy
Philip E. Batson, First results using the Nion third-order scanning transmission electron microscope corrector
Andrew L. Bleloch, Scanning transmission electron microscopy and electron energy loss spectroscopy: mapping materials atom by atom
Florent Houdellier, Martin Hÿtch, Florian Hüe and Etienne Snoeck, Aberration correction with the SACTEM-Toulouse: from imaging to diffraction
Bernd Kabius and Harald Rose, Novel aberration correction concepts
Angus I. Kirkland, Peter D Nellist, Lan-yun Chang and Sarah J. Haigh, Aberration-corrected imaging in conventional transmission electron microscopy and scanning transmission electron microscopy
Stephen J Pennycook, M.F. Chisholm, A.R. Lupini, M. Varela, K. van Benthem, A.Y. Borisevich, M.P. Oxley, W. Luo and S.T. Pantelides, Materials applications of aberration-corrected scanning transmission electron microscopy
Nobuo Tanaka, Spherical aberration-corrected transmission electron microscopy for nanomaterials
Knut Urban, Lothar Houben, Chun-lin Jia, Markus Lentzen, Shao-bo Mi, Andreas Thust and Karsten Tillmann, Atomic-resolution aberration-corrected transmission electron microscopy
Yimei Zhu and Joe Wall, Aberration-corrected electron microscopes at Brookhaven National Laboratory

1 Lists of the contents of volumes 100–149 are to be found in volume 150; the entire series can be searched on ScienceDirect.com

Volume 154

Henning F. Harmuth and Beate Meffert, Dirac's Difference Equation and the Physics of Finite Differences

Volume 155

Dmitry Greenfield and Mikhail Monastyrskiy, Selected Problems of Computational Charged Particle Optics

Volume 156

Vasileios Argyriou and Maria Petrou, Photometric stereo: an overview
Fred Brackx, Nele de Schepper and Frank Sommen, The Fourier transform in Clifford analysis
Niels de Jonge, Carbon nanotube electron sources for electron microscopes
Erasmo Recami and Michel Zamboni-Rached, Localized waves: a review

Volume 157

Mikhail I. Yavor, Optics of charged particle analyzers

Volume 158

Péter Dombi, Surface plasmon-enhanced photoemission and electron acceleration with ultrashort laser pulses
Brian J. Ford, Did physics matter to the pioneers of microscopy?
Jérôme Gilles, Image decomposition: theory, numerical schemes, and performance evaluation
Stina Svensson, The reverse fuzzy distance transform and its use when studying the shape of macromolecules from cryo-electron tomographic data
Marc van Droogenbroeck, Anchors of morphological operators and algebraic openings
Dong Yang, Shiva Kumar and Hao Wang, Temporal filtering technique using time lenses for optical transmission systems

Volume 159

Cold Field Emission and the Scanning Transmission Electron Microscope
Albert Victor Crewe, The work of Albert Victor Crewe on the scanning transmission electron microscope and related topics
Lyn W. Swanson and Gregory A. Schwind, A review of the cold-field electron cathode
Joseph S. Wall, Martha N. Simon and JamesF. Hainfeld, History of the STEM at Brookhaven National Laboratory
Hiromi Inada, Hiroshi Kakibayashi, Shigeto Isakozawa, Takahito Hashimoto, Toshie Yaguchi and Kuniyasu Nakamura, Hitach's development of cold-field emission scanning transmission electron microscopes
Peter W. Hawkes, Two commercial STEMs: the Siemens ST100F and the AEI STEM-1
Ian R.M. Wardell and Peter E. Bovey, A history of Vacuum Generators' 100-kV STEM
H. Sebastian von Harrach, Development of the 300-kV Vacuum Generators STEM (1985–1996)
Bernard Jouffrey, On the high-voltage STEM project in Toulouse (MEBATH)
Andreas Engel, Scanning transmission electron microscopy: biological applications
Kenneth C.A. Smith, STEM at Cambridge University: reminiscences and reflections from the 1950s and 1960s

Volume 160

Zofia Baranczuk, Joachim Giesen, Klaus Simon and Peter Zolliker, Gamut mapping
Adrian N. Evans, Color area morphology scale-spaces
Ye Pu, Chia-lung Hsieh, Rachel Grange and Demetri Psaltis, Harmonic holography
Gerhard X. Ritter and Gonzalo Urcid, Lattice algebra approach to endmember determination in hyperspectral imagery
Reinhold Rüdenberg, Origin and background of the invention of the electron microscope
H. Gunther Rudenberg and Paul G. Rudenberg, Origin and background of the invention of the electron microscope: commentary and expanded notes on Memoir of Reinhold Rüdenberg

Volume 161

Marian Mankos, Vassil Spasov and Eric Munro, Principles of dual-beam low-energy electron microscopy
Jorge D. Mendiola-Santibañez, Iván R. Terol-Villalobos and Israel M. Santillán-Méndez, Determination of adequate parameters for connected morphological contrast mappings through morphological contrast measures
Ignacio Moreno and Carlos Ferreira, Fractional Fourier transforms and geometrical optics
Vladan Velisavlevic, Martin Vetterli, Baltasar Berufell-Lozano and Pier Luigi Dragotti, Sparse image representation by directionlets
Michael H.F. Wilkinson and Georgios K. Ouzounis, Advances in connectivity and connected attribute filters

Volume 162

Kiyotaka Asakura, Hironobu Niimi and Makoto Kato, Energy-filtered x-ray photoemission electron microscopy (EXPEEM)
Eireann C. Cosgriff, Peter D. Nellist, Adrian J. d'Alfonso, Scott D. Findlay, Gavin Behan, Peng Wang, Leslie J. Allen and Angus I. Kirkland, Image contrast in aberration-corrected scanning confocal electron microscopy
Christopher J. Edgcombe, New dimensions for field emission: effects of structure in the emitting surface
Archontis Giannakidis and Maria Petrou, Conductivity imaging and generalised Radon transform: a review
Olivier Losson, Ludovic Macaire and Yanqin Yang, Comparison of color demosaicing methods

Volume 163

Wolfgang S. Bacsa, Optical interference near surfaces and its application in subwavelength microscopy
Ruy H.A. Farias and Erasmo Recami, Introduction of a quantum of time ("chronon"), and its consequences for the electron in quantum and classical physics
Andrew Neice, Methods and limitations of subwavelength imaging
A.Sever Škapin and P. Ropret, Identification of historical pigments in wall layers by combination of optical and scanning electron microscopy coupled with energy-dispersive spectroscopy
Markus E. Testorf and Michael A. Fiddy, Superresolution imaging–revisited

Volume 164

Amos Bardea and Ron Naaman, Magneto-lithography: from the bottom-up route to high throughput
Román Castañeda, The optics of spatial coherence wavelets
Junchang Li, Yanmei Wu and Yan Li, Common diffraction integral calculation based on a fast Fourier transform algorithm
Marcel Teschke and Stefan Sinzinger, A generalized approach to describe the interference contrast and phase contrast method
Dokkyu Yi and Booyong Choi, Nonlinear partial differential equations for noise problems
Henning F. Harmuth, Harmuth corrigenda

Volume 165

Natalie Baddour, Two-dimensional Fourier transforms in polar coordinates
Neil V. Budko, Superluminal, subluminal, and negative velocities in free-space electromagnetic propagation
Rowan Leary and Rik Brydson, Chromatic aberration correction: the next step in electron microscopy
Michele Marrocco, Methods for vectorial analysis and imaging in high-resolution laser microscopy
Tomoya Sakai, Masaki Narita, Takuto Komazaki, Haruhiko Nishiguchi and Atsushi Imiya, Image hierarchy in Gaussian scale space
Yusuf Ziya Umul, The theory of the boundary diffraction wave
Emil Wolf, History and solution of the phase problem in the theory of structure determination of crystals from x-ray diffraction measurements

Volume 166

Valeriy Syrovoy, Theory of Intense Beams of Charged Particles

Volume 167

Emmanuel de Chambost, A history of Cameca (1954–2009)
Johan Debayle and Jean-Charles Pinoli, Theory and applications of general adaptive neighborhood image processing
Mohamed ben Haj Rhouma, Mohamed Ali Khabou and Lotfi Hermi, Shape recognition based on eigenvalues of the Laplacian
Nicolas Loménie and Georges Stamon, Point set analysis
Leonid P. Yaroslavsky, Image recovery from sparse samples, discrete sampling theorem, and sharply bounded band-limited discrete signals

Volume 168

Luca Geretti and Antonio Abramo, The synthesis of a stochastic artificial neural network application using a genetic algorithm approach
Michel Jourlin, Josselin Breugnot, Frédéric Itthirad, Mohammed Bouabdellah and Brigitte Closs, Logarithmic image processing for color images
Rainer A. Leitgeb, Current technologies for high-speed and functional imaging with optical coherence tomography
Sergej A. Nepijko and Gerd Schönhense, Analysis of optical systems, contrast depth, and measurement of electric and magnetic field distribution on the object's surface in mirror electron microscopy
Chad M. Parish, Multivariate statistics applications in scanning transmission electron microscopy
Hidetaka Sawada, Fumio Hosokawa, Takeo Sasaki, Toshikatsu Kaneyama, Yukihito Kondo and Kazutomo Suenaga, Aberration correctors developed under the Triple C project
Tobias Schulz, Martin Albrecht and Klaus Irmscher, Spatially resolved thermoluminescence in a scanning electron microscope

Volume 169

Erchan Aptoula and Sébastien Lefèvre, Morphological texture description of grayscale and color images
Vera Guarrera and Herwig Ott, Electron microscopy of ultracold gases
Konstantinos Konstantinidis, Ioannis Andreadis and Georgios Ch. Sirakoulis, Application of artificial intelligence to content-based image retrieval
Xingwei Yang, Daniel B. Szyld and Longin Jan Latecki, Diffusion on a tensor product graph for semi-supervised learning and interactive image segmentation
S.A. Nepijko and G. Schönhense, Electron holography for electric and magnetic field measurement and its application for nanophysics

Volume 170

Alex S. Eggeman and Paul A. Midgley, Precession electron diffraction
Ray Hill, John A. Notte and Larry Scipione, Scanning helium ion microscopy
Hone-Ene Hwang and Pin Han, Signal reconstruction algorithm based on a single intensity in the Fresnel domain
Kazuhisa Sato, Toyohiko J. Konno and Yoshihiko Hirotsu, Electron microscpy studies on magnetic $L1_0$ FePd nanoparticles
D.A. Zanin, H. Cabrera, L. de Pietro, M. Pikulski, M. Goldmann, U. Ramsperger, D. Pescia and John P. Xanthakis, Fundamental aspects of near-field emission scanning electron microcopy

Volume 171

Gregor Esser, Wolfgang Becken, Werner Müller, Peter Baumbach, Josep Arasa and Dietmar Uttenweiler, Derivation of the reflection equations for higher order aberrations of local wavefronts by oblique incidence
Lila Iznita Izhar and Maria Petrou, Thermal imaging in medicine
Jean-Michel Tualle, Derivation of the radiative transfer equation in a medium with a spatially varying refractive index: a review
Kamlesh Shrivas and Mitsutoshi Setou, Imaging mass spectrometry. Sample preparation, instrumentation and applications
Robert T. Thompson and Steven A. Cummer, Transformation optics
Tobias Klein, Egbert Buhr and Carl Georg Frase, TSEM – a review of scanning electron

microscopy in transmission mode and its applications
Michel Jourlin, Maxime Carré, Josselin Breugnot and Mohamed Bouabdellah, Logarithmic image procesing: additive contrast, multiplicative contrast and associated metrics

Volume 172

Jay Theodore Cremer, Neutron and x-ray microscopy, Part 1

Volume 173

Jay Theodore Cremer, Neutron and x-ray microscopy, Part 2

Volume 174

Silicon-based Millimeter-wave Technology. Measurement, Modeling and Applications
M. Jamal Deen and Ognian Marinov, Measurement techniques and issues
Guennadi A. Kouzaev, M. Jamal Deen and Natalia K. Nikolova, Transmission lines and passive components
Mohamed H. Bakr and Mohamed H. Negm, Modeling and design of high-frequency structures using artificial neural networks and space mapping
Oana Moldovan, Antonio Lázaro, François Danneville, Rodrigo Picos, Bogdan Nae, Benjamin Iniguez and M. Jamal Deen, Nanoscale FETs
M. Daneshmand and R.R. Mansour, RF MEMS switches and switch matrices
Natalia K. Nikolova, Maryam Ravan and Reza K. Amineh, Substrate-integrated antennas on silicon

Volume 175

Jay Theodore Cremer, Small angle scatter with correlation, scatter and intermediate functions
Jay Theodore Cremer, Nuclear scatter of neutron spin states
Christian Dwyer, Atomic-resolution core-level spectroscopy in the scanning transmission electron microscope
Partha Pratim Mondal and Alberto Diaspro, Point spread function engineering for super-resolution single-photon and multi-photon fluorescence microscopy
Paul Murray and Stephen Marshall, A review of recent advances in the hit-or-miss transform
Stephen J. Sangwine, Perspectives on color image procesing by linear vector methods using projective geometric transformations

Volume 176

Katsushige Tsuno, Damaschin Ioanoviciu, Early History of Wien Filters
Damaschin Ioanoviciu, Katsushige Tsuno, Aberration Theory of the Wien Filter
Katsushige Tsuno, Damaschin Ioanoviciu, Wien Filter Instrumentation
Katsushige Tsuno, Damaschin Ioanoviciu, Simulation of Multipole Wien Filters
Damaschin Ioanoviciu, Katsushige Tsuno, Wien Filter Applications to Ions
Katsushige Tsuno, Damaschin Ioanoviciu, Application of Wien Filters to Electrons

Volume 177

Michel Jourlin , Josselin Breugnot, Bassam Abdallah, Joris Corvo, Enguerrand Couka, Maxime Carré, Image Segmentation in the Field of the Logarithmic Image Processing Model: Special Focus on the Hierarchical Ascendant Classification Techniques
Petros Maragos, Representations for Morphological Image Operators and Analogies with Linear Operators
Kenneth C. A. Smith Electron Microscopy at Cambridge University with Charles Oatley and Ellis Cosslett: Some Reminiscences and Recollections
Miguel José-Yacamán, Arturo Ponce, Sergio Mejía-Rosales, Francis Leonard Deepak, Advanced Methods of Electron Microscopy in Catalysis Research

Volume 178

Tony Lindeberg, Generalized Axiomatic Scale-Space Theory
Agnieszka Lisowska, Smoothlet Transform: Theory and Applications
Evgeniy M. Yakushev, Theory and Computation of Electron Mirrors: The Central Particle Method

Volume 179

Claude Daviau, Invariant Quantum Wave Equations and Double Space-Time
Niels de Jonge, In-Situ and Correlative Electron Microscopy
Vladimir P. Oleshko, James M. Howe, Electron Tweezers as a Tool for High-Precision Manipulation of Nanoobjects
Pilar Sobrevilla, Eduard Montseny, Aina Barcelo, Robustness Analysis of the Reduced Fuzzy Texture Spectrum and its Performance on Noisy Images
Arturo Tejada, Wouter Van den Broek, Arnold J. den Dekker, Measure-by-Wire (MBW): An Automatic Control Framework for High-Throughput Transmission Electron Microscopy

Volume 180

Mikhail Ya. Schelev, Mikhail A. Monastyrskiy, Nikolai S. Vorobiev, Sergei V. Garnov and Dmitriy E. Greenfield, Aspects of Streak Image Tube Photography

Volume 181

Ying Bai, Xiao Han, Jerry L. Prince, Octree Grid Topology-Preserving Geometric Deformable Model (OTGDM)
Maïtine Bergounioux, Second-order Variational Models for Image Texture Analysis
Victoria Klang, Nadejda B. Matsko, Electron Microscopy of Pharmaceutical Systems

Volume 182

Hans R. Gelderblom, Detlev H. Krüger, Helmut Ruska (1908–1973): His Role in the Evolution of Electron Microscopy in the Life Sciences, and Especially Virology
Hans R. Gelderblom, Detlev H. Krüger, Peter W. Hawkes Publications from the Düsseldorf University Institute for Biophysics and Electron Microscopy (Institut für Biophysikund Elektronenmikroskopie der Universität Düsseldorf) 1958-1973
Nebojsa Neškovič, P. Beličev, I. Telečki, S. Petrovič, Rainbow Lenses
Ben Adcock, Anders Hansen, Bogdan Roman, Gerd Teschke, Generalized Sampling: Stable Reconstructions, Inverse Problems and Compressed Sensing over the Continuum

Volume 183

M.M. El-Gomati, C.G.H. Walker, Toward Quantitative Scanning Electron Microscopy
Laurent Navarro, Guy Courbebaisse, Michel Jourlin, Logarithmic Wavelets
F. Lanusse , J.-L. Starck , A. Woiselle, M. J. Fadili, 3-D Sparse Representations

Volume 184

Anatoli A. Ischenko, Sergei A. Aseyev, Time-Resolved Electron Diffraction: for Chemistry, Biology and Materials Science

Volume 185

Pawel Berczynski, Slawomir Marczynski, Gaussian Beam Propagation in
Inhomogeneous Nonlinear Media. Description in Ordinary Differential Equations by Complex Geometrical Optics
David Agard, Yifan Cheng, Robert M. Glaeser, Sriram Subramaniam, Single-Particle Cryo-Electron Microscopy (Cryo-EM): Progress, Challenges, and Perspectives for Further Improvement
Martin Welk, Michael Breuß, Morphological Amoebas and Partial Differential Equations

Volume 186

Niels de Jonge, Marina Pfaff, Diana B. Peckys
Practical Aspects of Transmission
Electron Microscopy in Liquid
Jian-Jiun Ding, Soo-Chang Pei
Linear Canonical Transform
Andrey I. Denisyuk, Alexey V. Krasavin, Filipp E. Komissarenko, Ivan S. Mukhin
Mechanical, Electrostatic, and Electromagnetic Manipulation of Microobjects and Nanoobjects in Electron Microscopes

Volume 187

Ahmed Elgammal
Homeomorphic Manifold Analysis (HMA): Untangling Complex Manifolds
Teruo Kohashi
Spin-Polarized Scanning Electron Microscopy

Volume 188

Allen M. Carroll
Pattern Generators for Reflective Electron-Beam Lithography (REBL)
Frank Gunzer and Jürgen Grotemeyer
Recent Developments in Time-of-Flight Mass Spectrometry
Margit Pap
A Special Voice Transform, Analytic Wavelets, and Zernike Functions
Colin J.R. Sheppard, Shan S. Kou, and Jiao Lin
The Hankel Transform in n-dimensions and Its Applications in Optical Propagation and Imaging

INDEX

Note: Page numbers followed by "f" and "t" indicate figures and tables respectively

F

G

H

I

K

L

M

N

Y

Z

FSC
www.fsc.org
MIX
Papier | Fördert
gute Waldnutzung
FSC® C083411

Kolibri 360 GmbH
Ferdinand-Jühlke-Straße 7
99095 Erfurt, Deutschland
produktsicherheit@kolibri360.de